CONCISE ENCYCLOPEDIA
OF SYSTEM SAFETY

CONCISE ENCYCLOPEDIA OF SYSTEM SAFETY

DEFINITION OF TERMS AND CONCEPTS

Clifton A. Ericson II

A JOHN WILEY & SONS, INC., PUBLICATION

Copyright © 2011 by John Wiley & Sons, Inc. All rights reserved

Published by John Wiley & Sons, Inc., Hoboken, New Jersey
Published simultaneously in Canada

No part of this publication may be reproduced, stored in a retrieval system, or transmitted in any form or by any means, electronic, mechanical, photocopying, recording, scanning, or otherwise, except as permitted under Section 107 or 108 of the 1976 United States Copyright Act, without either the prior written permission of the Publisher, or authorization through payment of the appropriate per-copy fee to the Copyright Clearance Center, Inc., 222 Rosewood Drive, Danvers, MA 01923, (978) 750-8400, fax (978) 750-4470, or on the web at www.copyright.com. Requests to the Publisher for permission should be addressed to the Permissions Department, John Wiley & Sons, Inc., 111 River Street, Hoboken, NJ 07030, (201) 748-6011, fax (201) 748-6008, or online at http://www.wiley.com/go/permissions.

Limit of Liability/Disclaimer of Warranty: While the publisher and author have used their best efforts in preparing this book, they make no representations or warranties with respect to the accuracy or completeness of the contents of this book and specifically disclaim any implied warranties of merchantability or fitness for a particular purpose. No warranty may be created or extended by sales representatives or written sales materials. The advice and strategies contained herein may not be suitable for your situation. You should consult with a professional where appropriate. Neither the publisher nor author shall be liable for any loss of profit or any other commercial damages, including but not limited to special, incidental, consequential, or other damages.

For general information on our other products and services or for technical support, please contact our Customer Care Department within the United States at (800) 762-2974, outside the United States at (317) 572-3993 or fax (317) 572-4002.

Wiley also publishes its books in a variety of electronic formats. Some content that appears in print may not be available in electronic formats. For more information about Wiley products, visit our web site at www.wiley.com.

Library of Congress Cataloging-in-Publication Data

Ericson, Clifton A.
   Concise encyclopedia of system safety : definition of terms and concepts / Clifton A Ericson II.
     p. cm.
   Includes index.
   ISBN 978-0-470-92975-9 (cloth)
  1. System safety—Encyclopedias.   I. Title.
   TA169.7.E75 2011
   620.8'603—dc22
                                                         2010046368

Printed in Singapore

oBook ISBN: 978-1-118-02866-7
ePDF ISBN: 978-1-118-02864-3
ePub ISBN: 978-1-118-02865-0

10 9 8 7 6 5 4 3 2 1

CONTENTS

Preface	vii
Acknowledgments	ix
Author Biography	xi
List of Figures	xiii
List of Tables	xvii
Chapter 1 Introduction to System Safety	1
Chapter 2 System Safety Terms and Concepts	16
Chapter 3 System Safety Specialty Areas	456
Chapter 4 System Safety Acronyms	486
Index	507

PREFACE

System safety is an engineering discipline that is applied during system design and development of a product or system to identify and mitigate hazards, and in so doing eliminate or reduce the risk of potential mishaps and accidents. System safety is ultimately about saving lives. It is a proven technique that is currently applied on a diversity of systems, such as commercial aircraft, military aircraft, ships, trains, automobiles, nuclear power plants, weapon systems, chemical processing plants, mining, software, and medical devices.

The endeavor of system safety requires detailed knowledge and understanding of the tools, techniques, and processes involved in the system safety discipline. I wrote this book to fill a gap I perceived in the system safety body of knowledge, that is, the lack of a single source containing an explanation of the basic terms and concepts used in the system safety discipline. During my career as a system safety engineer, I discovered that many of the relevant terms are defined in numerous different documents. Some terms are learned only through experience and word of mouth, and some terms are actually defined ambiguously or incorrectly. This book attempts to meet the need to correctly explain the basic concepts and terms of system safety in a single volume, thereby serving as an everyday reference source for the definition of terms.

This book is intended for persons from various industries who are interested in making safer products and systems. It should be very useful to those new to the system safety discipline who would like to understand the basic terms used in the discipline. It is also intended for the system safety engineers who actually apply the system safety process in their daily jobs. I have written this book for engineers, analysts, and managers who are confronted with the many unique terms used in the system safety discipline as they interface with system safety engineers.

The lack of system safety costs millions of dollars in damages and loss of lives every year due to preventable accidents and mishaps. It is my greatest hope that the readers of this book can use the material contained herein to better understand and apply the system safety concept and thereby develop safe products and systems.

CLIFTON A. ERICSON II

ACKNOWLEDGMENTS

In a book of this undertaking there are naturally many people to acknowledge. This book reflects my life's journey through 45 years of engineering in the system safety discipline. My life has been touched and influenced by many people, far too many to adequately list and credit. To those whom I have left, out I apologize. But it seems that there are a few people who always remain in the forefront of one's memory. I would like to acknowledge and dedicate this book to the Boeing System Safety organization on the Minuteman Weapon System development program. This was the crucible where the experiment of system safety really started, and this is where I started my career in system safety engineering. This group has provided my most profound work-related memories and probably had the greatest influence on my life. It was led by Niel Classon, who was an early visionary and leader in the system safety field. Other people in this organization who helped in my development included Dave Haasl, Kaz Kanda, Dwight Leffingwell, Harvey Moon, Joe Muldoon, Bob Schroder, Hal Trettin, Gordon Willard, and Brad Wolfe. Later in my career, Perry D'Antonio of Sandia National Laboratories nudged me into holding various offices in the System Safety Society and to eventually become President of this international organization. My association with the System Safety Society, and the many exemplary members of the society, has helped me to expand both my knowledge and career in system safety. Danny Brunson, former Director of the Naval Ordnance Safety and Security Activity, while working on a project with me, stimulated my interest in the definition of system safety terms and the need for correct and consistent definitions.

AUTHOR BIOGRAPHY

Clifton A. Ericson II has had 45 years of experience in the field of systems safety, software safety, and fault tree analysis. He currently works for the URS Corporation (formerly EG&G Technical Services) in Dahlgren, Virginia. He provides technical analysis, consulting, oversight, and training on systems safety and software safety projects. He currently supports NAVAIR system safety on the UCAS and BAMS unmanned aircraft systems, and he is assisting in writing NAVAIR system safety policies and procedures. Prior to joining URS, Mr. Ericson worked at Applied Ordnance Technology (AOT), Inc., of Waldorf, Maryland, where he was a program manager of system and software safety. In this capacity he directed projects in system safety and software safety engineering. Prior to joining AOT, Mr. Ericson was employed as a Senior Principal Engineer for the Boeing Company for 35 years. At Boeing he worked in the fields of system safety, reliability, software engineering, and computer programming. Mr. Ericson has been involved in all aspects of system safety, including hazard analysis, fault tree analysis, software safety, safety certification, safety documentation, safety research, new business proposals, and safety training. He has worked on a diversity of projects, such as the Minuteman Missile System, SRAM missile system, ALCM missile system, Morgantown People Mover system, 757/767 aircraft, B-1A bomber, AWACS system, Boeing BOECOM system, EPRI solar power system, and the Apollo Technical Integration program.

Mr. Ericson has taught courses on software safety and fault tree analysis at the University of Washington. Mr. Ericson was President of the System Safety Society in 2001-2003, and served as Executive Vice President of the System Safety Society, and Co-Chairman of the 16th International System Safety Conference. He was the technical program chairman for the 1998 and 2005 system safety conferences. He is the founder of the Puget Sound chapter (Seattle) of the System Safety Society. In 2000 he won the Apollo Award for safety consulting work on the International Space Station, and the Boeing Achievement Award for developing the Boeing fault tree analysis course. Mr. Ericson won the System Safety Society's Presidents Achievement Award in 1998, 1999, and 2004 for outstanding work in the system safety field.

Mr. Ericson is author of *Hazard Analysis Techniques for System Safety*, published in 2005 by Wiley. He is also author of the *NAVSEA Weapon System Safety Guidelines Handbook*. He has prepared and presented training courses in system safety and software safety in the United States, Singapore, and Australia and has presented numerous technical papers at safety conferences. Mr. Ericson has published many technical articles in system and software safety and is currently editor of the *Journal of System Safety*, a publication of the System Safety Society.

LIST OF FIGURES

Figure 2.1   ALARP model   30
Figure 2.2   Barrier analysis model   37
Figure 2.3   Propane energy path with barriers   38
Figure 2.4   Battleshort model   42
Figure 2.5   Reliability bathtub curve   42
Figure 2.6   BPA concept   44
Figure 2.7   Bow-tie diagram   44
Figure 2.8   Cascading failure   52
Figure 2.9   Example of CCF   62
Figure 2.10   CCF versus CMF   63
Figure 2.11   FTA with CCF event   67
Figure 2.12   Example of CMF/CCF   68
Figure 2.13   FT with cut sets shown   84
Figure 2.14   Electromagnetic field wavelengths   115
Figure 2.15   The electromagnetic spectrum   116
Figure 2.16   Engineering development life cycle   125
Figure 2.17   Event sequence diagram (ESD) concept   132
Figure 2.18   Event tree analysis model   133
Figure 2.19   Event tree and hazard–mishap relationships   134
Figure 2.20   FMEA concept   146
Figure 2.21   Example FMEA worksheet   148
Figure 2.22   Example FMECA worksheet   149
Figure 2.23   Fault versus failure   150
Figure 2.24   Example fault hazard analysis worksheet   152
Figure 2.25   Fault tree analysis example   153
Figure 2.26   FT symbols for basic events, conditions, and transfers   154
Figure 2.27   FT gate symbols   155
Figure 2.28   Alternative FT symbols   155
Figure 2.29   Functional block diagram concept   166
Figure 2.30   Functional block diagram (FBD) of safety-critical function   166
Figure 2.31   Example FHA worksheet   168
Figure 2.32   Fuze system concept   172
Figure 2.33   Fuze S&A concept   173
Figure 2.34   Hazard/mishap relationship   178
Figure 2.35   Hazard triangle   179
Figure 2.36   Example of hazard components   181

xiv LIST OF FIGURES

Figure 2.37 HRI concept   195
Figure 2.38 The hazard tracking system (HTS)   199
Figure 2.39 Example HHA worksheet   203
Figure 2.40 Example independent protection layers (IPLs)   217
Figure 2.41 IPL evaluation using ETA   218
Figure 2.42 Mishap scenario breakdown   221
Figure 2.43 Interlock example   225
Figure 2.44 Missile launch interlock fault tree   226
Figure 2.45 Ishikawa diagram (or fish bone diagram)   229
Figure 2.46 Example JHA worksheet   232
Figure 2.47 Latency example   240
Figure 2.48 MORT top tiers   247
Figure 2.49 MA model for one component system with repair   249
Figure 2.50 MA model for two component parallel system with no repair   249
Figure 2.51 Master logic diagram (MLD) concept   251
Figure 2.52 Hazard–mishap relationship   256
Figure 2.53 Hazard–mishap example components   256
Figure 2.54 Example of CCFs and MOEs   264
Figure 2.55 Normal distribution curve   269
Figure 2.56 Example O&SHA worksheet   276
Figure 2.57 Example PN model with three transition states   286
Figure 2.58 Example PHA worksheet   292
Figure 2.59 Example PHL worksheet   294
Figure 2.60 PLOA, PLOC, and PLOM breakdown   300
Figure 2.61 Example of redundancy   311
Figure 2.62 RBD models of series and parallel systems   315
Figure 2.63 The requirements management process   322
Figure 2.64 Hazard/mishap risk   327
Figure 2.65 Example risk acceptance methods   331
Figure 2.66 Risk management model   333
Figure 2.67 Risk assessment summary   344
Figure 2.68 Bow-tie analysis of barriers   347
Figure 2.69 Elements of the safety case   349
Figure 2.70 Basic model for GSN   350
Figure 2.71 SCF thread for brake function   351
Figure 2.72 Load versus failure distribution   354
Figure 2.73 SIS concept   356
Figure 2.74 Safety precepts pyramid   365
Figure 2.75 SRCA methodology   367
Figure 2.76 Example SRCA requirements correlation matrix worksheet   368
Figure 2.77 FTA of single point failure (SPF)   373
Figure 2.78 CMM levels   376
Figure 2.79 SCL and LOR concept   378

Figure 2.80 Overview of the software safety process 389
Figure 2.81 Subsystem representation 397
Figure 2.82 Example SSHA worksheet 399
Figure 2.83 System representation 402
Figure 2.84 Example SHA worksheet 407
Figure 2.85 Major system life-cycle phases 410
Figure 2.86 Core process, closed-loop view 415
Figure 2.87 Core elements, task view 417
Figure 2.88 Core system safety process 421
Figure 2.89 Safety requirements pyramid 426
Figure 2.90 SETR process diagram 430
Figure 2.91 System types 431
Figure 2.92 Top-level mishap concept 438
Figure 2.93 Example what-if analysis worksheet 449

LIST OF TABLES

Table 2.1	Example Energy Sources, Targets, and Barriers	37
Table 2.2	Key Common Cause Attributes	65
Table 2.3	Hardware Design Assurance Level Definitions and Their Relationships to Systems Development Assurance Level	98
Table 2.4	System DAL Assignment	105
Table 2.5	Probable Effects of Shock	111
Table 2.6	Suitable Protection Measures	112
Table 2.7	Radio Frequency Bands	117
Table 2.8	EMR-Related Hazards	118
Table 2.9	Example Flash Point and Autoignition Temperatures	160
Table 2.10	Example Hazard Components	180
Table 2.11	Typical Human Error Causal Categories	208
Table 2.12	Interlock Influence on Fault Tree Top Probability	227
Table 2.13	Laser Classifications (Old System)	237
Table 2.14	Laser Classifications (New System)	238
Table 2.15	Typical Laser System Hazards	239
Table 2.16	Various Aspects of Requirements	319
Table 2.17	Characteristics of Good Requirements	320
Table 2.18	SILs Defined in IEC61508	358
Table 2.19	Software Control Categories (CC) from MIL-STD-882C	380
Table 2.20	Example LOR Task Table	381
Table 2.21	Software Levels from DO-178B	383
Table 2.22	Thermal Contact Limits from MIL-STD-1472	436
Table 2.23	Example TLMs for Different System Types	439

CHAPTER 1

Introduction to System Safety

INTRODUCTION

The endeavor for safety has been around as long as mankind; humans seem to have a predilection or natural instinct for self-preservation (i.e., safety). Prior to the advent of the system safety methodology, safety was generally achieved by accident … people did the best they could, and if an accident occurred, they merely made a fix to prevent a future occurrence and tried again. Often, this safety-by-chance approach would result in several accidents (and deaths) before the design was finally made safe. The next-generation safety approach was safety-by-prescription (compliance-based safety), where known good safety practices were prescribed for a particular product or system. As systems became larger and more techno-complex, accidents also became more complex, and knowing how to make a system safe was no longer a straightforward task. In addition, as the consequences of an accident became more drastic and more costly, it was no longer feasible or acceptable to allow for safety-by-chance or compliance. It became obvious that an intentional and proactive "systems" approach was needed. System safety was somewhat of a natural technological advancement, moving from the approach of haphazardly recovering from unexpected mishaps to deliberately anticipating and preventing mishaps. System safety is a design-for-safety concept; it is a deliberate, disciplined, and proactive approach for intentionally designing and building safety into a system from the very start of the system design. Overall, the objective of system safety is to prevent or significantly reduce the likelihood of potential mishaps in order to avoid injuries, deaths, damage, equipment loss, loss of trust, and lawsuits.

System safety as a formal discipline was originally developed and promulgated by the military–industrial complex to prevent mishaps that were costing lives, resources, and equipment loss. As the effectiveness of the discipline was observed by other industries, it was adopted and applied to these industries and technology fields, such as commercial aircraft, nuclear power, chemical

---

Concise Encyclopedia of System Safety: Definition of Terms and Concepts, First Edition. Clifton A. Ericson II.
© 2011 by John Wiley & Sons, Inc. Published 2011 by John Wiley & Sons, Inc.

processing, rail transportation, medical, Federal Aviation Administration (FAA), and National Aeronautics and Space Adminstration (NASA), just to name a few. System safety is an emergent property of a system, established by a system safety program (SSP) performed as a component of the systems engineering process. It should be noted that throughout this book, the terms "system" and "product" are interchangeable; system safety applies to both systems and products.

WHAT IS SAFETY?

In order to understand "System Safety," one must understand the related terms "safe" and "safety," which are closely intertwined, yet each term has different nuances such that they cannot be used interchangeably. In addition, the terms "hazard," "mishap," and "risk" must also be understood, as they are important components of the system safety process.

"Safe" is typically defined as freedom from danger or the risk of harm; secure from danger or loss. Safe is a state that is secure from the possibility of death, injury, or loss. A person is considered safe when there is little danger of harm threatening them. A system is considered safe when it presents low mishap risk (to users, bystanders, environment, etc.). Safe can be regarded as a *state* ... a state of low mishap risk (i.e., low danger); a state where the threat of harm or danger is nonexistent or minimal.

"Safety" is typically defined as the condition of being protected against physical harm or loss. Safety as defined in military standard MIL-STD-882D is "freedom from those conditions that can cause death, injury, occupational illness, damage to or loss of equipment or property, or damage to the environment." Since 100% freedom from risk is not possible, safety is effectively "freedom from conditions of *unacceptable* mishap *risk*." *Safety* is the "condition" of being protected against physical harm or loss (i.e., mishap). The term safety is often used in various casual manners, which can sometimes be confusing. For example, "the designers are working on aircraft safety" implies the designers are establishing the condition for a safe state in the aircraft design. Another example, "aircraft safety is developing a redundant design" implies an organizational branch of safety, "aircraft safety" that is endeavoring to develop safe system conditions. A "safety device" is a special device or mechanism used to specifically create safe conditions by mitigating an identified hazard.

The definitions for the terms *safe* and *safety* hinge around the terms *hazard*, *mishap*, and *risk*, which are closely entwined together. A mishap is an event that has occurred and has resulted in an outcome with undesired consequences. It should be noted that in system safety, the terms mishap and accident are synonymous. In order to make a system safe, the potential for mishaps must be reduced or eliminated. Risk is the measure of a potential future mishap event, expressed in terms of likelihood and consequence. Safety is measured by *mishap risk*, which is the likelihood of a potential mishap occurring multiplied by the potential severity of the losses expected when the mishap occurs. Hazards are the precursor to mishaps, and thus, potential mishaps are identified and evaluated via hazard identification and hazard risk assessment. Mishap risk provides a predictive measure that system safety uses to rate the safety significance of a hazard and the amount of improvement provided by hazard mitigation methods. In essence, mishap risk is a safety metric that characterizes the level of danger presented by a system design; potential mishap risk is caused by hazards that exist within the system design.

WHAT IS SYSTEM SAFETY?

System safety is a Design-for-Safety (DFS) process, discipline, and culture. It leans heavily on analysis—analysis of a proposed product, process, or system design that effectively anticipates potential safety problems through the identification of hazards in the design. Safety risk is calculated from the identified hazards, and risk is eliminated or reduced by eliminating or controlling the appropriate hazard causal factors. System safety also utilizes known safety requirements and guidelines for products and systems; however, it has been proven that compliance-based safety is insufficient alone for complex systems because compliance requirements do not cover subtle hazards created by system complexities. System safety begins early in the design process and continues throughout the life cycle of the product/system. System safety by necessity considers function, criticality, risk, performance, and cost parameters of the system.

System safety is often not fully appreciated for the contribution it can provide in creating safe systems that present minimal chance of deaths and serious injuries. System safety invokes and applies a planned and disciplined methodology for purposely designing safety into a system. A system can only be made safe when the system safety methodology (or equivalent) is consistently applied and followed. Safety is more than eliminating hardware failure modes; it involves designing the safe system interaction of hardware, software, humans, procedures, and the environment, under all normal and adverse failure conditions. Safety must consider the entirety of the problem, not just a portion of the problem; that is, a systems perspective is required for full safety coverage. System safety anticipates potential problems and either eliminates them, or reduces their risk potential, through the use of design safety mechanisms applied according to a safety order of precedence.

The basic interrelated goals of system safety are to:

- Proactively prevent product/system accidents and mishaps
- Protect the system and its users, the public, and the environment from mishaps
- Identify and eliminate/control hazards

- Design and develop a system presenting minimal mishap risk
- Create a safe system by intentionally designing safety into the overall system fabric

Since many systems and activities involve hazard sources that cannot be eliminated, zero mishap risk is often not possible. Therefore, the application of system safety becomes a necessity in order to reduce the likelihood of mishaps, thereby avoiding deaths, injuries, losses, and lawsuits. Safety must be designed intentionally and intelligently into the system design or system fabric; it cannot be left to chance or forced-in after the system is built. If the hazards in a system are not known, understood, and controlled, the potential mishap risk may be unacceptable, with the result being the occurrence of many mishaps.

Accidents and mishaps are the direct result of hazards that have been actuated. (Note: Accidents and mishaps are synonymous terms, and mishap has become the preferred term in system safety.) Mishaps happen because systems contain many inherent hazard sources (e.g., gasoline in an automobile), which cannot be eliminated since they are necessary for the objectives of the system. As systems increase in complexity, size, and technology, the inadvertent creation of system hazards is a natural consequence. Unless these hazards are controlled through design safety mechanisms, they will ultimately result in mishaps.

System safety is a process for conducting the intentional and planned application of management and engineering principles, criteria, and techniques for the purpose of developing a safe system. System safety applies to all phases of the system life cycle. The basic system safety process involves the following elements:

1. System safety program plan (SSPP)
2. Hazard identification
3. Risk assessment
4. Risk mitigation
5. Mitigation verification
6. Risk acceptance, and
7. Hazard/risk tracking

System safety is an intentional process, and when safety is intentionally designed into a system, mishap risk is significantly reduced. System safety is the discipline of identifying hazards, assessing potential mishap risk, and mitigating the risk presented by hazards to an acceptable level. Risk mitigation is achieved through a combination of design mechanisms, design features, warning devices, safety procedures, and safety training.

WHY SYSTEM SAFETY?

We live in a world surrounded by hazards and potential mishap risk; hazards, mishaps, and risks are a reality of daily life. One of the major reasons for hazards is the ubiquitous *system*; many hazards are the by-product of man-made systems, and we live in a world of systems and systems-of-systems. Systems are intended to improve our way of life, yet they also contain inherent capability to spawn many different hazards that present us with mishap risk. It is not that systems are intrinsically bad; it is that systems can go astray, and when they go astray, they typically result in mishaps. System safety is about determining how systems can go bad and is about implementing design safety mitigations to eliminate, correct, or work around safety imperfections in the system.

Systems go bad for various reasons. Many of these reasons cannot be eliminated, but they can be controlled when they are known and understood. Potential mishaps exist as hazards in system designs (see the definition of "hazard" in Chapter 2). Hazards are inadvertently designed-in to the systems we design, build, and operate. In order to make a system safe, we must first understand the nature of hazards in general and then identify hazards within a particular system. Hazards are predictable, and if they can be predicted, they can also be eliminated or controlled, thereby preventing mishaps.

Systems seem to have both a bright side and a dark side. The bright side is when the system works as intended and performs its intended function without a glitch. The dark side is when the system encounters hardware failures, software errors, human errors, and/or sneak circuit paths that lead to anything from a minor incident to a major mishap event. The following are some examples of the dark side of a system, which demonstrate different types and levels of safety vulnerability:

- A toaster overheats and the thermal electrical shutoff fails, allowing the toast to burn, resulting in flames catching low-hanging cabinets on fire, which in turn results in the house burning down.
- A dual engine aircraft has an operator-controlled switch for each engine that activates fuel cutoff and fire extinguishing in case of an engine fire. If the engine switches are erroneously cross-wired during manufacture or maintenance, the operational engine will be erroneously shut down while the other engine burns during an engine emergency.
- A missile system has several safety interlocks that must be intentionally closed by the operator in order to launch the missile; however, if all of the interlocks fail in the right mode and sequence, the system will launch the missile by itself.
- Three computers controlling a fly-by-wire aircraft all fail simultaneously due to a common cause failure, resulting in the pilot being unable to correctly maneuver the flight control surfaces and land the aircraft.
- A surgeon erroneously operates on the wrong knee of a patient due to the lack of safety procedures, checklists, training, and inspections in the hospital.

There are several "system laws" that essentially state that systems have a natural proclivity to fail. These laws create hazard existence factors which explain the various reasons why hazards exist within systems.

The system laws illuminating why hazards exist include:

- Systems must include and utilize components that are naturally hazardous.
- Physical items will always eventually fail.
- Humans do commit performance errors and always will.
- System components are often combined together with sneak fault paths and integration flaws.
- Systems are often designed with unintended functions that are not recognized.
- Environmental factors can influence safe functioning of components.
- Software is typically too complex to completely test for safety validation.

There is no getting around these system laws; they will happen, and they will shape the hazard risk presented by a system design. System safety must evaluate the potential impact of each of these system laws and determine if hazards will result, and if so, how the hazards can be eliminated or controlled to prevent mishaps. In other words, these system laws are hazard-shaping factors that must be dealt with during product/process/system design in order to develop a safe system. Since hazards are unique for each system design, safety compliance measures do not provide adequate safety coverage; system hazard analysis is thus necessary.

In order to achieve their desired objectives, systems are often forced to utilize hazardous sources in the system design, such as gasoline, nuclear material, high voltage or high pressure fluids. Hazard sources bring with them the potential for many different types of hazards which, if not properly controlled, can result in mishaps. In one sense, system safety is a specialized trade-off between *utility value* and *harm value*, where utility value refers to the benefit gained from using a hazard source, and harm value refers to the amount of harm or mishaps that can potentially occur from using the hazard source. For example, the explosives in a missile provide a utility value of destroying an intended enemy target; however, the same explosives also provide a harm value in the associated risk of inadvertent initiation of the explosives and the harm that would result. System safety is the process for balancing utility value and harm value through the use of design safety mechanisms. This process is often referred to as Design-for-Safety (DFS).

Systems have become a necessity for modern living, and each system spawns its own set of potential mishap risks. Systems have a trait of failing, malfunctioning, and/or being erroneously operated. System safety engineering is the discipline and process of developing systems that present reasonable and acceptable mishap risk, for both users and nearby nonparticipants.

To design systems that work correctly and safely, a safety analyst needs to understand and correct how things can go wrong. It is often not possible to completely eliminate potential hazards because a hazardous element is a necessary system component that is needed for the desired system functions, and the hazardous element is what spawns hazards. Therefore, system safety is essential for the identification and mitigation of these hazards. System safety identifies the unique interrelationship of events leading to an undesired event in order that they can be effectively mitigated through design safety features. To achieve this objective, system safety has developed a specialized set of tools to recognize hazards, assess potential mishap risk, control hazards, and reduce risk to an acceptable level.

A system is typically considered to be safe when it:

1. Operates without causing a system mishap under normal operation
2. Presents an acceptable level of mishap risk under abnormal operation

Normal operation means that no failures or errors are encountered during operation (fault free), whereas abnormal operation means that failures, sneak paths, unintended functions, and/or errors are encountered during operation. The failures and/or errors change the operating conditions from normal to abnormal. A system must be designed to operate without generating mishaps under normal operating conditions; that is, under normal operating conditions (no faults), a system must be mishap free. However, since many systems require the inclusion of various hazard sources, they are susceptible to potential mishaps under abnormal conditions, where abnormal conditions are caused by failures, errors, malfunctions, extreme environments, and combinations of these factors. Hazards can be triggered by a malfunction involving a hazard source. During normal operation, the design is such that hazard sources cannot be triggered. Normal operation relates directly to an inherent safe design without considering equipment or human failures. Abnormal operation relates to a fault-tolerant design that considers, and compensates for, the potential for malfunctions and errors combined with hazard sources.

WHEN SHOULD SYSTEM SAFETY BE APPLIED?

Essentially, every organization and program should always perform the system safety process on every product, process, or system. This is not only to make the system safe but also to prove and verify the system is safe. Safety cannot be achieved by chance. This concept makes obvious sense on large safety-critical systems, but what about small systems that seem naturally safe? Again, a system should be proven safe, not just assumed to be safe. An SSP can be tailored in size, cost, and effort through scaling, based on standards, common sense, and risk-based judgment, ranging from toasters to motorcycles to submarines to skyscrapers.

System safety should especially be applied to complex systems with safety-critical implications, such as nuclear power, underwater oil drilling rigs, commercial aircraft, computer-managed automobiles, and surgery. System safety should also be applied to integrated systems and systems of systems, such as automobiles within a traffic grid system within a highway system within a human habitat system.

The system safety process should particularly be invoked when a system can kill, injure, or maim humans. It should always be applied as good business practice, because the cost of safety can easily be cheaper than the costs of not doing safety (i.e., mishap costs). When system safety is not performed, system mishaps often result, and these mishaps generate associated costs in terms of deaths, injuries, system damage, system loss, lawsuits, and loss of reputation.

WHAT IS THE COST OF SYSTEM SAFETY?

System safety is a trade-off between risk, cost, and performance. The cost of safety is like a two-edged sword—there is a cost for performing system safety, and there is also a cost for not implementing system safety. An ineffective safety effort has a ripple effect; an unsafe system will continue to have mishaps until serious action is taken to correctly and completely fix the system. The cost of safety can be viewed as involving two competing components: the investment costs versus the penalty costs. These are the positive and the negative cost factors associated with safety (i.e., safety as opposed to un-safety). When evaluating the cost of safety, both components must be considered because there is a direct interrelationship. In general, more safety effort results in fewer mishaps, and less safety effort results in more mishaps. This is an inverse correlation; as safety increases, mishaps decrease. There is a counterbalance here; it takes money to make safety increase, but if safety is not increased, it will cost money to pay for mishap losses and to then make fixes that should have been done in the first place.

Investment costs are the costs associated with intentionally designing safety into the system. Safety should be viewed as an investment cost rather than just a cost of doing business because safety can save money in the long run. Safety investment costs are the actual amounts of money spent on a proactive safety program to design, test, and build the system such that the likelihood of future potential mishaps occurring is reduced to an acceptably low level of probability or risk. This expenditure is made as an investment in the future, because as the system is designed to be inherently safe, potential future mishaps are eliminated or controlled such that they are not likely to occur during the life of the system. This investment should eliminate or cancel potential mishap penalty costs that could be incurred due to an unsafe system, thus saving money. One reason decision makers like to avoid the necessary investment cost of safety is because the results of the investment expenditure are usually not apparent or visible; they tend to be an intangible commodity (that is until a mishap occurs). Penalty costs are the costs associated with the occurrence of a mishap or mishaps during the life of the system. Penalty costs should be viewed as the un-safety costs incurred due to mishaps that occur during operation of the system.

For example, consider the case where an SSP is conducted during the development of a new toaster system. Hazard analyses show that there are certain over-temperature hazards that could result in a toaster fire, which could in turn result in a house fire and possible death or injury of the occupants. The SSP recognizes that the risk of the new toaster system is too high and recommends that certain safety features be incorporated, such as an over-temperature sensor with automatic system shutdown. These safety features will prevent potential fires, along with preventing the penalty costs associated with the predicted mishaps. However, the total number of house fires actually prevented by the new design might not ever be known or appreciated (thus, safety has an intangible value).

The cost of safety is ultimately determined by how much system developers are willing to invest or pay. Are they willing to pay little or nothing and take a higher risk, or pay a reasonable value and take a lower risk? There are many stakeholders involved, yet some of the stakeholders, such as the final user, often have nothing to say in the final decision for how much to spend on safety or the risk accepted. The user should receive the commensurate level of safety he or she expects. One problem is that quite often, the system developer is a different organization than the system user. The developer can save money by not properly investing in safety, which results in un-safety penalty costs being transferred to the system user, rather than the developer.

To some degree, ethics is an influential factor in the cost of safety because ethics sometimes controls how much safety is applied during system development. Is it unethical for a system developer to *not* make a system as reasonably safe as possible, or practical? Is it unethical for a system developer to make a larger profit by not properly investing in safety, and passing the risk and penalty costs on to the user?

Safety cost revolves around the age-old question of "how safe is safe." The system should be made as safe as possible, and the safety investment cost should be less than the potential penalty costs. Perhaps the best way to determine how much an SSP should cost is to add up the potential penalty costs and use some percentage of that amount. Unfortunately, this approach requires a significant amount of effort in analysis and future speculation that most programs are not willing to expend.

Another advantage of performing a good SSP is liability protection. Quite often, courts are awarding zero or limited liability following a mishap if it can be shown that a reasonable SSP was implemented and followed. Or, if it can be shown that an SSP was not implemented, the defense against liability is significantly damaged.

There is no magic number for how much to invest in system safety or to spend on an SSP. Complex, safety-critical, and high-consequence systems should understandably receive more system safety budget than simple and benign systems. Risk versus mishap costs and losses should be the determining factor.

THE HISTORY OF SYSTEM SAFETY

From the beginning of mankind, safety seems to have been an inherent human genetic element or force. The Babylonian Code of Hammurabi states that if a house falls on its occupants and kills them, the builder shall be put to death. The Bible established a set of rules for eating certain foods, primarily because these foods were not always safe to eat, given the sanitary conditions of the day. In 1943, the psychologist Abraham Maslow proposed a five-level hierarchy of basic human needs, and safety was number two on this list. System safety is a specialized and formalized extension of our inherent drive for safety.

The system safety concept was not the invention of any one person, but rather a call from the engineering community, contractors, and the military, to design and build safer systems and equipment by applying a formal proactive approach. This new safety philosophy involved utilizing safety engineering technology, combined with lessons learned. It was an outgrowth of the general dissatisfaction with the fly-fix-fly, or safety-by-accident, approach to design (i.e., fix safety problems after a mishap has occurred) prevalent at that time. System safety as we know it today began as a grass roots movement that was introduced in the 1940s, gained momentum during the 1950s, became established in the 1960s, and formalized its place in the acquisition process in the 1970s.

The first formal presentation of system safety appears to be by Amos L. Wood at the Fourteenth Annual Meeting of the Institute of Aeronautical Sciences (IAS) in New York in January 1946. In a paper titled "The Organization of an Aircraft Manufacturer's Air Safety Program," Wood emphasized such new and revolutionary concepts as:

- Continuous focus of safety in design
- Advance analysis and postaccident analysis
- Safety education
- Accident preventive design to minimize personnel errors
- Statistical control of postaccident analysis

Wood's paper was referenced in another landmark safety paper by William I. Stieglitz titled "Engineering for Safety," presented in September 1946 at a special meeting of the IAS and finally printed in the IAS Aeronautical Engineering Review in February 1948. Mr. Stieglitz's farsighted views on system safety are evidenced by the following quotations from his paper:

Safety must be designed and built into airplanes, just as are performance, stability, and structural integrity. A safety group must be just as important a part of a manufacturer's organization as a stress, aerodynamics, or a weights group. ...

Safety is a specialized subject just as are aerodynamics and structures. Every engineer cannot be expected to be thoroughly familiar with all developments in the field of safety any more than he can be expected to be an expert aerodynamicist.

The evaluation of safety work in positive terms is extremely difficult. When an accident does not occur, it is impossible to prove that some particular design feature prevented it.

The need for system safety was motivated through the analysis and recommendations resulting from different accident investigations. For example, on May 22, 1958, the Army experienced a major accident at a NIKE-AJAX air defense site near Middletown, New Jersey, which resulted in extensive property damage and loss of lives to Army personnel. The accident review committee recommended that safety controls through independent reviews and a balanced technical check be established, and that an authoritative safety organization be established to review missile weapon systems design. Based on these recommendations, a formal system safety organization was established at Redstone Arsenal in July 1960, and Army Regulation 385-15, "System Safety," was published in 1963.

As a result of numerous Air Force aircraft and missile mishaps, the Air Force also became an early leader in the development of system safety. In 1950, the United States Air Force (USAF) Directorate of Flight Safety Research (DFSR) was formed at Norton Air Force Base (AFB), California. It was followed by the establishment of safety centers for the Navy in 1955 and for the Army in 1957. In 1954, the DFSR began sponsoring Air Force-industry conferences to address safety issues of various aircraft subsystems by technical and safety specialists. In 1958, the first quantitative system safety analysis effort was undertaken on the Dyna-Soar X-20 manned space glider.

The early 1960s saw many new developments in system safety. In July 1960, an Office of System Safety was established at the USAF Ballistic Missile Division (BMD) at Inglewood, California. BMD facilitated both the pace and direction of system safety efforts when in April 1962 it published the first system-wide safety specification titled Ballistic System Division (BSD) Exhibit 62-41, "System Safety Engineering: Military Specification for the Development of Air Force Ballistic Missiles." The Naval Aviation Safety Center was among the first to become active in promoting an inter-service system safety specification for aircraft, BSD Exhibit 62-82, modeled after BSD Exhibit 62-41. In the fall of 1962, the Air Force Minuteman Program Director, in another system safety first, identified system safety as a contract deliverable item in accordance with BSD Exhibit 62-82.

The first formal SSPP for an active acquisition program was developed by the Boeing Company in December of 1960 for the Minuteman Program. The first military specification (Mil-Spec) for safety design requirements, missile specification MIL-S-23069, "Safety Requirements, Minimum, Air Launched Guided Missiles," was issued by the Bureau of Naval Weapons on October 31, 1961. In 1963, the Aerospace System Safety Society (now the System Safety Society) was founded in the Los Angeles area. In 1964, the University of Southern California's Aerospace Safety Division began a master's degree program in Aerospace Operations Management from which specific system safety graduate courses were developed. In 1965, the University of Washington and the Boeing Company jointly held the first official System Safety Conference in Seattle, Washington. By this time, system safety had become fully recognized and institutionalized.

For many years, the primary reference for system safety has been MIL-STD-882, which was developed for Department of Defense (DoD) systems. It evolved from BSD Exhibit 62-41 and MIL-S-38130. BSD Exhibit 62-41 was initially published in April 1962 and again in October 1962; it first introduced the basic principles of safety, but was narrow in scope. The document applied only to ballistic missile systems, and its procedures were limited to the conceptual and developmental phases from initial design to and including installation or assembly and checkout. However, for the most part, BSD Exhibit 62-41 was very thorough; it defined requirements for systematic analysis and classification of hazards and the design safety order of precedence used today. In addition to engineering requirements, BSD Exhibit 62-41 also identified the importance of management techniques to control the system safety effort. The use of a system safety engineering plan and the concept that managerial and technical procedures used by the contractor were subject to approval by the procuring authority were two key elements in defining these management techniques.

In September 1963, the USAF released MIL-S-38130. This specification broadened the scope of the system safety effort to include "aeronautical, missile, space, and electronic systems." This increase of applicable systems and the concept's growth to a formal Mil-Spec were important elements in the growth of system safety during this phase of evolution. Additionally, MIL-S-38130 refined the definitions of hazard analysis. These refinements included system safety analyses: system integration safety analyses, system failure mode analyses, and operational safety analyses. These analyses resulted in the same classification of hazards, but the procuring activity was given specific direction to address catastrophic and critical hazards.

In June 1966, MIL-S-38130 was revised. Revision A to the specification once again expanded the scope of the SSP by adding a system modernization and retrofit phase to the defined life-cycle phases. This revision further refined the objectives of an SSP by introducing the concept of "maximum safety consistent with operational requirements." On the engineering side, MIL-S-38130A also added another safety analysis: the Gross Hazard Study (now known as the Preliminary Hazard Analysis). This comprehensive qualitative hazard analysis was an attempt to focus attention on hazards and safety requirements early in the concept phase and was a break from other mathematical precedence. But changes were not just limited to introducing new analyses; the scope of existing analyses was expanded as well. One example of this was the operating safety analyses, which would now include system transportation and logistics support requirements as well. The engineering changes in this revision were not the only significant changes. Management considerations were highlighted by emphasizing management's responsibility to define the functional relationships and lines of authority required to "assure optimum safety and to preclude the degradation of inherent safety." This was the beginning of a clear focus on management control of the SSP.

MIL-S-38130A served the DoD well, allowing the Minuteman Program to continue to prove the worth of the system safety concept. By August 1967, a tri-service review of MIL-S-38130A began to propose a new standard that would clarify and formalize the existing specification as well as provide additional guidance to industry. By changing the specification to a standard, there would be increased program emphasis and accountability, resulting in improved industry response to SSP requirements. Some specific objectives of this rewrite were: obtain a system safety engineering program plan early in the contract definition phase, and maintain a comprehensive hazard analysis throughout the system's life cycle.

In July 1969, MIL-STD-882 was published, entitled "System Safety Program for Systems and Associated Subsystems and Equipment: Requirements for." This landmark document continued the emphasis on management and continued to expand the scope to apply to all military services in the DoD. The full life-cycle approach to system safety was also introduced at this time. The expansion in scope required a reworking of the system safety requirements. The result was a phase-oriented program that tied safety program requirements to the various phases consistent with program development. This approach to program requirements was a marked contrast to earlier guidance, and the detail provided to the contractor was greatly expanded. Since MIL-STD-882 applied to both large and small programs, the concept of tailoring was introduced, thus allowing the procuring authority some latitude in relieving the burden of the increased number and scope of hazard analyses. Since its advent, MIL-STD-882 has been the primary reference document for system safety.

The basic version of MIL-STD-882 lasted until June 1977, when MIL-STD-882A was released. The major contribution of MIL-STD-882A centered on the concept of risk acceptance as a criterion for SSPs. This evolution required introduction of hazard probability and established categories for frequency of occurrence to accommodate the long-standing hazard severity categories. In addition to these engineering developments, the management side was also affected. The responsibilities of the managing activity became more specific as more emphasis was placed on contract definition.

In March 1984, MIL-STD-882B was published, reflecting a major reorganization of the A version. Again, the evolution of detailed guidance in both engineering and management requirements was evident. The task of sorting through these requirements was becoming complex, and more discussion on tailoring and risk acceptance was expanded. More emphasis on facilities and off-the-shelf acquisition was added, and software was addressed in some detail for the first time. The addition of Notice 1 to MIL-STD-882B in July 1987 expanded software tasks and the scope of the treatment of software by system safety. With the publication in January 1993 of MIL-STD-882C, hardware and software were integrated into system safety efforts. The individual software tasks were removed so that a safety analysis would include identifying the hardware and software tasks together as an integrated systems approach.

The mid-1990s brought the DoD acquisition reform movement which included the Military Specifications and Standards Reform (MSSR) initiative. Under acquisition reform, program managers are to specify system performance requirements and leave the specific design details up to the contractor. In addition, the use of Mil-Specs and standards would be kept to a minimum. Only performance-oriented military documents would be permitted. Other documents, such as contractual item descriptions and industry standards, are now used for program details. Because of its importance, MIL-STD-882 was allowed to continue as a MIL-STD, as long as it was converted to a performance-oriented MIL-STD practice. This was achieved in MIL-STD-882D, which was published as a DoD Standard Practice in February 2000.

SYSTEM SAFETY GUIDANCE

System safety is a process that is formally recognized internationally and is used to develop safe systems in many countries throughout the world. MIL-STD-882 has long been the bedrock of system safety procedures and processes; the discipline tended to grow and improve with each improvement in MIL-STD-882. In 1996, the Society of Automotive Engineers (SAE) established Aerospace Recommended Practice ARP-4754, Certification Considerations for Highly-Integrated or Complex Aircraft Systems. This standard emulates the system safety process for the FAA certification of commercial aircraft. In 2009, the system safety process was formally documented in an American National Standards Institute (ANSI) Standard, ANSI/ Government Electronics and Information Technology Association GEIA-STD-0010-2009, Standard Best Practices for System Safety Program Development and Execution, February 12, 2009.

SYNOPSIS

We live in a perilous world composed of many different hazards that present risk for potential mishaps. Hazards and risk are inevitable; one cannot live life without exposure to hazards. However, mishaps are not inevitable—they can be controlled. This perilous world we live in is a world composed of technological systems. When viewed from an engineering perspective, most aspects of life involve interfacing with systems of one type or another. For example, consider the following types of systems we encounter in daily life: toasters, television sets, homes, electrical power, electrical power grid, and hydroelectric power plant. Commercial aircraft are systems that operate within a larger national transportation system, which in turn operate within a worldwide airspace control system. The automobile is a system that interfaces with other systems, such as other vehicles, fuel filling stations, highway systems, bridge systems. Everything can be viewed as a system at some level, and the unique interconnectedness and complexity of each system presents special challenges for safety. Hazards tend to revolve around systems and processes within these systems; system components fail and wear out in many diverse ways; humans are susceptible to performing erroneous system actions.

Systems can, and will, fail in various different key modes, thereby contributing to hazards. Harm, damage, and losses result from actualized hazards that become accidents (mishaps). Since many systems and activities involve hazard sources that cannot be eliminated, zero mishap risk is often not possible. System safety was established as a systems approach to safety, where safety is applied to an entire integrated system design as opposed to a single component. System safety takes a sum of the parts view rather than an individual component view. It has been demonstrated over the years that the best way to develop and operate safe systems is to apply the system safety methodology to the design and operation of the complete system as a whole, as opposed to applying safety to isolated individual components. A structured disciplined approach is necessary to develop system designs that can counter and tolerate failures and errors in safety-related situations.

At the micro-safety level, system safety is a trade-off between safety cost and risk for eliminating or controlling known hazards. However, at the macro-safety level, system safety is more than just cost versus risk; it is also a matter of safety culture, integrity, and ethics. Should an organization decide how much risk they are willing to pay for and then pass that risk to the user, or are they obligated to provide risk acceptable to the user?

Is system safety worth the cost and effort? Typically, the cost of safety is much less than the cost of not making the system/product safe, as the mishaps that may result can be quite expensive. *Safety must be earned* through the system safety process; it cannot be achieved by accident, chance, or luck. Safety is not free, but it costs less than the direct, indirect, and hidden costs of mishaps. System safety is necessary because hazards almost always exist within product and system designs. For various (and valid) reasons, some hazards can be eliminated when recognized, while others must be allowed to persist; these residual hazards, however, can be reduced to an acceptable level of risk.

CHAPTER 2

System Safety Terms and Concepts

ABORT

An abort is the premature termination of the mission, operation, function, procedure, task, and so on before intended completion. An abort is generally necessary because something has unexpectedly gone wrong, and it is necessary to abort in order to avoid a potentially impending mishap, an unsafe state, or a high-risk state. Depending upon the system conditions at the time, an abort will effect transition from an unsafe system state to a safe state. If transition directly to a safe state is not possible, then it may be necessary to continue operation in a reduced capacity mode and switch to alternate contingency plans. Abort termination is an intentional decision and should be based on preestablished criteria.

In system safety, special consideration must be given to the possible need for an abort, and the mechanisms implementing an abort, when designing system operations. In other words, safe system operation necessitates that abort contingencies be considered and planned in advance of actual operation, particularly for safety-critical functions (SCFs) and operations. This involves identifying potential abort scenarios: the functions that may require an abort, the factors that cause the need for an abort, and the methodology for initiating and safely conducting the abort. Timing of events may be a critical safety factor (SF). Warning, escape, and protection mechanisms are also important safety considerations when developing abort contingency plans.

An abort should only be set into motion by the issuance of a valid abort command from an "authorized entity." This requirement is particularly germane to unmanned systems (UMSs). In a manned system, the authorized entity is generally the official operator; however, in a UMS much planning must go into determining what constitutes a valid command and a valid controller (authorized entity) because it may be possible for an unauthorized entity to take control of the UMS.

Concise Encyclopedia of System Safety: Definition of Terms and Concepts, First Edition. Clifton A. Ericson II.
© 2011 by John Wiley & Sons, Inc. Published 2011 by John Wiley & Sons, Inc.

Example 1: The pilot decided to abort the intended mission and return to base after one of the two engines on the aircraft failed. In this case, aircraft safety is reduced, and preestablished contingency plans require the pilot to terminate the mission rather than expose the aircraft to higher mishap risk and loss of the aircraft. In this situation, the pilot is the authorized entity.

Example 2: An aircraft weapon system enters the abort state if any unsafe store conditions are detected in other states, and/or if any abnormal conditions that preclude completion of the normal initialization and release sequence occur. An abort command is issued to the store, and all station power is subsequently removed. If transition to the abort state occurs after the launch state has been entered (and irreversible functions have been initiated), power is removed from the store interface, and no further attempts to operate the store are conducted during the ongoing mission. In this case, the system itself is the authorized entity.

ABNORMAL OPERATION

Abnormal operation (of a system) is system behavior which is not in accordance with the documented requirements and expectations under normally conceivable conditions. Abnormal system operation is when a disturbance, or disturbances, causes the system to deviate from its normal operating state. The effects of abnormal operation can be minimal or catastrophic.

System safety views abnormal operation as the result of failures and/or errors experienced during operations. Failures and/or errors change system operating conditions from normal to abnormal, thereby causing system disturbances and perturbations. These factors typically involve hazards and mishaps. Since many systems require the inclusion of various hazard sources, they are naturally susceptible to potential mishaps under abnormal (fault) operating conditions, where the abnormal conditions are caused by failures, errors, malfunctions, extreme environments, and combinations of these factors.

Factors that can cause operational disturbances and perturbations include:

- Human error
- Hardware failures
- Software errors
- Secondary effects from other sources, such as from electromagnetic radiation (EMR)
- Sneak circuit paths resulting from design flaws

ABOVE GROUND LEVEL (AGL)

In aviation an altitude is said to be AGL when it is measured with respect to the underlying ground surface. This is as opposed to above mean sea level (AMSL). The expressions AGL and AMSL indicate where the "zero level" or

18   SYSTEM SAFETY TERMS AND CONCEPTS

"reference altitude" is located. A pilot flying a plane under instrument flight rules must rely on its altimeter to decide when to deploy the aircraft landing gear, which means the pilot needs reliable information on the altitude of the plane with respect to the airfield. The altimeter, which is normally a barometer calibrated in units of distance instead of atmospheric pressure, must therefore be set in such a way as to indicate the altitude of the craft above ground. This is done by communicating with the control tower of the airport (to get the current surface pressure) and setting the altimeter so as to report a null altitude AGL on the ground of that airport. Confusing AGL with AMSL, or improperly calibrating the altimeter may result in controlled flight into terrain (CFIT).

ACCEPTABLE RISK

Acceptable risk is risk that is known, understood, and judged to be acceptable by an individual or a group. There are many different risk types that can be evaluated and accepted, such as cost risk, schedule risk, investment risk, programmatic risk, and safety risk. It should be noted that if risk is not identified or communicated, it is accepted by default without knowledge. An accepted risk does not necessarily mean that the risk has been made as low as potentially possible.

Typically, acceptable risk refers to the risk presented by a hazard that is considered to be acceptable by a risk decision authority without further risk reduction (i.e., residual risk). It is the accepted level of risk based on preestablished criteria that delineate risk levels, acceptance decision criteria, and decision authority levels. Accepted risk does not necessarily equate to the lowest risk possible; it may be a high risk that must be accepted for various program reasons or mission needs. The system, system user, equipment, and environment are consciously exposed to this risk. Accepting residual risk is a difficult, yet necessary task, which is often negotiated or tempered with competing factors such as cost, benefit, schedule, and operational effectiveness.

System safety is a mishap risk management process, whereby mishap risk is identified through hazards, and if the risk does not meet the established level of acceptability, design action is taken to reduce the risk to an acceptable level. For various reasons, it is often impossible to eliminate mishap risk in many systems. System safety should be involved in establishing the criteria and constraints for acceptable risk for system programs. Military standard (MIL-STD)-882 identifies the criteria for four levels of risk: high, serious, medium, and low, each of which is accepted by a different level of decision authority.

Stating that mishap risk for a particular hazard is acceptable can be misleading if not thoroughly defined. If a high-ranking authority accepts a high-risk hazard because a lower-ranking person cannot, does that really make the system suitably safe, or does it discount the risk? It may be more ethical and cost-effective to state that the potential mishap risk presented by a particular hazard has been reduced to the lowest level possible, or reasonably practical. Some questions that should be answered include:

- Is the risk known and accepted by the actual system/product user?
- Is the amount of accepted risk appropriate for the operations involved?
- Does allowing a higher-level authority accept a higher risk provide an ethical level of safety?

See *As Low as Reasonably Practicable (ALARP)*, *Hazard*, *Mishap Risk*, *Residual Risk*, and *Risk* for additional related information.

ACCEPTANCE TEST

An acceptance test is a single test, or suite of tests, performed on a product to determine its performance acceptability. A product can be accepted or rejected based on how well the product's performance compares with preestablished acceptance criteria. An acceptance test is the validation process that demonstrates that hardware and/or software is acceptable for use. It provides the basis for delivery of an item under terms of a contract, where the contract should specify acceptance criteria. It also serves as a quality control screen to detect deficiencies.

System safety is generally involved in the acceptance test process. Safety may review test results to ensure safety concerns are adequately resolved and that design safety requirements are met. Safety may also review acceptance test procedures to ensure that no test hazards exist and that the potential mishap risk of a test is acceptable. It may be necessary for safety to be present as a witness to the conduct of certain acceptance tests, generally on safety-critical components.

ACCIDENT

Many dictionaries define accident as "an event occurring by chance or unintentionally." An accident is also defined as "an unplanned event that results in a harmful outcome; for example, death, injury, occupational illness, or major damage to or loss of property" (System Safety Handbook: Practices and Guidelines for Conducting System Safety Engineering and Management, Federal Aviation Administration [FAA], December 2000). An accident event is undesired, unintentional, and results in negative consequences.

An accident is an unplanned event, or events, that results in an outcome culminating in death, injury, damage, harm, and/or loss. An accident is the event and associated outcome that results from an actualized hazard. There is a direct relationship between hazards and accidents. For example, the crash of an aircraft is an accident event, with death of the occupants and loss of the aircraft the resulting outcome. This accident could be the result of a hazard such as insufficient fuel during flight to allow for a safe landing.

In system safety, the terms *accident* and *mishap* are synonymous, and no distinction should be made between the two terms. Under the definition of a mishap, MIL-STD-882C equates accident and mishap as synonymous. In system safety parlance, the term *mishap* has become the preferred term (rather than *accident*) primarily to maintain a consistency of terminology when discussing mishap risk. The term *incident* is related to accident and mishap but has a slightly different meaning.

See *Hazard*, *Incident*, *Mishap*, and *Risk* for additional related information.

ACCIDENT CAUSE

Accident cause refers to the causal factors resulting in an accident or mishap. To better understand accident causes, see the definition for *Hazard*, which describes the components of a hazard. Accidents and mishaps cannot occur unless a hazard preexists and the hazard components establish the associated mishap causal factors.

See *Hazard* for additional related information.

ACCIDENT COST

Accident costs are the direct and indirect costs associated with an accident or mishap. These costs can be in terms of deaths, injuries, dollars, equipment loss, and environmental damage. Direct costs include costs such as medical costs, absenteeism costs, equipment costs, and legal costs. Indirect costs include costs such as product devaluation and loss of reputation.

ACCIDENT INVESTIGATION

Accident investigation is the determination by qualified personnel as to the specific causing for a particular accident or mishap. Causal factor considerations include management errors, technical design, hardware failures, procedural errors, and so on. An accident investigation can be conducted by a formal board or by an informal analysis performed by one or more individuals.

ACCIDENT SCENARIO

An accident scenario is the aggregation of factors, conditions, and events that together ultimately lead to the occurrence of an accident (mishap) event. It may be a combination of failures, errors, events, and/or conditions that occur together to finally cause a resultant accident. The sequence of events begins with an initiating event (IE) and is (usually) followed by one or more pivotal events (PEs) that lead to the undesired end accident event. Since an accident (or mishap) is an actualized hazard, an accident scenario can be defined by a hazard description. IE and PEs are part of the hazard and should be described in the hazard description. In some situations, the specific accident outcome can vary depending upon which PEs occur and which do not. These lead to the concept of a family of hazards, whereby the base hazard is the same, but the scenario and outcome come vary depending on which PEs actually occur. The term accident scenario is used in event tree analysis (ETA).

See *Event Tree Analysis (ETA)*, *Initiating Event (IE)*, and *Pivotal Event (PE)* for additional related information.

AMERICAN STANDARD CODE FOR INFORMATION INTERCHANGE (ASCII)

The ASCII is a character-encoding scheme based on the ordering of the English alphabet. ASCII codes represent text in computers, communications equipment, and other devices that use text. ASCII includes definitions for 128 characters: 33 are nonprinting control characters that affect how text and space is processed, 94 are printable characters, and the space character is considered an invisible graphic. For example, the ASCII code for the letter "A" is 65 and for the letter "a" is 97.

ACTION

An "action" is a product of an audit; it is an assignment to an organization or person, with a date for completion, to correct a finding identified during the audit. Typical categories of items resulting from an audit include: compliance, finding, observation, issue, and action.

See *Audit* for additional related information.

ACTUATOR

An actuator is an electromechanical device that is a mechanism used to effect or produce motion. It is a power mechanism that typically converts electrical, hydraulic, or pneumatic energy into longitudinal or rotary motion. Actuators are used for many purposes, such as moving flight control surfaces and opening/closing valves. Whenever actuators are used in a system design, they should always be considered as potential hazard sources or hazard initiating mechanisms (IMs), which require further safety evaluation. Safety consideration should be given to failure modes such as fails to operate, operates incorrectly, and operated inadvertently. Failure or incorrect operation of an actuator could

22 SYSTEM SAFETY TERMS AND CONCEPTS

be the IM for a particular hazard where correct actuator operation is necessary for a particular critical system function, such as flap control. For example, an aircraft flight control actuator provides the force and direction to physically move the flight control surface, such as an aileron. If the actuator fails or functions incorrectly, this will cause loss of flight control and possible loss of the aircraft.

ADA

Ada is a high-level computer programming language designed by a team under contract to the United States Department of Defense (DoD). It was developed from 1977 through 1983 and was named after Ada Lovelace, who is often credited with being the first computer programmer. The intent of the DoD was to standardize all projects on one single language rather than supporting the many already in DoD usage. It was also the goal of the DoD to develop a modern language that would be correct and consistent through the special validation of Ada compilers. Validation provides for safety and reliability in mission-critical applications, such as avionics software.

Notable features of Ada include strong typing, modularity mechanisms (packages), run-time checking, parallel processing (tasks), exception handling, and generics. Ada 95 added support for object-oriented programming, including dynamic dispatch. Ada supports run-time checks in order to protect against access to unallocated memory, buffer overflow errors, off-by-one errors, array access errors, and other avoidable bugs. These checks can be disabled in the interest of run-time efficiency but can often be compiled efficiently. It also includes facilities to help program verification. For these reasons, Ada is widely used in critical systems, where any anomaly might lead to very serious consequences, that is, accidental death or injury. Examples of systems where Ada is used include aircraft avionics, train systems, weapons, and spacecraft.

There are some safe subsets of Ada, such as Spark Ada, which are more restricted for special safety applications and use on safety-critical systems. These subsets restrict some of the standard language functions and syntax that could be somewhat risky because they are difficult to verify, such as multithread processing, dynamic memory allocation, and recursion.

AEROSPACE

Aerospace denotes applications that are designed to operate in both airborne and space environments.

AIRBORNE

Airborne denotes applications peculiar to aircraft and missiles and other systems designed to operate within the Earth's atmosphere environment.

AIRCRAFT AIRWORTHINESS

Aircraft airworthiness is the demonstrated capability of an aircraft, aircraft subsystem, or aircraft component to function satisfactorily when used within prescribed limits on the aircraft. Generally, flight clearances are prepared and issued to indicate an aircraft is deemed airworthy for testing.

See *Airworthiness* and *Flight Clearance* for additional related information.

AIRCRAFT AIRWORTHINESS AUTHORITY

Aircraft Airworthiness Authority is a term used to describe the organization responsible for determining the safety suitability and effectiveness of parts that go into aviation (aircraft) systems. This organization determines if the aircraft, and/or it parts, are suitable for safe flight (i.e., are they airworthy). System safety is usually involved in evaluating airworthiness for systems and equipment from a safety perspective and in ensuring that the Aircraft Airworthiness Authority's safety standards are satisfied. The Airworthiness Authority for commercial aircraft in the United States is the FAA. The Civil Aviation Authority (CAA) is the Airworthiness Authority for commercial aircraft in Europe. The Air Force, Army, and Navy have their own internal airworthiness authorities.

AIRCRAFT SAFETY CARD

An aircraft safety card is a card instructing airline passengers about the procedures for dealing with various emergency conditions that might arise during the flight. The safety cards are usually provided by airlines on all commercial flights, usually located in the back of the seat in front of each passenger. The contents are usually in the form of pictures, graphically illustrating such procedures as buckling the seatbelts, bracing for impact in an airplane crash, dealing with depressurization, opening the emergency exit door, or inflating life rafts in the event of a water landing. The graphic representation allows the cards to be understood regardless of language or reading ability.

AIRWORTHINESS

Airworthiness is generally known as the capability for "fitness to fly." Airworthiness is the demonstrated capability of an aircraft, aircraft subsystem, or aircraft component to function satisfactorily when used within prescribed limits on the aircraft. Generally, flight clearances are prepared and issued for hardware and software documenting their level of airworthiness, which allows (or disallows) for flight with or without limitations. The demonstrated condition is usually achieved by a combination of analysis and testing. An airworthiness flight clearance certificate states that an item is approved for flight and there is evidence that it will operate in a safe manner to accomplish its intended function. It should be noted that test and support equipment also fall into the airworthiness flight clearance certification process when used aboard an aircraft, particularly test support equipment.

Military Handbook (MIL-HDBK)-516B, Airworthiness Certification, DoD Handbook, delineates the U.S. military aircraft airworthiness certification process. It also contains a list of the typical certification data required for airworthiness certification, which leads to a flight clearance. It includes a section on system safety, a system safety program (SSP), and a software safety program (SwSP). System safety generally provides the safety evaluation of hardware and software for airworthiness flight clearance recommendations. System safety is involved in ensuring that the Aircraft Airworthiness Authority's standards are satisfied for the airworthiness authority involved.

See *Flight Clearance* and *Safety of Flight (SOF)* for additional information.

ALL-UP-ROUND (AUR)

An AUR is a completely assembled ammunition intended for delivery to a target or configured to accomplish its intended mission. This term is identical to the term "all-up-weapon." System safety is concerned with weapons and munitions in the final AUR configuration (i.e., as a system). The AUR configuration provides a system architecture and design that can be analyzed for hazards and potential mishap risk.

ALGORITHM

An algorithm is a prescribed set of well-defined rules or procedures for solving a problem. Algorithms are typically used in computer programs. For example, there are several different algorithms for sorting a list of numbers or characters, such as the Bubble sort, Insertion sort, Shell sort, Merge sort, Heapsort, and Quicksort. Each sort algorithm has advantages and disadvantages.

AMMUNITION

Ammunition is an item containing one or more projectiles, together with propellant needed to impart velocity to the projectile(s) which are propelled from a reusable launcher. The projectiles may be an inert or contain a high explosive, smoke generator, or other energetic composition. The launcher may be a gun. The North Atlantic Treaty Organization (NATO) and U.S. term *ammunition* is covered by the term *munitions*; ammunition is a subset of munitions. Ammunition, or ammo, is defined as a device charged with explosives, propellants, pyrotechnics, initiating elements, compositions, riot control agents, chemical herbicides, smoke, or flame used in connection with defense or offense, demolition, training, ceremonial, or nonoperational purposes. Ammo includes cartridges, bombs, projectiles, grenades, mines, pyrotechnics, bullets, primers, propellants, fuzes, detonators, torpedoes, mines, initiators, and propelling charges.

An SSP for a system containing ammunition must always take special safety precautions and include an element of explosives safety. Ammunition and explosives (A&Es) are a primary hazard source that should always be considered in a hazard analysis (HA).

See *Explosives Safety* and *Munitions* for additional related information.

AMMUNITION AND EXPLOSIVES (A&Es)

A&E covers any non-nuclear ordnance, ammunition, explosive, or explosive material/item/device/hazardous waste classed or being developed to be classed as a United Nations Organization (UNO) Class 1, Divisions 1 through 6 item. A&E does not constitute devices filled with chemical or biological material. One of the means to describe or identify A&E is to categorize an item by its intended use. Usage is described as service (wartime), practice or training, exercise (recoverable items such as torpedoes and mines), inert items used for demonstration handling drills or displays, drill items (shaped like service items but nonexplosive), empty/used (being returned for reuse), and non-service (shapes). A&E materials are colored coded according to their use as described in Naval Sea Systems Command (NAVSEA) Ordnance Pamphlet (OP) 2238/ NAVAIR Manual 11-1-117.

ANALYSIS OF VARIANCE (ANOVA)

ANOVA is a statistical method for segregating the sources of variability affecting a set of observations to determine if certain factors associated with each variable contributed to the variance. ANOVA is a collection of statistical models, and their associated procedures, in which the observed variance is partitioned into components due to different explanatory variables. In its simplest form, ANOVA provides a statistical test of whether or not the means of several groups are all equal, and therefore generalizes to more than two groups. ANOVA typically provides a measure of the variance due to each of the sources of variation.

ANALYSIS TECHNIQUE

Analysis technique refers to the various safety analysis techniques or methodologies available for safety analysis and HA.

See *Safety Analysis Technique* for further information.

ANALYSIS TYPE

Analysis type refers to the various safety analysis types available for safety analysis and HA.
See *Safety Analysis Type* for further information.

AND GATE

An AND gate is a logic gate used in Fault Tree Analysis (FTA). The AND gate logic states that a gate output only occurs when all of the gate inputs occur together simultaneously. See Figure 2.27 below for an example of an AND gate.

ANOMALOUS BEHAVIOR

Anomalous behavior is deviation from a general rule or regime; it is inconsistent behavior deviating from what is usual, normal, or expected. In systems applications, anomalous behavior is behavior of an item that is not in accordance with the documented requirements or specifications for the item. It is unexpected behavior, and it is generally undesired behavior. System safety is concerned with anomalous behavior because it can typically be the cause of a mishap, or be part of a sequence of events leading to a mishap. Safety HA typically looks for anomalous behavior that is safety-essential or safety-critical in outcome. When performing an HA, potential anomalous behavior should be considered in hardware actions, hardware functions, software functions, and human tasks. For example, inadvertent rudder movement on an aircraft, without a valid command to move, is considered anomalous behavior with safety significance.

ANOMALY

An anomaly is a deviation from the expected behavior; it is anomalous behavior. In systems theory, an anomaly is a system action, state, or condition that is not expected or intended. It may or may not be hazardous, but it is generally the result of a failure, error, or combinations of failures and errors. For example, uncommanded rudder movement on an aircraft is an anomaly that is referred to as anomalous behavior. Identifying and evaluating potential anomalies and anomalous system behavior is one of the basic steps in HA. It is important to know and understand potential anomalous behavior and the risk presented.
See *Anomalous Behavior* for additional related information.

ANTHROPOMETRICS

Anthropometrics involves the quantitative descriptions and measurements of the physical body variations in people. These metrics are useful in human factors design and human engineering. System designs that do not appropriately account for anthropometric considerations can often lead to safety problems.

APERTURE

An aperture is an opening in the protective housing, shielding enclosure of a product that controls the amount of light admitted. In a laser system, the aperture controls the laser and collateral radiation that is emitted, which also allows human exposure to the radiation.

APPLICATIONS SOFTWARE

Application software is the software written to get the machine to perform a particular task. The way it is written and what it can do will depend on the design of the software. Most application software is developed as part of a system development program. Some applications software is provided by commercial off-the-shelf (COTS) software.

APPLICATION-SPECIFIC INTEGRATED CIRCUIT (ASIC)

An ASIC is an integrated circuit (IC) customized for a particular use, rather than intended for general-purpose use.

ARCHITECTURE

Architecture is the organizational structure of a system or software design, identifying its components, their interfaces, and concept of execution among them. Architecture is a unifying or coherent form or structure of something, such as a system or a building; the manner in which the components of a system or building are organized and integrated. In systems theory, architecture is the unique manner in which the components of a system are organized and integrated. It is the structure, in terms of components, connections, and constraints, of a system, product, or process. It includes all elements of a system that determine its form and function. By definition, all systems have an architecture, explicit or implicit, which can often be viewed from different perspectives. Architecture is the organizational structure of a system that identifies its components, their interfaces, and the concept of execution between them. Architecture is effectively a specially developed system design.

System safety performs HAs and risk assessment on system architectures in order to identify hazards and potential mishap risk. A system hazard analysis (SHA) is essentially an HA of the overall system architecture, including its components and functions. System safety influences system architecture design by requiring the implementation of safety design features, such as redundancy, to reduce the overall mishap risk potential.

See *System Hazard Analysis (SHA)* for additional related information.

ARM

In systems theory, to "arm" a device or function is to facilitate the ability of the device or function to perform its intended operation. The item that requires arming is typically a potential hazardous device or function. Interlocks are placed in its functional path to prevent inadvertent operation, and arming removes these interlocks just prior to intended operation. Arm is a general term that implies the energizing of electronic and electrical circuitry, which in turn controls power sources or other components used to initiate explosives. The arming operation completes all steps preparatory to electrical initiation of explosives except the actual fire signal. A design utilizing an arm function is also used for command and control functions, as well as other hazardous functions or operations. In some system designs, the arm function is on a timer, which removes the arm capability if the fire function is not performed within the specified time limit. For example, a high-energy laser on a helicopter must be armed before it can be fired. Arming and firing is a two-stage sequential process. In addition, the arm function may only be valid for 30s (example) before it is placed back into the unarmed state (safing). This helps to prevent operator error resulting in unintended operation.

See *Armed*, *Arming Device*, and *Safing* for additional related information.

ARMED

"Armed" is the state of a device, subsystem, or system when all safety interlocks and switches have been made ineffective with the exception of the single function which would initiate the intended operation. Armed typically means that the arm function has been activated and the device is ready to be fired. For example, a missile is technically "armed" and ready to fire after the "arm" button has been pressed by the aircraft pilot; it is functionally armed when the arm mechanism has functioned. The pilot then presses the "fire" button to launch the missile. If the missile has not been armed first, it will not launch when the fire button is pressed.

See *Arm*, *Arming Device*, *Safe Separation*, and *Safing* for additional related information.

ARMING DEVICE

An arming device is a special device used to intentionally arm something, such as an explosive detonator or an electronic circuit. It is considered to be a safety device that prevents the intended function from occurring until the arming criteria have been met. Typical arming devices also provide a visual mode for observing if the device is in the armed or safe state. For example, a common warhead arming device criteria requires that two separate independent and mechanical environments are satisfied, such as pressure and velocity, before the device goes into the armed state. MIL-STD-1316E provides numerical probability requirements that must be met for fuze arming designs used for arming weapons. System safety must perform FTA to verify that these numerical requirements are met by the design.

See *Fuze* for additional related information.

ARTIFICIAL INTELLIGENCE (AI)

AI is the concept and application of giving machines decision-making capability that simulates human cognition. It involves the programming and ability of a robot to perform functions that are normally associated with human intelligence, such as reasoning, planning, problem solving, pattern recognition, perception, cognition, understanding, learning, speech recognition, and creative response. AI is an important factor in the design of autonomous UMSs and robotic systems. System safety is concerned with AI because if it is designed incorrectly or malfunctions, it can be the cause of a mishap, or be part of a sequence of events leading to a mishap. Safety HA typically looks for potential anomalous AI behavior that is safety-essential or safety-critical in outcome.

AS LOW AS REASONABLY PRACTICABLE (ALARP)

ALARP is an acronym for "as low as reasonably practicable." If a given risk can be shown to have been reduced to as low a level as is reasonably practicable, taking into consideration the costs and benefits of reducing it further, then it is said to be a tolerable risk. As the risk is further reduced, the less proportionately it is necessary to spend to reduce it further. The concept of diminishing risk proportion is shown by the triangle in Figure 2.1.

See *Acceptable Risk* for additional related information.

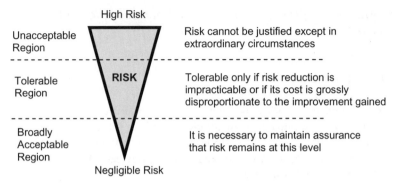

Figure 2.1 ALARP model.

ASSEMBLER OR ASSEMBLY LANGUAGE

Assembler is a computer programming language referred to as an assembly language. It is known as a low-level type language that corresponds closely to the instruction set of a given computer, allows symbolic naming of operations and addresses, and generally results in a one-to-one translation of program instructions into machine instructions. Assembly language implements a symbolic representation of the numeric machine codes and other constants needed to program a particular central processing unit (CPU) architecture. This representation is usually defined by the hardware manufacturer, and is based on abbreviations (called mnemonics) that help the programmer remember individual instructions, registers, and so on. A particular assembly language is specific to a physical or virtual computer architecture (as opposed to most high-level languages, most of which are portable).

Assembly languages were first developed in the 1950s, and were referred to as second-generation programming languages. They eliminated much of the error prone and time-consuming first-generation programming needed with the earliest computers, freeing the programmer from tedium such as remembering numeric codes and calculating addresses. However, by the 1980s, their use had largely been supplanted by high-level languages, in the search for improved programming productivity. Today, assembly language is used primarily for direct hardware manipulation, or to address critical performance issues. Typical uses are device drivers, low-level embedded systems, and real-time systems.

A program called an assembler is used to translate assembly language statements into the target computer's binary machine code. The assembler performs a more or less one-to-one mapping from mnemonic statements into machine instructions and data. Some newer assemblers have incorporated structured programming elements, such as IF/ELSE/ENDIF and similar control flow blocks. This was a way to reduce or eliminate the use of GOTO operations in assembly code, one of the main factors causing spaghetti code.

ASSEMBLY

An assembly is an integrated set of components and/or subassemblies that comprise a defined part of a subsystem, for example, the pilot's radar display console or the fuel injection assembly of an aircraft propulsion subsystem. An assembly can also be composed of a number of parts, subassemblies, or any combination thereof, joined together to perform a specific function and which can be disassembled without destruction of designed use (i.e., they can be reassembled). Typical examples of assemblies are power supplies, memory boards, switching devices, and so on. An assembly would be reflected as a specific level in a system hierarchy.

See *System Hierarchy* for additional related information.

ATTRIBUTE

An attribute is a quantitative or qualitative characteristic of a system or system element. For example, safety, reliability, and quality are system attributes that result as an emergent property of the final system design and manufacture.

AUTOMATIC TEST EQUIPMENT (ATE)

ATE is equipment that is designed to automatically conduct analysis of functional or static parameters and to evaluate the degree of unit under test (UUT) performance degradation, and may be used to perform fault isolation (FI) of UUT malfunctions. The decision making, control, and evaluative functions are conducted with minimum reliance on human intervention and are usually done under computer control.

AUDIT

An audit is an independent examination of a work product or set of work products to assess compliance with specifications, standards, contractual agreements, or other criteria. There are many different types of audits, such as functional configuration audit (FCA), physical configuration audit (PCA), software audit, and safety audit. The purpose is to conduct an independent review and examination of system records and activities in order to determine the adequacy and effectiveness of the work performed, to ensure compliance with established policy and operational procedures, and to recommend any necessary changes. An audit can be either a formal or informal review of a program, to determine if objectives and requirements have been met. An audit also involves identifying deficiencies, problems, and issues.

The following are typical categories of items resulting from an audit:

1. Compliance—A compliance is the complete satisfaction of an objective or requirement.
2. Finding—A finding is the identification of a failure to show compliance to one or more of the objectives or requirements. A finding may involve an error, deficiency, or other inadequacy. A finding might also be the identification of the nonperformance of a required task or activity.
3. Observation—An observation is the identification of a potential improvement. An observation is not a compliance issue and does not need to be addressed before approval.
4. Issue—An issue is a concern; it may not be specific to compliance or process improvement but may be a safety, system, program management, organizational, or other concern that is detected during the audit review.
5. Action—An action is an assignment to an organization or person, with a date for completion, to correct a finding identified during the audit.

In system safety an audit is typically performed on an SSP. The purpose of the audit is to determine if the SSP is on track and if all contractual requirements are being satisfied and that mishap risk is being properly identified, assessed, controlled, and accepted. A safety audit is typically conducted according to a preplanned schedule that is established in the System Safety Program Plan (SSPP). A safety audit typically consists of a review of the system developer, contractor, or subcontractor's documentation, hardware, and/or software to verify that it complies with project requirements and contractual requirements. A safety audit should be performed by someone not working on the program under review, that is, someone independent from the program. An audit trail is a chronological record of audit evidence that is sufficient to enable the reconstruction of the final results by an independent person or group of people.

See *Safety Audit* for additional related information.

AUTHORIZED ENTITY

An authorized entity is the individual operator or control element authorized to direct or control system or mission functions. It is the individual, or system control element, authorized for making command and control decisions. Manned systems typically have an operator that is authorized to make real-time operational decisions. However, for UMSs, there may not be an operator in some control elements; therefore, decisions must be made by internal authorized elements of the system itself. This is decision logic that must be programmed into the system which cannot be counterfeited by an external source. This is to ensure that an outside source cannot take over control of the system.

Authorization rules should be designed into the system. As UMSs evolve and increase in their level of autonomy, a system operator or human controller may not be a valid assumption; control may be completely relinquished to the UMS. Systems may use man-to-machine or machine-to-machine control. In this context, the term authorized entity is used to denote the entity which, by design, exercises immediate control over the UMS. The term is used to denote two methods of valid control: human or machine. A system safety design goal should be that the system, manned or unmanned, shall be designed to only respond to valid commands from the authorized entity or entities in order to preclude intrusion by hackers or counterfeiters.

AUTOIGNITION

Autoignition refers to the self-ignition of a substance or material. A system safety concern is the autoignition of safety-critical components, such as an explosive, solid rocket fuel, or combustible fuel mixtures. The autoignition temperature or kindling point of a substance is the lowest temperature at which it will spontaneously ignite in a normal atmosphere without an external source of ignition, such as a flame or spark. This temperature is required to supply the activation energy needed for combustion. The temperature at which a chemical will ignite decreases as the pressure increases or the oxygen concentration increases. Autoignition temperature estimates seem to vary in the literature and should only be used as estimates. Factors which may cause variation include partial pressure of oxygen, altitude, humidity, and amount of time required for ignition. Some example autoignition temperatures include:

- Gasoline: 246–280°C (475–536°F)
- Diesel or Jet A-1: 210°C (410°F)
- White phosphorus: 34°C (93°F)
- Carbon disulfide: 90°C (194°F)
- Diethyl ether: 160°C (320°F)
- Butane: 405°C (761°F)
- Magnesium: 473°C (883°F)
- Hydrogen: 536°C (997°F)

AUTOMATIC MODE

Automatic mode is the mode of operation where the system is operating itself without human interaction. This is primarily accomplished via a control system that operates in accordance with a preprogrammed set of instructions (i.e., a computer program) and control laws. For example, the autopilot mode of an aircraft flight control system operates in an automatic mode. Automatic modes of operation are of system safety concern due to the possibility that they could fail to operate or operate incorrectly, particularly without detection or warning. Any data or information input into a control system is also of safety concern as data corruption could cause the system to operate in an unsafe manner.

See *Autonomous Operation* for additional related information.

AUTOMATIC OPERATION

Automatic operation is when the system is in an operational mode or state, in which the system is automatically executing its preprogrammed task or set of tasks. For example, most aircraft have an automatic flight control system that automatically flies the aircraft according to a set of preprogrammed instructions and control laws. Automatic modes of operation are of system safety concern due to the possibility that they could fail to operate or operate incorrectly, particularly without detection or warning. Any data or information input into a control system is also of safety concern as data corruption could cause the system to operate in an unsafe manner.

See *Autonomous System* for additional related information.

AUTONOMOUS

In systems theory, autonomous refers to the autonomous operation of a system or subsystem. This means a system operates without external influence, such as a robotic system or an autonomous unmanned aircraft (UA).

AUTONOMOUS OPERATION

In systems theory, autonomous operation refers to the autonomous operation of a system or subsystem. This means a system operates without external influence, such as a robotic system or an autonomous UA.

AUTONOMOUS SYSTEM

Autonomous is typically defined as the quality or state of being self-governing; undertaken or carried on without outside control; existing or capable of existing independently. In systems theory, autonomous refers to the autonomous operation of a system or subsystem, where the system is operating itself without human interaction. This characteristic requires specific design considerations and methods to correctly build-in the capability for automatic operation, considering factors such as the ability to sense, perceive, analyze, communicate, and make decisions.

UMSs and robots are primarily designed for autonomous operation. Autonomy is characterized into levels of autonomy by factors, such as mission complexity, environmental difficulty, and level of human–robot interaction to accomplish the missions. The two primary levels of autonomy are fully autonomous and semiautonomous. Finer-grained autonomy level designations can also be established and applied as necessary. The two levels of autonomy are defined as follows:

- Fully autonomous—A mode of operation wherein the UMS is expected to accomplish its mission, within a defined scope, without human intervention.
- Semiautonomous—A mode of operation wherein the UMS accomplishes its mission with some level of human interaction.

System safety is a prime consideration in the design and operation of autonomous systems. The autonomous mode of operation of a system can be safety-critical, should it fail to operate or operate incorrectly, particularly without detection or warning. Any data or information input into the control system is also of safety concern as data corruption could cause the system to operate in an unsafe manner.

AUTONOMY

In systems theory, autonomy refers to the autonomous operation of a system or subsystem. This means a system operates without external influence, such as a robotic system or an autonomous UA. A system may have several different levels of autonomy.

AVAILABILITY

Availability is a measure of the degree to which an item is ready to operate when called for use at an unknown (random) time. Availability is the probability that a system is not failed or undergoing repair action when it needs to be used. It is a performance criterion for repairable systems that accounts for both the reliability and maintainability properties of a component or system.

AVERAGE

In mathematics, an "average" is a general statistical term applied to the measure of central tendency of a data set; it is a measure of the "middle" or "expected" value of the data set. An average is a single value that is meant to typify a list of values. If all the numbers in the list are the same, then this number should be used. If the numbers are not the same, an easy way to obtain a representative value from a list is to randomly pick any number from the list. However, the word "average" is usually reserved for more sophisticated methods that are generally found to be more useful. In the latter case, the average is calculated by combining the values from the set in a specific way and computing a single number as being the average of the set.

There are many different descriptive statistics that can be chosen as a measurement of the central tendency of a data set. These include arithmetic mean, the median, and the mode. Other statistical measures such as the *standard deviation* and the *range* are called measures of spread and describe how spread out the data are.

Three commonly used averages include:

- Arithmetic mean = $(X_1 + X_2 + \ldots + X_N)/N$
- Median—The middle value that separates the higher half from the lower half of the data set.
- Mode—The most frequent value in the data set.

BACKOUT AND RECOVERY

Backout and recovery involves the action(s) necessary in a contingency or emergency to restore normal safety conditions in order to avoid a potential accident. Typically, this involves repeating certain already performed steps in reverse order to achieve a safe state. System safety should evaluate SCFs and operations in order to anticipate hazards and develop contingency plans for backout and safe recovery. Without preestablished contingency plans, it may not be possible to avoid certain mishaps.

BARRIER (OR SAFETY BARRIER)

A safety barrier is a safety mechanism or device designed and implemented to control hazards and mitigate risk. Barrier safety is based on the theory that when hazardous energy sources exist within a system, they pose a hazardous threat to certain targets. Placing barriers between the energy source and the target can mitigate the threat to targets. This concept is illustrated in Figure 2.2.

Barriers can be designed to be passive or active in nature. Passive safety barriers provide a natural fixed roadblock, such as a tank pit or a firewall. Active barriers provide a response to certain states or conditions; they involve a sequence of detection, diagnosis, and action (also referred to as detect–diagnose–deflect). Both physical and nonphysical barriers are utilized and applied in hazard control and risk mitigation. Anything used to control,

BARRIER (OR SAFETY BARRIER)

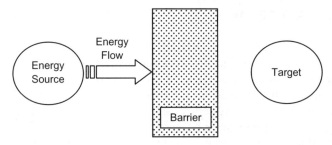

Figure 2.2 Barrier analysis model.

TABLE 2.1 Example Energy Sources, Targets, and Barriers

Energy Sources	Barriers	Targets
Electrical	Walls	Personnel
Thermal	Barricades	System
Mechanical	Guard rails	Product
Chemical	Shields	Equipment
Hydraulic	Lockouts	Environment
Pneumatic	Procedures	Processes
Kinetic	Interlocks	Functions
	Switches	
	Relays	
	Hard hats	

prevent, or impede unwanted adverse energy flow. Also, anything used to control, prevent, or impede unwanted event flow.

Barriers should be designed such that they cannot be bypassed. Safety consideration must be given to the potential failure modes of the elements in a barrier. Barriers typically involve physical devices; however, they can also be implemented by a separation through time and space, warning devices, procedures/work processes, knowledge and skill, and so on. A barrier diagram is a graphical depiction of the evolution of unwanted events (IEs or conditions) through different system states depending on the functioning of the safety barriers intended to prevent this evolution. A barrier diagram represents possible (accident) scenarios. Example energy sources, targets, and barriers include, but are not limited to, the examples shown in Table 2.1.

In essence, a safety barrier is any design or administrative method that prevents a hazardous energy source from reaching a potential target in sufficient magnitude to cause damage or injury. Barriers separate the target from the source by various means involving time or space. Barriers can take many forms, such as physical barriers, distance barriers, timing barriers, procedural barriers, and the like.

See *Barrier Analysis (BA)* and *Safety Barrier Diagram* for additional related information.

38 SYSTEM SAFETY TERMS AND CONCEPTS

BARRIER ANALYSIS (BA)

BA is a system safety analysis technique for identifying and mitigating hazards specifically associated with hazardous energy sources. BA provides a tool to evaluate the unwanted flow of hazardous energy to targets, such as personnel or equipment, and the evaluation of barriers preventing or reducing the hazardous energy flow. Many system designs cannot eliminate energy sources from the system since they are a necessary part of the system. The purpose of BA is to evaluate these energy sources and determine if potential hazards in the design have been adequately mitigated through the use of energy barriers.

BA involves the meticulous tracing of energy flows through the system. BA is based on the premise that a mishap is produced by unwanted energy exchanges associated with energy flows through barriers into exposed targets. The BA process begins with the identification of energy sources within the system design. Diagrams are then generated tracing the energy flow from its source to its potential target. The diagram should show barriers that are in place to prevent damage or injury. If no barriers are in place, then safety design requirements must be generated to establish and implement effective barriers.

BA is a powerful and efficient system safety analysis tool for the discovery of hazards associated with energy sources. The sequentially structured procedures of BA produce consistent, logically reasoned, and less subjective judgments about hazards and controls than many other analysis methods available. It should be noted that BA alone is not comprehensive enough to serve as the sole HA of a system, as it may miss critical human errors or hardware failures not directly associated with energy sources.

The simple concept of BA and its graphical portrayal of accident causation is a powerful analysis tool. It should be noted that an unwanted energy source from a single source may attack multiple targets. Also, in some situations, multiple barriers may be required for optimum safety and risk mitigation. Figure 2.3 depicts an example BA diagram for a hypothetical water heating

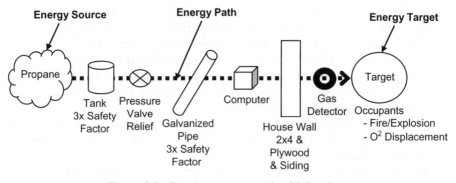

Figure 2.3 Propane energy path with barriers.

system. This diagram shows the energy source on the left and the target on the right. Along the energy flow path there are several different types of barriers implemented into the system design.

BA is an HA technique that is important and essential to system safety. For more detailed information on the BA technique, see Clifton A. Ericson II, *Hazard Analysis Techniques for System Safety* (2005), chapter 19.

See *Barrier* and *Safety Barrier Diagram* for additional related information.

BARRIER FUNCTION

A barrier function is a special safety function designed to prevent, control, or mitigate the propagation of failures or energy into an undesired event (UE) or mishap. A safety barrier can be a series of elements that together implement a barrier function, each element consisting of a technical system or human action. For example, series of interlocks are designed into a missile fire control system to prevent the inadvertent and unauthorized outcome of inadvertent missile launch function. The set of interlocks are safety barriers that implement a barrier function.

See *Interlock* for additional related information.

BARRIER GUARD

A barrier guard is a safety device designed to protect machine operators from hazard points on the machinery or equipment. Some example barrier guard types include enclosures, gates, fences, and interlocks.

BEGINNER'S ALL-PURPOSE SYMBOLIC INSTRUCTION CODE (BASIC)

BASIC is a computer programming language. It is known as a high-level language or high-order language (HOL). BASIC is an acronym for Beginner's All-purpose Symbolic Instruction Code. The original BASIC was designed in 1964 by John Kemeny and Thomas Kurtz at Dartmouth University in New Hampshire, USA, to provide computer access to nonscience students. At the time, nearly all use of computers required writing custom software, which was primarily limited to scientists and mathematicians. The language and its variants became widespread on microcomputers in the late 1970s and 1980s. BASIC remains popular to this day in a handful of highly modified dialects and new languages influenced by BASIC such as Microsoft Visual Basic.

See *High-Order Language (HOL)* for additional related information.

BATTERY SAFETY

Battery types can be separated into two categories: primary batteries and secondary batteries. A primary battery is one that cannot be recharged and is typically thrown away after the battery is completely discharged. Examples of these types of batteries include alkaline batteries and lithium metal batteries. A secondary battery is one that can be recharged and used again and again. They do eventually "die," but most can be charged and discharged many times. Examples of these battery types include nickel metal hydride, nickel cadmium, and lithium ion.

Lithium primary batteries offer high voltage outputs, high energy density, flat discharge characteristics, wide operating temperature ranges, and one of the best storage lives. This translates into power systems that are lighter weight, are lower cost per unit of energy, and have higher and more stable voltages. Therefore, lithium primary batteries enjoy a wide usage in man-portable electronics. One of the unfortunate attributes of lithium primary batteries is that they occasionally vent gas violently. During violent venting, pressures of several hundred pounds per square inch (psi) are capable of being generated within the battery enclosure. These pressures are capable of turning a benign piece of electronics into a "grenade" with potentially harmful flying debris. Therefore, it is imperative that electronics using lithium primary batteries be designed to safely vent gases when necessary.

The guidelines for designing a safe lithium battery enclosure fall into two categories: electrical and mechanical. Electrical design guidelines attempt to prevent the conditions that may lead to a violent battery venting. Mechanical design guidelines are aimed at minimizing the effects of a violent venting once it occurs, and considerations include increasing free space volume, proper material selection, predetermining the direction of gas venting through the enclosure cover, and the use of blow-out plugs. Proper material selection is also very important. The goal of a successful lithium battery enclosure design is not to contain the pressure generated by a violent venting but rather to safely vent the gas. Containment of the gas may be possible, but the size and weight penalties would be so great as to cause the equipment to be no longer portable from a practical standpoint. In the construction of battery enclosures, pliable materials are preferred over brittle materials.

Some typical battery safety guidelines include:

- Provide casing around lithium batteries to protect against explosion with lethal projectiles
- Isolate lithium batteries from external power source to prevent inadvertent recharging
- Provide safety measures for sulfuric acid in lead acid batteries
- Lead acid batteries under charge gives off the flammable gas hydrogen; therefore, keep away from naked flame or source of ignition

- Ensure there is adequate ventilation when charging lead acid batteries to dissipate hazardous gasses that vent from the battery when charging
- If the following conditions are violated, the result may be a battery of lower capacity going into voltage reversal when discharged, causing the battery to vent:
  - Never mix primary (non-rechargeable) and secondary (rechargeable) batteries in equipment at the same time
  - Never mix different types of rechargeable batteries (i.e., lithium ion and nickel cadmium [NiCAD]) batteries in equipment at the same time
  - Never mix new and used lithium ion batteries in the same equipment

BATTLESHORT

A battleshort is a design feature that provides the capability to short or bypass certain safety features in a system to ensure completion of a mission without interruption that would normally result from the triggering of a safety interlock. Battleshorts are typically manual switches that bypass designed-in safety interlocks during emergency situations; a battleshort is a un-safety design feature.

In a particular system design, safety interlocks may have been designed into a function to shutdown the function if certain hazardous conditions or parameters occur, such as over temperature. The intent is to shut down the system so that the high temperature does not severely damage the equipment or cause a mishap. However, the system may be in a survival mode whereby the system function is more important for survival than the risks from over temperature; thus, the need to bypass the safety interlocks to prevent shutdown. For example, a ship's self-defense system may be tracking an incoming missile during a battle whereby the ship has already suffered damage. The damage may be causing system over temperature and when the preset conditions are met, the interlocks will shut down the missile firing system. The battleshort gives the crew the option to prevent the automatic safety shutdown in order to fire missiles in order to protect the ship.

When implementing battleshorts, care must be taken to ensure that they do not become a potential single point failure (SPF) that could bypass several safety interlocks during normal operation. System safety should perform a battleshort analysis, which is a special HA on the battleshort and interlock design to ensure they are safely implemented. Figure 2.4 shows an example of a battleshort model and how the battleshort is an SPF that bypasses several interlocks, where the interlocks serve as a barrier-type design safety feature (DSF). A battleshort implementation should always include an indicator light to warn users when the battleshort has been activated (normally or due to failures).

See *Interlock* for additional related information.

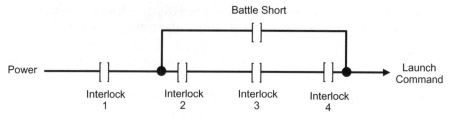

Figure 2.4 Battleshort model.

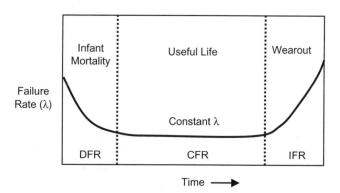

Figure 2.5 Reliability bathtub curve.

BATHTUB CURVE

The reliability bathtub curve, shown in Figure 2.5, represents the change in probability of failure over time of a component. The bathtub can be divided into three regions: burn-in period, useful life period, and wearout period.

The bathtub curve has proven to be particularly appropriate for electronic equipment and systems. It depicts the characteristic pattern over period of decreasing failure rate (DFR) followed by a period of constant failure rate (CFR), followed by a period of increasing failure rate (IFR). The DFR region is the infant mortality (DFR) period characterized by an initially high failure rate. This is normally the result of poor design, the use of substandard components, or lack of adequate controls in the manufacturing process. When these mistakes are not caught by quality control inspections, an early failure is likely to result. Early failures can be eliminated by "burn-in" whereby the equipment is operated at stress levels equal to the intended actual operating conditions. The equipment is then released for actual use only when it has successfully passed through the "burn-in" period.

The CFR region is the useful life period, which is characterized by an essentially constant failure rate. This is the period dominated by chance failures. Chance failures are those failures that result from strictly random or chance causes. They cannot be eliminated by either lengthy burn-in periods or good preventive maintenance practices. Equipment is designed to operate under certain conditions and up to certain stress levels. When these stress levels are exceeded due to random unforeseen or unknown events, a chance failure will occur. While reliability theory and practice is concerned with all three types of failures, its primary concern is with chance failures, since they occur during the useful life period of the equipment. The time when a chance failure will occur cannot be predicted; however, the likelihood or probability that one will occur during a given period of time within the useful life can be determined by analyzing the equipment design. If the probability of chance failure is too great, either design changes must be introduced or the operating environment must be made less severe.

The IFR region is the wearout period, which is characterized by an increasing rate of failure as a result of equipment deterioration due to age or use. For example, mechanical components such as transmission bearings will eventually wear out and fail, regardless of how well they are made. Wearout failures can be postponed, and the useful life of equipment can be extended by good maintenance practices. The only way to prevent failure due to wearout is to replace or repair the deteriorating component before it fails.

BENT PIN ANALYSIS (BPA)

BPA is a system safety analysis technique for identifying hazards caused by bent pins within cable connectors. It is possible to improperly attach two connectors together and have one or more pins in the male connector bend sideways and make contact with other pins within the connector. If this should occur, it is possible to cause open circuits and/or short circuits to +/− voltages, which may be hazardous in certain system designs. For example, a certain cable may contain a specific wire carrying the fire command signal (voltage) for a missile. This fire command wire may be contained within a long wire that passes through many connectors. If a connector pin in the fire command wire should happen to bend and make a short circuit with another connector pin containing +28 VDC the missile fire command may be inadvertently generated. BPA is a tool for evaluating all of the potential bent pin combinations within a connector to determine if a potential safety hazard exists.

The use of this technique is recommended for the identification of system hazards resulting from potential bent connector pins or wire shorts. BPA should always be considered for systems involving safety-critical circuits with connectors. The value of BPA is determining the potential safety effect of one or more pins bending inside a connector and making contact with other pins or the casing. If a safety-critical circuit were to be short-circuited to another circuit containing positive or negative voltage, the overall effect might be catastrophic. BPA is a tool for evaluating all of the potential single pin-to-pin bent pin combinations within a connector to determine if a potential safety hazard exists, given that a bent pin occurs. BPA diagrams provide a pictorial aid for

44 SYSTEM SAFETY TERMS AND CONCEPTS

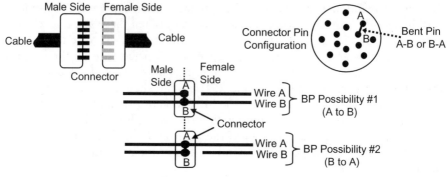

Figure 2.6 BPA concept.

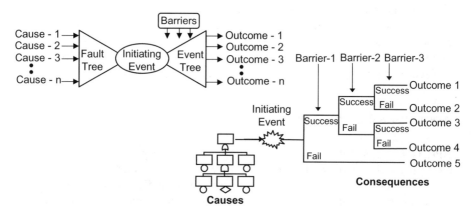

Figure 2.7 Bow-tie diagram.

analysts to visualize pin-to-pin or wire-to-wire short hazards. These types of hazards might be overlooked by other analysis techniques.

Figure 2.6 provides an overview of the BPA concept.

Note in Figure 2.7 that pins A and B are close enough in proximity that they can make physical contact if one of them bends in the right direction, and that pins C and D are too far apart to make physical contact should one of them bend. BPA would evaluate the two scenarios of the pins A–B combination; pin A is bent such that it makes physical contact with pin B, or, pin B is bent such that it makes physical contact with pin A. Each of these two bent pin (BP) possibilities presents two different possible outcomes, for a total of four possible outcomes.

In this A–B and B–A scenario, there are four possible outcomes:

1. Pin A makes contact with pin B; wires A and B become a short circuit and the content of wire A has an upstream or downstream effect on wire B (depending upon the contents of both A and B).

2. Pin A makes contact with pin B; wire A becomes an open circuit after the short (downstream).
3. Pin B makes contact with pin A; wires A and B become a short circuit and the content of wire B has an upstream or downstream effect on wire A (depending upon the contents of both A and B).
4. Pin B makes contact with pin A; wire B becomes an open circuit after the short (downstream).

BPA is an HA technique that is important and essential to system safety. For more detailed information on the BPA technique, see Clifton A. Ericson II, *Hazard Analysis Techniques for System Safety* (2005), Chapter 20.

BEYOND-LINE-OF-SIGHT (BLOS)

In radio frequency (RF) communications, BLOS refers to communications beyond visual distance or beyond-line-of-sight. High-frequency (HF) communications in the 2–30 MHz band provide BLOS communication. Maintaining communications with an autonomous UMS has safety implications; loss of communications or incorrect communications is a hazard source.

BINGO

Bingo is an order (for an aircraft) to proceed to land at a specified field utilizing a maximum range profile. A bingo aircraft is in an emergency situation. The range profile is based on aircraft configuration, distance headwind, weight, and so on. For example, bingo fuel means an aircraft has just enough fuel to return to the nearest landing field.

BINGO FUEL

The term "bingo fuel" is referred to as the amount of fuel necessary to have on board to divert to an alternate landing site (bingo field) if needed. Therefore, when the crew is doing mission planning, they will establish a bingo fuel load, and they will not fly below this minimum in case they must divert.

BIOHAZARD

A biohazard is a hazard where the hazard source component is an organism or a product of organisms.

BLACK BOX

Black box refers to having visibility of only the externally visible performance and interfaces of an item, unit, subsystem, and so on.

BLACK BOX TESTING

Black box and white box testing are terms used to describe the point of view a test engineer takes when designing test cases; black box testing takes an external view of the test object, while white box testing takes an internal view. Both methods have advantages and disadvantages, but it is only when black box and white box testing methodologies are combined that comprehensive test coverage is achieved.

Black box testing uses external descriptions of the software, including specifications, requirements, and design to derive test cases. These tests can be functional or nonfunctional, though usually functional. The test designer selects valid and invalid inputs and determines the correct output. The tester has no knowledge of the test object's internal structure. This method of test design is applicable to all levels of software testing: unit, integration, functional testing, system, and acceptance. The larger and more complex the software, the more one is forced to use black box testing in order to simplify the process. While this method can uncover unimplemented parts of the specification, one cannot be sure that all existent paths are tested.

Typical black box test techniques include:

- Requirements testing
- Functional testing
- Stress testing
- Decision table testing
- State transition testing
- Boundary value analysis testing
- Regression testing
- Reliability testing

Black box testing looks at the available inputs for an application and compares the results to the expected outputs that should result from each input. It is not concerned with the inner workings of the application, the process that the application undertakes to achieve a particular output, or any other internal aspect of the application that may be involved in the transformation of an input into an output. An example of a black box system would be a search engine, where text is entered and then the "search" key is pressed and results are returned. In this a case, the specific process that is being employed to obtain the search results is not known, the tester simply provides an input, a search term, and then receives an output which is checked for correctness.

See *White Box Testing* for additional related information.

BLASTING CAP

A blasting cap is a device intended to cause initiation of explosives. It is composed of a metallic tube, closed at one end, containing a charge of one or more detonating compounds. It is typically initiated by a spark, aflame, or an electrical current.

BOOSTER EXPLOSIVE

Booster and lead explosives are compounds and formulations that are used to transmit and augment the detonation reaction. They are typically used in the beginning stage of an explosive train, for example, in a Fuze.

BOUNDARY

A boundary is a parameter that limits the extent of something. It may be a physical, conceptual, or a mental parameter. In system design, a boundary is a parametric condition, sometimes vague and subjectively stated, which delimits and defines a system and sets it apart from its environment. It can also be a parameter that separates components or subsystems within a system. A boundary is a point of demarcation to provide a needed distinction between parts. In system design, a boundary can also be a design constraint. These design constraints may be externally imposed (e.g., safety, environmental) or internally imposed as a result of prior decisions which limit subsequent design alternatives. Examples of these constraints include: form, fit, function, interface, technology, material, standardization, cost, and time. A boundary interface is a functional, physical, electrical, electronic, mechanical, hydraulic, pneumatic, optical, software, or similar characteristic that defines a common boundary between two or more systems, products, or components.

See *Interface* and *System* for additional related information.

BOW-TIE ANALYSIS

Bow-tie analysis is an analysis technique that combines FTA and ETA together to evaluate multiple possible outcomes from an undesired IE. The various outcomes result from the operation or failure of barriers intended to prevent a mishap. The analysis begins with identification of the IE of concern in the center. An FTA is performed to identify the causal factors and probability of this event. Then an ETA is performed on all the barriers associated with the IE and the possibilities of each barrier function failing. The various different failure combinations provide the various outcomes possible, along with the probability of each outcome. Bow-tie analysis is also referred to as X-tree analysis. Figure 2.7 shows an example bow-tie diagram and demonstrates how the tool received its name.

See *Event Tree Analysis (ETA)*, *Safety Barrier Diagram*, and *X-Tree Analysis* for additional related information.

BUILD

Build refers to a version of software that meets a specified subset of the requirements that the completed software will meet. Typically during program development, several software builds are developed to incorporate changes, fixes, updates, and new requirements. Different builds are differentiated by a "build number," which is established by the program's software development process.

BUILT-IN-TEST (BIT)

BIT is an integral capability of the mission equipment which provides an onboard, automated test capability, consisting of software or hardware (or both) components, to detect, diagnose, or isolate product (system) failures. The fault detection and, possibly, isolation capability, is used for periodic or continuous monitoring of a system's operational health, and for observation and, possibly, diagnosis as a prelude to maintenance action.

BUILT-IN-TEST EQUIPMENT (BITE)

BITE is any device permanently mounted in the prime product or item and used for the express purpose of testing the product or item, either independently or in association with external test equipment.

BURN

A burn is the result of contact with a thermal, chemical, or physical agent causing skin injury or damage. First degree burns show redness of the unbroken skin, second degree burns show skin blisters and some breaking of the skin, and third degree burns show destruction of the skin and underlying tissues with possible charring.

BURN-IN

Burn-in is the process of operating components or devices for a predetermined length of time in order to identify and eliminate components susceptible to infant mortality. Early failures can be eliminated by "burn-in" whereby the equipment is operated at stress levels equal to the intended actual operating conditions. The equipment is then released for actual use only when it has successfully passed through the "burn-in" period. Not to be confused with debugging.

BURNING

With regard to explosives, this is the least violent type of explosive event. The energetic material ignites and burns, non-propulsively. The case may open, melt, or weaken sufficiently to rupture nonviolently, allowing mild release of combustion gases. Debris stays mainly within the area of the fire. The debris is not expected to cause fatal wounds to personnel or be a hazardous fragment beyond 15 m (49 feet).
   See *Explosive Event* for additional related information.

C

C is a computer programming language. It is known as a high-level type language or HOL.
   See *High-Order Language (HOL)* for additional related information.

C++

C++ is a computer programming language, which is an object oriented implementation of the original C language. It is known as a high-level type language or HOL.
   See *High-Order Language (HOL)* for additional related information.

CALIBRATION

Calibration is a comparison of a measuring device with a known standard and a subsequent adjustment to eliminate any differences. Not to be confused with alignment. Calibration of safety-critical items (SCIs) is a safety concern, and may be established as a critical safety item (CSI).

CAPABILITY MATURITY MODEL (CMM)

CMM is a development model created by Carnegie Mellon University (CMU) for software development. The CMM was created by the Software Engineering

Institute (SEI), which was established by CMU. The maturity model can be described as a structured collection of elements that describe certain aspects of maturity in an organization. When it is applied to an existing organization's software development processes, it allows an effective approach toward developing software and improving the process.

There are five levels defined along the CMM continuum. Predictability, effectiveness, and control of an organization's processes are believed to improve as the organization moves up these five levels. The five levels are

Level 1—Initial (chaotic)

This is the starting point for use of a new process. At this level the process is driven in an ad hoc, uncontrolled, and reactive manner by users or events and typically involves individual heroics. Processes and methods are typically undocumented and/or poorly controlled, which provides a chaotic or unstable environment for the processes.

Level 2—Repeatable

At this level the process is managed according to the metrics. It is characteristic of processes at this level that some processes are repeatable, possibly with consistent results. Process discipline is unlikely to be rigorous, but where it exists, it may help to ensure that existing processes are maintained during times of stress.

Level 3—Defined

At this level the process is defined and confirmed as a standard business process. It is characteristic of processes at this level that there are sets of defined and documented standard processes established and subject to some degree of improvement over time. These standard processes are in place and are used to establish consistency of process performance across the organization.

Level 4—Managed

At this level the process is measured and controlled. It is characteristic of processes at this level that, using process metrics, management can effectively control the process. In particular, management can identify ways to adjust and adapt the process to particular projects without measurable losses of quality or deviations from specifications. Process capability is established from this level.

Level 5—Optimized

At this level the process involves deliberate process optimization and improvement; the focus is on continually improving process performance through both incremental and innovative technological changes/

improvements. Processes are concerned with addressing statistical common causes of process variation and changing the process to improve process performance.

The CMM model can be applied to other processes, such as the Capability Maturity Model Integration (CMMI), which has been developed as a process improvement approach that helps organizations improve their performance. CMMI can be used to guide process improvement across a project, a division, or an entire organization.

See *Software Capability Maturity Model (CMM)* for additional related information.

CAPABILITY MATURITY MODEL INTEGRATION (CMMI)

CMMI is the application of the CMM process to organizational performance. There are five CMMI levels, and at each higher level, the organization is able to operate more effectively. CMMI can be used to guide process improvement across a project, a division, or an entire organization. CMMI helps integrate traditionally separate organizational functions, set process improvement goals and priorities, provide guidance for quality processes, and provide a point of reference for appraising current processes.

CASCADING FAILURE

A cascading failure is the failure scenario whereby the failure of an item directly causes another item to fail. A failure in a system of interconnected parts, where the service provided depends on the operation of a preceding part, and the failure of a preceding part can trigger the failure of successive parts. Typically, the items are linked together in a chain-like fashion, and the first item causes the second item to fail, similar to a domino effect. This can be considered to be a failure for which the probability of occurrence is substantially increased by the existence of a previous failure (i.e., a dependent failure). Cascading failures also cause cascading fault states, whereby a sequence of normally "off" modes can be initiated to the "on" fault state by a component failure. For example, the simple failure of a switch can result in prematurely (or inadvertently) applying power to a missile launch system. The switch failure results in providing the normally expected power to the launch system, except at the wrong time. This is considered a "command fault" state.

Figure 2.8 demonstrates the cascading common cause failure (CCF) scenario. In this example, several items are connected in series, with an operational interdependency between the items. If item A should fail, it places a heavier load on item B, exceeding item B's design load limits. As a result, item B either fails, or passes a heavier load onto item C, and so on. For example, a steel beam plate may have seven rivets holding two beams together. If two

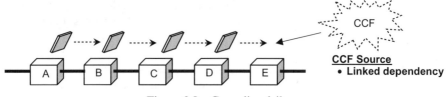

Figure 2.8 Cascading failure.

rivets fail, the load weakens the plate–beam connection causing the remaining rivets to fail. Or, several electrical components may be connected in series. If one component fails, it may pass a higher than designed-for current to the next component, resulting in a cascading effect on the circuit. The scenario is similar to a series of falling dominos/each domino knocking down the next one.

In this cascading failure model, the first failure in the series (A in this example) effectively causes the other components to fail as they are each overstressed by the previous failure. In this example, the root cause of the final outcome is the failure of component A, and the coupling factor is the design vulnerability of the components to a stress level higher than the design limit. From another viewpoint, the first failure can also command the other components to operate prematurely. When component A fails in a certain mode, it provides its input to component B prematurely, which causes B to perform as intended (designed), except prematurely and unintended. This is often referred to as the command mode failure sequence.

See *Dependent Failure* for additional related information.

CATALYST

A catalyst is a substance that alters the speed of a chemical reaction without changing its composition. It can increase or retard the reaction speed.

CATASTROPHE

A catastrophe is an event resulting in great loss and/or misfortune; a state of extreme damage and irreparable ruin; an event resulting in losses of extraordinarily large proportions. A mishap resulting in extreme loss, damage, and deaths would be considered a catastrophe.

CATASTROPHIC HAZARD

A catastrophic hazard is a hazard that has a Category I (Catastrophic) severity level, as defined by the hazard severity criteria in MIL-STD-882.

See *Hazard Severity Levels* for additional information.

CAUTION

A caution is an operating procedure, practice, or condition, that may result in injury or damage to equipment if not carefully observed or followed. The intent of a caution is to prevent a potential mishap. A caution is less critical than a warning. The need for cautions are typically identified through HAs, and the cautions are incorporated into the appropriate manuals and procedures for operation, maintenance, repair, testing, and so on, of a system or product.

See *Warning* for additional related information.

CERTIFICATION

Certification is the confirmation that some fact or statement is true through the use of documentary evidence. Certification follows testing to verify that the certified person, product, or activity is in accordance with specifications and has demonstrated compliance with documented requirements. It is a statement of adequacy provided by a responsible agency for a specific area of concern in support of the validation process.

CHAIN REACTION

A chain reaction is a chemical or nuclear process in which some of the products (or energy released) of the process help to continue the process.

CHANGE CONTROL BOARD (CCB)

During the system development process, a CCB is a group delegated to make decisions regarding whether or not proposed changes to a project should be implemented. The CCB is constituted of project stakeholders or their representatives. The authority of the CCB may vary from project to project, but decisions reached by the CCB are often accepted as final and binding. The CCB is part of the configuration control process and is formally documented as part of the configuration control plan.

Typically, system safety is a member of the CCB in order to evaluate all changes and proposed changes for safety impact. A proposed change may make an existing safe design unsafe by compromising existing safety features, or it may introduce new hazards into the design. The CCB works with engineering change proposals (ECPs) for each change. ECPs should have a safety box on the form to ensure safety assessment of the ECP. ECPs should only be evaluated for safety by an inexperienced system safety analyst.

See *Configuration Control* and *Engineering Change Proposal (ECP)* for additional related information.

CLASS A ACCIDENT

Accidents (mishaps) are classified into three categories as delineated by the Department of Defense Instruction (DoDINST) 6055.7. These categories are Class A, Class B, and Class C, which are defined as follows:

- Class A accident—The resulting total cost of damages to government and other property in an amount of $1 million or more; a DoD aircraft is destroyed; or an injury and/or occupational illness results in a fatality or a permanent total disability.
- Class B accident—The resulting total cost of damage is $200,000 or more, but less than $1 million. An injury and/or occupational illness results in permanent partial disability, or when three or more personnel are hospitalized for in-patient care (which, for accident reporting purposes only, does not include just observation and/or diagnostic care) as a result of a single accident.
- Class C accident—The resulting total cost of property damage is $20,000 or more, but less than $200,000; a nonfatal injury that causes any loss of time from work beyond the day or shift on which it occurred, or a nonfatal occupational illness or disability that causes loss of time from work or disability at any time.

CLASS B ACCIDENT

Accidents (mishaps) are classified into three categories as defined by DoDINST 6055.7. These categories are Class A, Class B, and Class C.
See *Class A Accident* for additional related information.

CLASS C ACCIDENT

Accidents (mishaps) are classified into three categories as defined by DoDINST 6055.7. These categories are Class A, Class B, and Class C.
See *Class A Accident* for additional related information.

CLASS DESK

In the Naval Air Systems Command (NAVAIR) branch of the U.S. Navy, the class desk is the program manager (PM) for systems engineering for an acquisition program. In an operational program, the class desk is the systems engineering PM involved in system upgrades, replacement components, and safety problems that arise in the fleet.

CLOSED SYSTEM

A closed system is a system that is isolated from its surrounding environment and does not interact with its environment. In general, no system can be completely closed; there are only varying degrees of closure.

See *Open System* for additional related information.

CODE COVERAGE

Code coverage is a measure of the degree to which a source code has been tested. It inspects the code directly and is therefore a form of white box testing. It has been extended to include hardware description languages (HDLs) as well as functional languages. Two common forms of code coverage are statement (or line) coverage and path (or edge) coverage. Line coverage reports on the execution footprint in terms of which lines of code were executed. Path or edge coverage reports on which branches or code decision points were executed. They both report a coverage metric, measured as a percentage; the message depends on which code coverage has been used, as 67% path coverage is more comprehensive than 67% statement coverage.

A condition is a Boolean expression (e.g., IF). A decision is a Boolean expression composed of conditions and Boolean operators (e.g., AND, OR). A decision without a Boolean operator is a condition. Typical code coverage criteria include the following:

- Function coverage
    Has each function (or subroutine) in the program been called?
- Statement coverage
    Has each statement in the program been executed?
- Entry/exit coverage
    Has every possible call and return of the function been executed?
- Condition coverage
    Every condition in a decision in the program has taken all possible outcomes at least once.
- Decision coverage
    Every point of entry and exit in the program has been invoked at least once, and every decision in the program has taken all possible outcomes at least once.
- Condition/decision coverage
    Every point of entry and exit in the program has been invoked at least once, every condition in a decision in the program has taken all possible outcomes at least once, and every decision in the program has taken all possible outcomes at least once.

- Modified condition/decision coverage (MC/DC)
    Every point of entry and exit in the program has been invoked at least once, every condition in a decision in the program has taken on all possible outcomes at least once, and each condition has been shown to affect that decision outcome independently. A condition is shown to affect a decision's outcome independently by varying just that condition while holding fixed all other possible conditions. The condition/decision criterion does not guarantee the coverage of all conditions in the module because in many test cases, some conditions of a decision are masked by the other conditions. Using the MC/DC criterion, each condition must be shown to be able to act on the decision outcome by itself, everything else being held fixed. The MC/DC criterion is thus much stronger than condition/decision coverage.

The amount of code coverage testing that should be performed depends on the safety certification level desired for the software. The most stringent coverage, MC/DC testing (or equivalent), is required by the Radio Technical Commission for Aeronautics (RTCA)/DO-178B for Level A software certification.

COLLATERAL DAMAGE

Collateral damage is damage that is unintended or incidental to the intended outcome. The term originated in the military, but it has since expanded into broader use. System safety may identify collateral damage that results from a mishap.

COLLATERAL RADIATION

Collateral radiation is the extraneous radiation (such as secondary beams from optics, flash lamp light, RF radiation, and X-rays) that is not the intended laser beam as a result of the operation of the product or any of its components. System indicator lights would not normally be considered sources of collateral radiation.

COMBUSTIBLE LIQUID

A combustible liquid is any liquid that has a flash point at, or above, 100 degrees Fahrenheit (37.7 degrees Celsius).

COMBUSTION

Combustion is a chemical process that involves oxidation sufficient to produce light or heat.

COMBUSTION PRODUCTS

Combustion products consist of the gases and solid particulates and residues evolved or remaining from a combustion process. They can be a potential hazard source for many different types of hazards.

COMMAND MODE FAILURE SEQUENCE

A command mode failure is a cascading failure scenario whereby the failure of an item directly causes another item to fail, or to operate prematurely. It is a failure in a system of interconnected parts, where the service provided depends on the operation of a preceding part, and the failure of a preceding part can trigger the failure of successive parts. Typically, the items are linked together in a chain-like fashion, and the first item causes the second item to fail, similar to a domino effect. From this viewpoint, the first failure commands the other components to operate prematurely. When component A fails in a certain mode, it provides its input to component B prematurely, which causes B to perform as intended (designed), except prematurely and unintended. This is often referred to as the command mode failure sequence. This concept is used quite effectively in developing FTs for an FTA.

For example, in an example system, a motor-operated valve (MOV) is waiting for an electrical input signal to open the valve and allow liquid nitrogen to flow through the system. If a failure occurs in the electrical input circuit such that the input signal is inadvertently generated, the result is that the MOV is *commanded* to operate as designed. The MOV is normally expected to operate to an open position when it receives an input signal. However, in this situation, there is no intent, and the MOV was falsely commanded to function, thereby allowing the liquid nitrogen to move into the system unexpectedly, with potential harmful consequence. Command fault sequences can cause subsystems to operate prematurely (unintended; inadvertently) or to fail after a long sequence of events.

See *Cascading Failure* for additional related information.

COMMERCIAL OFF-THE-SHELF (COTS)

COTS refers to items that are commercially available from existing inventory sources, that is, off-the-shelf. COTS items are generally commercially developed items that require no development, unique modifications, or maintenance over the life cycle of the product to meet the needs of the system. COTS items are purchased by a program for as-is use in the system, as opposed to being designed and developed as part of the program. In some cases, COTS items are modified for use in the system, while in other cases COTS items cannot be modified. An example of COTS hardware would be the items that are purchased from a catalog, such as electronic power supplies, pipes, and valves. An example of COTS software would be the different operating systems that can be purchased for computer systems. The term COTS has become a generic term that includes COTS, nondevelopmental item (NDI), and government-furnished equipment (GFE).

Some advantages of COTS include:

- Cost savings (no development costs)
- Rapid insertion of new technology
- Proven product/process
- Possible broad user base
- Possible technical and logistics support

Some disadvantages of COTS include:

- Unknown development history (standards, quality assurance, test, analysis, failure history, etc.)
- Design and test data unavailable (drawings, test results, etc.)
- Proprietary design (prohibits information)
- Unable to modify
- Unknown limitations (operational, environmental, stress, etc.)
- May not be supported (configuration control, technical support, updates, etc.)
- May include extra unnecessary capabilities
- Unknown part obsolescence factor
- Safety analyses unavailable or not applicable
- May require increased test and analysis for safety verification

Projects must understand that often, the use of COTS items is cheaper only because standard development tasks have not been performed and, when the costs to perform these necessary tasks to fully evaluate the COTS items in the system application may prove COTS not to be the best alternative. Specifically, hazards must be identified, risks must be assessed, and the risk must be made acceptable regardless of how the component/function is developed. The decision to use COTS items *does not* negate system safety requirements (SSRs).

See *Commercial Off-the-Shelf (COTS) Safety* for additional related information.

COMMERCIAL OFF-THE-SHELF (COTS) SAFETY

The use of COTS items presents many concerns and problems for system safety. There is an increasing trend to use COTS items because the financial benefits appear attractive. The use of COTS items promises to save on development costs, schedule reduction, programmatic risk, and supportability. However, in reality, these promises may be difficult to keep when there is a significant safety impact.

Most of the safety concerns and problems associated with COTS items are a result of the particular characteristics involved, and their impact on the new system design and operating environment. COTS item data, or specifically the lack of it, is the prime contributor to safety problems and concerns. Similarly, the lack of design, test and qualification information, and knowledge makes the safety assessment process more difficult. If a COTS item is used in an SCF, a considerable amount of design and test data on the COTS item is required to assist the system safety engineer in assessing COTS mishap risk potential. The environment for which the COTS item was originally designed must be known, in order to determine if it can meet the new system operational environment.

There are many aspects to the safety of a COTS item, but in general, safety issues revolve around several key questions. As these particular questions are answered, the safety steps to be taken in a COTS item safety program will naturally evolve. These key questions are:

1. Does the COTS item meet all system development requirements involving safety?
2. Does the COTS item have any inherent hazards?
3. Can the COTS item contribute to any system hazards when integrated into the system?
4. Can the COTS item be changed or modified without safety review?
5. Is the COTS item part of an SCF?
6. Does the COTS item contain unneeded functionality?

In order for the system to meet all design SSRs, it is necessary to know what requirements the COTS item was designed and built to meet. This includes requirements for design, development testing, and qualification testing. If the COTS item was not designed or tested to meet a specific system requirement of the program utilizing the COTS item, then the program has identified a safety concern that must be resolved. If none of the COTS item development requirements are known, then the safety problem is even greater.

For example, say that a COTS ultra-high frequency (UHF) radio is going to be used in a new military aircraft refueling tanker. This new tanker aircraft has a safety requirement that all cockpit equipment be designed and tested to meet certain levels of electromagnetic interference (EMI). The SSP must determine if the UHF radio was designed and tested for EMI and if it was tested to the specific EMI levels required. If this requirement is unknown or not met, then the program must make some critical decisions, such as use a different UHF radio, perform extra tests on the radio, waive the requirement, or accept the mishap risk. Knowing the commercial standards the COTS item was built to is a key element in the safety effort.

It is necessary to determine if the COTS item has any inherent hazards. This involves obtaining and reviewing previous safety studies performed on the COTS item. It would also involve evaluating historical usage experience data on the COTS item to determine what types of problems have already been experienced. If the existing inherent safety of the COTS item is unknown, then system safety must be conservative and assume the worst. Perhaps the product will require special safety testing, or reverse engineering, to evaluate inherent safety.

The most critical question that must be answered is, will the COTS item contribute to any system hazards when integrated into the new system? The COTS item must be an integral part of all HAs performed on the system to determine if it can contribute to causing any significant hazards. This means that design information, reliability information, and previous HAs data must be available. If this information is not available to the safety analyst, then a comprehensive HA of the system cannot be performed, and total system mishap risk is not fully known.

Another important safety aspect of COTS items is the assurance that the SSP reviews all changes to the COTS item by the COTS supplier, and that the PM accepts the changes. Programs must know that when replacement COTS items are provided, they do not contain any internal changes since the last safety analysis and acceptance of the item. The capability of the COTS vendor to maintain accurate configuration management (CM) and provide update and version control information is a key aspect to COTS item safety. This would be a good example of technology refresh without safety visibility. Many COTS item safety concerns reduce down to vendor support. Good vendor support can make the COTS item safety integration and verification process much simpler. Types of vendor support that are beneficial include providing design information, reliability data, problem reports, design changes, obsolescence information, maintenance data, and manuals.

A key safety aspect of a COTS item is whether or not it is used in a safety-critical system function. If a COTS item is part of an SCF, then it becomes a key system player and all of its design, operation, and test attributes must be known. If a COTS item is part of a safety-related (SR) function, then its safety sensitivity is a little lower (than an SCF). And, if a COTS item is part of a non-SR function, then there will likely be very little safety concern about the item; however, this does not waive the need for COTS safety analysis. Another COTS item safety concern is unneeded functionality. If the COTS item has more functionality than is needed by the new system, it is possible that this functionality may be the source of safety problems or hazards in the new system design. This is an aspect that must be ruled out by safety analysis and/or testing.

A very important rule-of-thumb for COTS safety is that the use of COTS does not eliminate the requirement for a COTS safety analysis and risk assessment. COTS and NDI must be evaluated for safety as an integral part of the entire system. COTS items are *not* exempt from the system safety process. The decision to use COTS items does not negate the need for SSRs and processes for COTS applications.

See *Commercial Off-the-Shelf (COTS)*, *Government-Furnished Equipment (GFI)*, and *Nondevelopmental Item (NDI)* for additional related information.

COMMERCIAL OFF-THE-SHELF (COTS) SOFTWARE

COTS software refers to software that is purchased off-the-shelf or is reused from another application. The intent of using COTS software is to use already developed software and thereby avoid development costs and schedules that might otherwise be incurred. COTS software can be proprietary or open source. COTS software can present many problems for safety, particularly when used in SCFs or applications.

See *Commercial Off-the-Shelf (COTS)*, *Commercial Off-the-Shelf (COTS) Safety*, and *Software Reuse* for additional related information.

COMMON CAUSE FAILURE (CCF)

A CCF is the simultaneous failure of multiple redundant components due to a common or shared cause. For example, a CCF occurs when two redundant electrical motors become inoperable simultaneously due to a common circuit breaker failure that provides power to both motors. In this example, the common circuit breaker provides the CCF event. CCFs can include causes other than just design dependencies, such as environmental factors, and human error. Ignoring the effects of CCFs in the system design can result in overestimation of the level of safety and reliability (i.e., the level of safety appears better than it actually is).

For system safety considerations, a CCF event consists of item or component failures that meet the following criteria:

1. Two or more individual redundant components fail or are degraded such that they cannot be used when needed, or used safely if still operational
2. The component failures result from a single shared cause (coupling mechanism)

CCFs create the subtlest type of hazards because they are not always obvious, and they can be difficult to identify. The potential for this type of event exists in any system design that relies on redundancy or uses identical components or software in multiple subsystems. CCF vulnerability results from system failure dependencies inadvertently designed into the system. Another example of a CCF would be the forced failure of two independent, and redundant, hydraulic flight control systems due to a failure event that cuts both hydraulic lines simultaneously. A safe design would physically separate the hydraulic lines by a sufficient distance in order that this type of event could not happen.

CCFs can be caused from a variety of sources or coupling factors, such as:

1. Common weakness in design redundancy
2. The use of identical components in multiple subsystems
3. Common software design
4. Common manufacturing errors
5. Common requirements errors
6. Common production process errors
7. Common maintenance errors
8. Common installation errors
9. Common environmental factor vulnerabilities

Figure 2.9 is an example of a CCF situation. In this example, a CCF causes loss of the common power source to both computers, causing simultaneous failure of both redundant computers, which effectively eliminates the intended redundancy.

CCFs and common mode failures (CMFs) are similar in nature in that they are both involved with the simultaneous loss of redundant equipment to a single shared cause. However, they differ by the type of the single shared causal event that causes the redundant items to fail simultaneously. A CCF is caused by an external event, whereas the CMF is caused by an identical failure internal to each item. CMFs normally fail in the same functional mode. Quite often, CMFs are (erroneously) referred to as CCFs. Although it is reasonable to include CMFs under the CCF umbrella, CCFs are much larger in scope and coverage. Figure 2.10 shows this conceptual difference between CCF and CMF. Note that the boxes represent redundant system elements, and the redundancy is effectively shunted by the CMFs and CCFs. Redundancy is the key for identifying CCFs and CMFs.

CCFs and CMFs are a form of dependent failures, created by dependencies built into the system design. In essence, CCF and CMF modes are SPFs that result in the elimination of intended design redundancy. Design diversity in redundant components can eliminate some CCF and CMF scenarios.

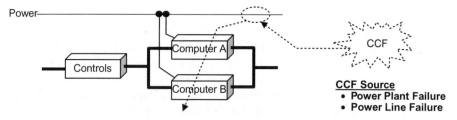

Figure 2.9 Example of CCF.

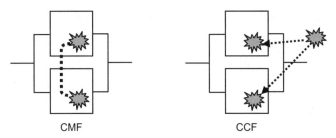

Figure 2.10 CCF versus CMF.

See *Common Cause Failure Analysis (CCFA)* and *Common Mode Failure (CMF)* for additional related information.

COMMON CAUSE FAILURE ANALYSIS (CCFA)

CCFA is an analysis methodology for identifying CCFs in a system, where a CCF is an SPF that eliminates independent redundant designs. The objective of CCFA is to discover common cause vulnerabilities in the system design that can result in the common failure of redundant subsystems and to develop design strategies to mitigate these types of hazards.

Many systems utilize subsystem design redundancy to ensure that a specific function occurs upon demand. The idea is that two separate and independent subsystems are much less likely to fail from independent failures than a single independent subsystem. System designs have become so complex, however, that occasionally, a dependency is inadvertently built into the redundancy design. One form of dependency is the CCF event that can cause failure of both redundant subsystems. A CCF is effectively an SPF that nullifies independent redundant subsystem designs. For example, a DC-10 crash occurred when an engine exploded and a fan blade from the engine cut two independent and separated hydraulic lines running along the top of the aircraft. Aircraft control depended upon system hydraulics, and the design, therefore, intentionally had two independent and redundant hydraulic systems. Though the redundant hydraulic lines were physically separated by a large distance, the engine exploding resulting in uncontained shrapnel was the common cause SPF resulting in the loss of both critical hydraulic subsystems.

The purpose of CCFA is to identify CCF vulnerabilities in the system design that eliminate or bypass design redundancy, where such redundancy is necessary for safe and reliable operation. Once CCFs are identified and evaluated for risk, defense strategies mitigating critical CCFs can be established and implemented. CCFA also provides a methodology for determining the quantitative risk presented by CCFs. CCFA can be applied to any type of system, but it is particularly useful for safety-critical systems using design redundancy. CCFs create the subtlest type of hazards because they are not always obvious, making them difficult to identify. The potential for this type of event exists in any system design that relies on redundancy or uses identical components or software in multiple subsystems. CCF vulnerability results from system failure dependencies inadvertently designed into the system. If a CCF is overlooked, total system risk is understated because the probability of this hazard is not included in the total risk calculation. If the common cause dependency is in a critical subsystem, CCFs could contribute to a significant impact in overall system risk.

Applying CCFA to the analysis of a system design is not a trivial process. It is more difficult than an analysis technique such as preliminary hazard analysis (PHA), primarily because it requires a thorough and detailed understanding of the system design, along with extensive data collection and analysis of CCFA components. In general, the basic steps in CCFA include the following:

1. Understand the system—Examine the system and define the critical system redundancies.
2. Develop initial system logic model—Develop an initial component level system logic model using FTA that identifies major contributing components.
3. Screening—Screen system design and data for identification of CCF vulnerabilities and CCF events.
4. Detailed CCF logic analysis—Place CCF components in FTA and perform a qualitative and quantitative analysis to assess CCF risk.
5. Evaluate outcome risk—Evaluate the outcome risk of each CCF event and determine if the risk is acceptable.
6. Corrective action—If the outcome risk is not acceptable, develop design strategies to countermeasure CCF effect and change system risk.

The most efficient approach to identifying CCF system susceptibilities is to focus on identifying coupling factors, regardless of defenses that might be in place. A CCF coupling factor is a characteristic of a group of components as they are situated in the system that identifies them as susceptible to the same causal mechanisms of failure. Such factors include similarity in design, location, environment, mission and operational, maintenance, and test procedures. The resulting list will be a conservative assessment of the system susceptibilities to CCFs. A coupling mechanism is the factor that distinguishes CCFs from multiple independent failures. Coupling mechanisms are suspected to exist when two or more component failures exhibit similar characteristics, both in the cause and in the actual failure mechanism. The analyst, therefore, should focus on identifying those components of the system which share common characteristics.

A list of common factors is provided in Table 2.2. This list helps to identify the presence of identical components in the system and most commonly observed coupling factors. Any group of components that share similarities in

TABLE 2.2 Key Common Cause Attributes

Characteristic	Description
Same design	The use of the same design in multiple subsystems can be the source of a CCF coupling factor vulnerabilities. This is particularly true of software design.
Same hardware	The use of identical components in multiple subsystems resulting in a vulnerability of multiple subsystems.
Same function	When the same function is used in multiple places, it may require identical or similar hardware that provides CCF vulnerabilities.
Same staff	Items are vulnerable to the same installation, maintenance, test, or operations staff that can make common errors.
Same procedures	Items are vulnerable to the same installation, maintenance, test, or operations procedures, which may have common errors.
Redundancy	When redundant items are identical, they are vulnerable to the same failure modes, failure rates, and CCF coupling factors.
Same location	Items are located in the same physical location, making them vulnerable to the same undesired conditions (fire, water, shock, etc.).
Same environment	Items are vulnerable to the same undesired environmental conditions (fire, water, shock, electromagnetic radiation, dust, salt, etc.).
Same manufacturer	Components have the same manufacturer, making all components vulnerable to the same failure modes and failure rates.
Common requirements	Common requirements for items or functions may contain common errors that generate CCF vulnerabilities.
Common energy sources	Items with common energy sources (e.g., electrical, chemical, and hydraulic) generate CCF vulnerabilities.
Common data sources	Items with common data sources generate CCF vulnerabilities, particularly in software design.
Common boundaries	Items that share a common boundary (physical, functional, logical, etc.) may have CCF vulnerabilities.

one or more of these characteristics is a potential point of vulnerability to CCF.

Coupling factors can be divided into four major classes:

- Hardware based
- Operation based
- Environment based
- Software based

Hardware-based coupling factors are factors that propagate a failure mechanism among several components due to identical physical characteristics. An example of hardware-based coupling factors is failure of several residual heat removal pumps because of the failure of identical pump air deflectors. There are two subcategories of hardware-based coupling factors: (1) hardware design and (2) hardware quality (manufacturing and installation). Hardware design coupling factors result from common characteristics among components determined at the design level. There are two groups of design-related hardware couplings: system level and component level. System-level coupling factors include features of the system or groups of components external to the components that can cause propagation of failures to multiple components. Features within the boundaries of each component cause -level coupling factors.

The development of a system-level FTA is a key step in CCFA process. The initial system FT logic model is developed as a basic system model that identifies the primary contributing components fault events leading to the undesired top-level event. The initial FT is developed around basic independent failure events, which provides a first approximation of cut sets (CSs) and probability. Many component failure dependencies among the components are not accounted for explicitly in the first approximation FT model, resulting in an underestimation of the risk of the FT's top-level event. As CCF events are identified, the analyst expands the FT model in to include identified CCFs. The FT model is much more complete and accurate with CCF events included. The FT is re-computed to reevaluate the criticality, sensitivity and probability of the CCF within the FT. The revised FT probability risk estimate that includes CCFs provides a correct risk estimate over the first approximation FT. Figure 2.11 contains two example FTs for a 3-redundant element system; one FT without consideration of CCF and a second FT that includes a CCF event.

Note that in Figure 2.11, the initial FTA produces only one CS, which is a 3 order CS, indicating that the probability of system failure should be small. Note also that the updated FTA produces two CSs; the original 3 order CS and an SPF CS. Depending upon the probability of the SPF CCF event, the overall FTA probability could be much higher than the initial FTA probability, as demonstrated in this example.

There are three methods of defense against a CCF:

1. defend against the CCF root cause
2. defend against the CCF coupling factor
3. defend against both items 1 and 2

Developing a defense strategy against a CCF root cause is usually somewhat difficult because components generally have an inherent set of failure modes that cannot be eliminated. However, the failure modes can be protected via redundancy, diversity, and barriers. A defense strategy for coupling factors typically includes diversity, barriers, location, personnel training, and staggered

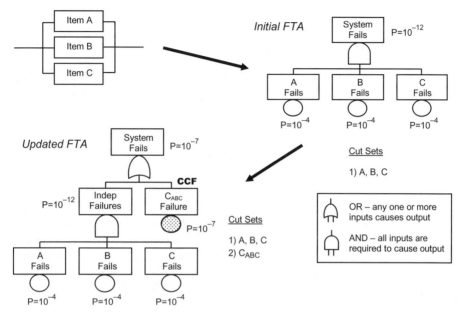

Figure 2.11 FTA with CCF event.

testing and maintenance. It is recommended that the SSP implement CCFA in order to support the goal of identifying and mitigating system risk produced by CCFs. CCFA is also recommended as part of a probabilistic risk assessment (PRA), particularly in order to obtain a truer view of system risk.

CCFA is an analysis technique that is important and essential to reliability engineering and system safety. For more detailed information on the CCFA technique, see Clifton A. Ericson II, *Hazard Analysis Techniques for System Safety* (2005), chapter 23.

See *Common Cause Failure (CCF)* for additional related information.

COMMON MODE FAILURE (CMF)

A CMF is the failure of multiple similar components in the same mode, for various reasons. A CMF is an event or failure that simultaneously affects a number of elements otherwise considered as being independent. For example, a set of identical resistors from the same manufacturer may all fail in the same mode (and exposure time) due to a common manufacturing flaw. The term CMF, which was used in the early safety literature and is still used by some practitioners, is more indicative of the most common symptom of the CCF, but it is not a precise term for describing all of the different dependency situations that can result in a CCF event. A CMF is a special case or a subset of a CCF. Typically, the term CCF is used to include both CCFs and CMFs.

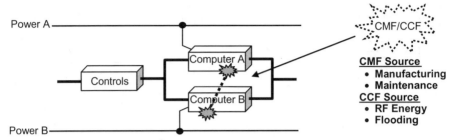

Figure 2.12 Example of CMF/CCF.

CMFs/CCFs create the subtlest type of hazards because they are not always obvious, and they can be difficult to identify. The potential for this type of event exists in any system design that relies on redundancy or uses identical components or software in multiple subsystems. CMF/CCF vulnerability results from system failure dependencies inadvertently designed into the system. An example of a CMF/CCF would be the forced failure of two independent, and redundant, flight control computers due to the failure of a common circuit breaker in the system design providing electrical power.

If a critical CMF/CCF is overlooked, total system risk is understated because the probability of this hazard is not included in the total risk calculation. If the common cause dependency is in a critical subsystem, CMFs/CCFs could contribute to a significant impact in overall system risk. It is imperative to perform CCFA in safety-critical applications. Figure 2.12 is an example of a CMF/CCF situation. In this example, a CMF/CCF causes simultaneous failure of both redundant computers, which effectively eliminates the intended redundancy.

CCFs and CMFs are similar in nature in that they are both involved with the simultaneous loss of redundant equipment to a single shared cause. However, they differ by the type of the single shared causal event that causes the redundant items to fail simultaneously. A CCF is caused by an external event, whereas the CMF is caused by an identical failure internal to each item. CMFs normally fail in the same functional mode. Quite often, CMFs are (erroneously) referred to as CCFs. Although it is reasonable to include CMFs under the CCF umbrella, CCFs are much larger in scope and coverage. Figure 2.10 showed the conceptual difference between CCF and CMF; note that the boxes represent redundant system elements, and the redundancy is effectively shunted by the CMFs and CCFs. Redundancy is the key for identifying CCFs and CMFs.

CCFs and CMFs are a form of dependent failures, created by dependencies built into the system design. In essence, CCF and CMF modes are SPFs that result in the elimination of intended design redundancy. Design diversity in redundant components can eliminate some CCF and CMF scenarios. A CMF occurs when events are not failure independent or statistically independent, that is, when one event causes multiple items or systems to fail. For example, when all of the pumps for a fire sprinkler system are located in one room, if the room becomes too hot for the pumps to operate, they will all fail essentially at the same time, from one cause (heat). The principle of redundancy states that when the failure events of two components are statistically independent, the probabilities of their joint occurrence multiply. For example, if the probability of failure of a component of a system is 10^{-3} per year, the probability of the joint failure of two of them is 10^{-6} per year, provided that the two events are statistically independent. This principle favors the design strategy of component redundancy; however, the beneficial effect of redundancy is weakened if there is CCF/CMF susceptibility.

As an example of CMF, consider a computer system where reliability is often achieved by using redundant hard drives. If one drive fails, data are still available from the backup drive. Even in this simple design concept, there can be many common modes which defeat the redundancy, such as:

- The disks are likely to be from the same manufacturer and are of the same model; therefore, they share the same design flaws.
- The disks are likely to have similar serial numbers; thus, they may share any manufacturing flaws affecting production of the same batch.
- The disks are likely to have been shipped at the same time; thus, they are likely to have suffered from the same transportation damage.
- As installed both disks are attached to the same power supply, making them vulnerable to the same power supply issues.
- As installed both disks are in the same case, making them vulnerable to the same overheating events.
- They will be both attached to the same card or motherboard, and driven by the same software, which may have the same bugs.
- Because of the very nature of the design, both disks will be subjected to the same workload and very closely similar access patterns, stressing them in the same way.

See *Common Cause Failure (CCF)* and *Common Cause Failure Analysis (CCFA)* for additional related information.

COMPILER

A compiler is a computer program that translates high-level language statements into directly executable machine code. A single high-level statement generally results in many machine instructions. The compiler translates the machine code into binary code (0's and 1's) for the computer to understand.

See *High-Order Language (HOL)* for additional related information.

COMPLEXITY

Complexity is the degree of intricacy of a system so interconnected as to make analysis difficult and complicated. Complexity is an attribute of systems or items which makes their operation difficult to comprehend. Increased system complexity is often caused by such items as sophisticated components and multiple interrelationships. Complexity is a system characteristic denoted by a large number of densely connected parts with multiple levels of embeddedness and entanglement. Complexity should not be confused with complicatedness, which is a design, architecture, or situation that is not easy to understand, regardless of its degree of complexity.

COMPLIANCE

Compliance is the complete satisfaction of an objective or requirement. In a system development program, the design must be in compliance with the requirements, and the program must be in compliance with contractual requirements. Safety compliance means the system design is in full compliance with applicable safety standards, requirements, and guidelines. A "compliance" is also a product of an audit; it is the complete satisfaction of an objective or requirement evaluated during the audit. Typical categories of items resulting from an audit include: compliance, finding, observation, issue, and action.

See *Audit* and *Compliance Based Safety* for additional related information.

COMPLIANCE-BASED SAFETY

Compliance-based safety is a prescriptive approach to safety based on laws, regulations, guidance standards, and so on. In this approach, the contractor or managing authority must follow prescriptive design requirements or guidance and show evidence of compliance. Prescribed design measures must be implemented into the system design. In the case of prescribed requirements, there is usually no room for different design options, as the prescribed requirement must be implemented as stated. In the case of guidance standards, the guidance is a little more general, allowing for implementation via different design options.

The compliance-based safety approach is effective and useful; however, its major drawback is that it does not ensure that all system potential mishap risk is reduced to the lowest effective value practical. Prescriptive safety provides a known basic level of safety, but it may fall short if further safety features are necessary for a particular system. Prescribed safety requirements only address a known set of hazards. Even for a compliance-based safety program, it is still necessary to perform HA to ensure that all hazards have been identified and the risk mitigated to the lowest level practical.

See *Prescriptive Safety* and *Risk-Based Safety* for additional related information.

COMPONENT

A component is a cleanly identified item composed of multiple parts. A component is one element in a system typically composed of a number of components, parts, or any combination thereof, joined together to perform a specific function. A component is generally the second smallest, most specific element that can be identified as a discrete item. A typical example of a component would be a cathode ray tube or the earpiece of the pilot's radio headset. A component is an artifact that is one of the individual composite entities in a system.

In system safety, parts and components are of prime interest because it is often their unique failure modes, within unique system architectures, that provide the IM for certain hazards within a system design. Failure mode and effects analysis (FMEA) and FTA typically deal with the system at the part or component level in order to determine the risk presented by a particular hazard. When an FTA is performed to determine the causal factors for a particular hazard or UE, the FTA is generally conducted to the part level. Failure rates can be obtained for parts, which can be used in the FTA to generate a quantitative result.

See *System Hierarchy* for additional related information.

COMPUTER-AIDED SOFTWARE ENGINEERING (CASE)

CASE is the use of software tools to assist in the development and maintenance of software. All aspects of the software development life cycle can be supported by software tools; therefore, the use of tools from across the spectrum can be described as CASE, from project management software through tools for business and functional analysis, system design, code storage, compilers, translation tools, test software, and so on.

Some typical CASE tools include:

- Code generation tools
- Data modeling tools
- Unified modeling language (UML) tools
- Refactoring tools
- Model transformation tools
- CM tools including revision control
- Data flow tools
- Data models in the form of entity-relationship diagrams
- Code test and verification tools

COMPUTER SOFTWARE COMPONENT (CSC)

A CSC is a functional or logically distinct part of a Computer Software Configuration Item (CSCI) or Software Configuration Item (SCI). A CSC is typically an aggregate of two or more computer software units (CSU). CSCs may be further decomposed into other CSCs and CSUs.

COMPUTER SOFTWARE CONFIGURATION ITEM (CSCI)

A CSCI is an aggregate of software that is designated for CM and is treated as a single entity in the CM process. A CSCI is also referred to as a software item (SI) or SCI.

COMPUTER SOFTWARE UNIT (CSU)

A CSU is the smallest subdivision of a CSCI for the purposes of engineering management. CSUs are typically separately compiled and are testable pieces of code.

CONCEPT OF OPERATIONS (CONOPs)

CONOPs is a statement, in broad outline, of a commander's assumptions or intent in regard to an operation or series of operations. The CONOPs frequently is embodied in campaign plans and operation plans; in the latter case, particularly when the plans cover a series of connected operations to be carried out simultaneously or in succession. The concept is designed to give an overall picture of the operation. CONOPs typically defines how a system will be operated and in what environments.

CONCURRENT DEVELOPMENT MODEL

This method performs several of the standard development tasks concurrently, in an attempt to save development time. For example, production may begin before final design and test are complete. This method has a higher probability for technical risk problems since some items are in preproduction before full development and testing.

See *Engineering Development Model* and *System Life-Cycle Model* for additional related information.

CONDITION

A condition is a state at a particular time (e.g., a condition of disrepair). A condition is also a statement of what is required as part of an agreement or part of an event. A condition can be an assumption on which rests the validity or effect of something else. In system safety, a hazard is based on certain conditions being present. For example, a hazard is a potential or existing condition, or conditions, that can result in a mishap when the condition or conditions are actuated or fulfilled.

CONFIGURATION

The functional and physical characteristics of a product or system and all of its integral parts, assemblies, and systems that are capable of fulfilling the fit, form, and functional requirements defined by performance specifications and engineering drawings.

CONFIGURATION CONTROL

Configuration control is the planned and disciplined control over the physical and functional design of a system or product. It involves the systematic evaluation, coordination, and formal approval/disapproval of proposed changes and implementation of all approved changes to the design and production of an item, the configuration of which has been formally approved by the contractor or by the purchaser, or both. Configuration control is accomplished through the CM process. System safety should be actively involved as a team member in the configuration control process and should be involved in assessing the safety impact of proposed and actual design changes.

See *Configuration Management (CM)* for additional related information.

CONFIGURATION ITEM (CI)

A CI is an item, or aggregate of items, designated for formal CM. A CI can be hardware, software, or firmware, or an aggregation of these items. A CI is typically an aggregation of system elements that satisfies an end use function and is designated for separate CM. CIs may vary widely in complexity, size, and type, from an aircraft, electronic, or ship system to a test meter or round of ammunition. Generally, any item required for logistics support and designated for separate procurement is a CI.

See *Configuration Management (CM)* for additional related information.

CONFIGURATION MANAGEMENT (CM)

CM is the formal management process that focuses on establishing and maintaining consistency of a system's or product's performance and its functional and physical attributes with its requirements, design, and operational information throughout its life cycle. CM can be defined as the management of system or product design through control of changes made to hardware, software, firmware, documentation, test, test fixtures, and test documentation throughout the life cycle of the system. CM involves formal management and control of CIs.

CM involves the systematic control and evaluation of all changes to baseline documentation and subsequent changes to that documentation which define the original scope of effort to be accomplished and the systematic control, identification, status accounting, and verification of all CIs. Typically, a formal CM plan is developed for a project, which specifies and documents all aspects of the process, including system safety.

System safety should be actively involved as a team member in the CM process. All ECPs must be evaluated by an experienced system safety analyst to determine if any hazards are associated with the proposed change, the associated risk, and the safety impact of the ECP on the existing system. The program should be notified when an ECP will decrease the level of safety of the existing system. It is important that only qualified safety engineers make the safety determination on proposed ECPs because they have the most experience and knowledge in regard to system hazards and risk assessments.

See *Engineering Change Proposal (ECP)* for additional related information.

CONFLAGRATION

A conflagration is a fire extending over a considerable area.

CONFORMANCE

Conformance is when an item is established as correct with reference to a standard, specification, or drawing. For example, a design verified to meet applicable requirements is considered to be in conformance with those requirements. Nonconformance may generate a potential safety issue.

CONTAMINATION

Contamination occurs when there is presence of unwanted material in an item that degrades the performance of the item or changes its desired characteristics. Contamination can be of molecular or particulate nature. Contamination can be the act of contaminating or polluting (either intentionally or accidentally) with unwanted substances or factors.

CONTINGENCY

A contingency is an emergency situation requiring special plans, rapid response, and special procedures to avoid an impending mishap and ensure the safety of personnel, equipment, and facilities. It is an emergency situation caused by an unexpected event or events. Due to the uncertainty of the situation, contingencies require plans, rapid response, and special procedures to ensure the safety and readiness of personnel, installations, and equipment. Safety is contingent on execution of the contingency plans.

CONTINGENCY ANALYSIS

Contingency analysis is a type of analysis conducted to determine procedures, equipment, and materials required to prevent a contingency from deteriorating into an accident. This could be considered analogous to an HA. The resulting contingency plans are the mitigations for the hazard or contingency.

CONTINUOUS WAVE (CW)

A CW or continuous waveform is an electromagnetic (EM) wave of constant amplitude and frequency, and in mathematical analysis, of infinite duration. CW is also the name given to an early method of radio transmission, in which a carrier wave is switched on and off. Information is carried in the varying duration of the on and off periods of the signal. In radio transmission, CWs are also known as "undamped waves," to distinguish this method from damped wave transmission. In laser physics and electrical engineering, the term "CW" refers to a laser which produces a continuous output beam, sometimes referred to as "free-running." This is as opposed to a q-switched, gain-switched, or mode-locked laser, which produces pulses of light. A CW is when the output of a laser is operated in a continuous (duration > 0.25 s) rather than a pulsed mode.

CONTRACT

A contract is a legal agreement between two or more competent parties. In the systems acquisition process, a contract is typically between the government and a contractor, whereby the contractor develops and produces a product or system for the government. In addition to specified performance requirements, contract requirements include those requirements defined in the statement of work (SOW); specifications, standards, and related documents; the Contract Data Requirements List (CDRL); management systems; and contract terms and conditions.

CONTRACTING PROCESS

When the government desires to procure a product or system, there is typically a formal process that is followed. The process is initiated with the government's request for proposal (RFP) along with the government SOW and/or statement of objectives (SOOs). The contracting process concludes when the government Source Selection Evaluation Board (SSEB) selects a winning contractor and formalizes the contract.

Developing the system safety approach, conveying it to the contract bidders, and selecting the best safety program is an integral part of the total acquisition contracting process. Each step involves system safety input, from both the government and the contractor. Both the government and the contractor must have expertise in system safety in order to objectively and effectively establish an SSP for the product/system.

CONTRACTOR

A contractor is a private sector enterprise or corporation engaged to provide services or products specified within a contract. Typically, the contractor provides goods or services to the government, which is typically known as the managing activity (MA). Occasionally, the contractor may be an organizational element of the DoD or any other government agency.

See *Managing Activity (MA)* for additional related information.

CONTRACTOR DATA REQUIREMENTS LIST (CDRL)

A CDRL is a DD Form 1423 list of contract data requirements that are authorized for a specific acquisition and made a part of the contract. All data delivered under contract must be specified by CDRL. The managing authority prepares the CDRL to require that the data package will be prepared and forwarded through the procurement agency, in sufficient copies, at the appropriate date to support program milestones. Many CDRLs refer to an existing Data Item Description (DID) for data format and content.

See *Data Item Description (DID)* for additional related information.

CONTROL ELEMENT

A control element is an electromechanical element of a UMS designed to control the UMS with decision-making capability. A computer would be considered an electromechanical control element. This term was created to support the nonhuman aspect of an authorized entity controlling the operation of a system.

See *Authorized Entity* for additional related information.

CONTROL ENTITY

Control entity is an individual operator or control element directing or controlling system functions for a mission. This term was created to support the nonhuman aspect of an authorized entity controlling the operation of a system.
See *Authorized Entity* for additional related information.

CONTROLLED AREA

A controlled area is an area where the occupancy and activity of personnel within is subject to control and supervision of protection from hazards, such as radiation or laser hazards.

CONTROLLED FLIGHT INTO TERRAIN (CFIT)

CFIT describes a mishap in which an airworthy aircraft, under pilot control, is unintentionally flown into the ground, a mountain, water, or an obstacle. The pilots are generally unaware of the danger until it is too late. There are many reasons for CFIT, including bad weather, imprecise navigation, and pilot error. Pilot error is the single biggest factor leading to a CFIT incident. Pilot fatigue, loss of situational awareness, or disorientation may play a role. CFIT incidents often involve a collision with significantly raised terrain such as hills or mountains, and may occur in conditions of clouds or otherwise reduced visibility. CFIT often occurs during aircraft descent to landing, near an airport. CFIT may be associated with subtle equipment malfunctions. If the malfunction occurs in a piece of navigational equipment and it is not detected by the crew, it may mislead the crew into improperly guiding the aircraft despite other information received from all properly functioning equipment, or despite clear sky visibility that should have allowed the crew to easily notice ground proximity.

CORRECTIVE MAINTENANCE (CM)

CM refers to actions performed as a result of failure, to restore an item to a specified condition. CM can include any or all of the following steps: localization, isolation, disassembly, interchange, reassembly, alignment, and checkout.

CORRELATION

Correlation, or track correlation, is the process of combining one track with another track. Only one track retains its track number, and the other is dropped. This is in reference to radar tracking from one or more tracking systems watching an incoming enemy system of some type.

CORROSION

Corrosion is a physical change in a material brought about by chemical or electrochemical action. Corrosion typically has a deteriorating or destructive effect. Note that erosion is caused by mechanical action.

COUNTERMEASURE

A countermeasure, in military applications, is a system designed to prevent sensor-based weapons from acquiring and/or destroying a target. A countermeasure is any device or technique used to prevent or counter a recognized threat or hazard. It is intentionally implemented in the design of a system for use during system operation. For example, military aircraft often employ electronic countermeasures or chaff to prevent being hit by an incoming missile.

In system safety, a countermeasure to a hazard is often referred to as a hazard countermeasure, DSF, design safety measure, or design safety mechanism. In this capacity, a safety mechanism is designed into the system to counter or mitigate an identified hazard. For example, two redundant flight control computers may be employed in the system design because a single flight control computer presents unacceptable mishap risk. A hazard countermeasure may not be necessary for system function, but it is necessary for safety (i.e., risk reduction). A hazard countermeasure can be any technique, device, or method designed to eliminate or reduce the risk presented by a hazard, and should be developed following the safety order of precedence (SOOP).

See *Design Safety Feature, Design Safety Measure, Design Safety Mechanism,* and *Hazard Countermeasure* for additional information.

CPU

The CPU or the processor is the portion of a computer system that carries out the instructions of a computer program, and is the primary element carrying out the computer's functions. Typically, it is composed of a single computer chip. It is also known as a microprocessor. The fundamental operation of a CPU is to execute a sequence of stored instructions called a computer program (i.e., the software). There are four steps that nearly all CPUs use in their operation: fetch, decode, execute, and writeback. The fetch step involves retrieving an instruction (which is represented by a number or sequence of numbers) from program memory. The location in program memory is determined by a program counter (PC), which stores a number that identifies the current position in the program. The PC keeps track of the CPU's place in the current program. After an instruction is fetched, the PC is incremented by the length of the instruction word in terms of memory units. The instruction that the CPU fetches from memory is used to determine what the CPU is to do. In the decode step, the instruction is broken up into parts that have significance to other portions of the CPU. The way in which the numerical instruction value is interpreted is defined by the CPU's instruction set architecture (ISA). Often, one group of numbers in the instruction, called the opcode, indicates which operation to perform. The remaining parts of the number usually provide information required for that instruction, such as operands for an addition operation. In the execute step, various portions of the CPU are connected so they can perform the desired operation. For example, if an addition operation was requested, an arithmetic logic unit (ALU) will be connected to a set of inputs and a set of outputs. The inputs provide the numbers to be added, and the outputs will contain the final sum. The ALU contains the circuitry to perform simple arithmetic and logical operations on the inputs. If the addition operation produces a result too large for the CPU to handle, an arithmetic overflow flag in a flags register may also be set. Writeback simply "writes back" the results of the execute step to some form of memory. After the execution of the instruction and writeback of the resulting data, the entire process repeats, with the next instruction cycle normally fetching the next instruction in the sequence after the PC has been incremented. An HA may look for erroneous output from a CPU, but it seldom delves into the internal failure modes of the CPU due to complexity.

CRASHWORTHINESS

Crashworthiness is the capability of a vehicle to protect its occupants against an accidental impact such that they receive no serious injury.

CREDIBLE ENVIRONMENT

A credible environment is an identified potential environment that is reasonable and realistic based on the best information available. A credible environment is one that a device or system may be exposed to during its life cycle (manufacturing to tactical employment or eventual demilitarization). These environments include extremes of temperature and humidity, EM effects, line voltages, lightning, and so on. Combinations of environments that can be reasonably expected to occur must also be considered within the context of credible environments. System safety should consider all credible environments when conducting an HA to identify system hazards.

CREDIBLE EVENT

A credible event is an identified potential event that is reasonable and realistic based on the best information available. The causal factors are plausible and convincing. A credible event is typically the most credible outcome expected from the occurrence of a mishap. Credible events can be identified through HA.

CREDIBLE FAILURE MODE

A credible failure mode is a failure mode resulting from the failure of either a single component, or the combination of multiple components, which has a reasonable probability of occurring during the systems life cycle. A credible failure mode is an identified potential failure that is reasonable and realistic based on the best information available. System safety should consider all credible failure modes when conducting an HA to identify system hazards.

CREDIBLE HAZARD

A credible hazard is an identified hazard that is reasonable and realistic based on the best design information available. The hazard causal factors (HCFs) are plausible and convincing. When the hazard or its causal factors can be proven to not exist, then the hazard is no longer credible. If a hazard has an extremely small probability of occurrence that does not mean it is not credible; credible refers to the fact that the hazard is possible, regardless of probability.

CRITICAL CHARACTERISTIC

Critical characteristic is any feature throughout the life cycle of a critical item, such as dimension, tolerance, finish, material or assembly, manufacturing or inspection process, operation, field maintenance, or depot overhaul requirement that if nonconforming, missing, or degraded may cause the failure or malfunction of the critical item. The term "critical characteristic" is synonymous with "critical safety characteristic."

CRITICAL DESIGN REVIEW (CDR)

The CDR is part of the Systems Development Life Cycle (SDLC) process of systems engineering and is one of the major design reviews in this process. The CDR demonstrates that the maturity of the design is appropriate to support proceeding with full-scale fabrication, assembly, integration, and test. CDR determines that the technical effort is on track to complete the flight and ground system development and mission operations, meeting mission performance requirements within the identified cost and schedule constraints. The CDR follows the preliminary design review (PDR). A rough rule of thumb is that at CDR, the design should be at least 85% complete. Many programs use drawing release as a metric for measuring design completion.

System safety should be involved in the PDR, typically making a presentation summarizing the safety effort to date, the DSFs in the system design, and the current level of mishap risk which the design presents. System safety provides, as a minimum, a subsystem hazard analysis (SSHA), SHA, safety assessment report (SAR), and final safety requirements for this review.

See *Preliminary Design Review (PDR)*, *System Requirements Review (SRR)*, and *Systems Engineering Technical Reviews (SETR)* for additional related information.

CRITICAL FAILURE

A critical failure is a failure that will directly cause death or serious injury when it occurs. It can also be defined as a failure that may cause a mission abort, mission failure, personal injury, or equipment damage. Typically, it is the most important failure mechanism in a hazard. A critical failure is a failure that will cause a Category I (Catastrophic) or Category II (Critical) hazard as these severity categories are defined in MIL-STD-882. It is not a failure that would cause a Category III (Marginal) or Category IV (Negligible) hazard.

CRITICAL FEW

Critical few refers to the general principle that a minority of the system elements will cause a majority of the system hazards. This philosophy follows the Pareto principle, which states that for many events, roughly 80% of the effects come from 20% of the causes. For example, 80% of a product's sales will typically come from 20% of the clients.

CRITICAL HAZARD

A critical hazard is a hazard that has a Category II (Critical) severity level, as defined by the hazard severity criteria in MIL-STD-882.

See *Hazard Severity Levels* for additional related information.

CRITICAL ITEM (CI)

In the case of safety a critical item is an item whose failure to operate, or incorrect operation, could lead or contribute to an accident (mishap). It can be a part, assembly, installation, or production process with one or more critical characteristics which will cause an unacceptable degradation of performance that would affect safety. A CI is typically referred to as an SCI or CSI. It should be noted that although somewhat related, there is a difference between SCI and CSI items.

See *Critical Safety Item (CSI)* and *Safety-Critical Item (SCI)* for additional related information.

CRITICAL ITEM LIST (CIL)

The CIL is a list of items that is considered critical for reliable and/or safe operation of the system. The list is generated from the FMEA and the failure modes and effects and criticality analysis (FMECA). Although it is primarily a reliability product, it can be used for system safety purposes.

CRITICAL SAFETY CHARACTERISTIC

Critical safety characteristic is any feature, such as tolerance, finish, material composition, manufacturing, assembly, or inspection process or product, which if nonconforming or missing could cause the failure or malfunction of a CSI.

See *Critical Safety Item (CSI)* for additional related information.

CRITICAL SAFETY ITEM (CSI)

CSI is a part, assembly, installation, subsystem, or system with one or more critical safety characteristics that, if missing or not conforming to the design requirements, quality requirements, or overhaul, and maintenance requirements, would result in an unsafe condition leading to death or serious injury. It is typically a part that must be manufactured to specific tolerances and maintained within these tolerances for safe system operation. A CSI part is a controlled part where certain safety characteristics are critical and are therefore tightly controlled.

A CSI is essentially the same as an SCI except that systems required to identify CSIs have additional statutory and regulatory requirements that the contractor must meet in supplying those CSIs to the government. For systems required to have a CSI list, HA and mishap risk assessment is used to develop that list. The determining factor in CSIs is the consequence of failure, not the probability that the failure or consequence would occur. CSIs include items determined to be "life-limited," "fracture critical," "fatigue-sensitive," and so on. Unsafe conditions relate to hazard severity categories I and II of MIL-STD-882. A CSI is also identified as a part, subassembly, assembly, subsystem, installation equipment, or support equipment for a system that contains a characteristic, failure mode, malfunction, or absence of which could result in a Class A or Class B accident as defined by DoDINST 6055.7.

For systems required to have a CSI list, HA and mishap risk assessment is used to develop that list. A CSI is essentially the same as an SCI except that systems required to identify CSIs have additional statutory and regulatory requirements that the contractor must meet in supplying those CSIs. An item can be classified as a CSI even after mitigation to an acceptable level of risk because maintaining quality control of the CSI characteristic is necessary as part of the mitigation. What constitutes a CSI is a critical characteristic of the item, which if defective, will cause the item to fail and thereby make the hazard mitigation ineffective. The critical characteristic may be a special process, finish, tolerance, material composition, and so on. And, the critical characteristic is difficult to control during manufacture or use, thereby necessitating special procedures, quality inspections, and so on, to ensure the critical characteristic is achieved and maintained (with a high probability) throughout the life of the system.

U.S. Army document AMC-R-702-32, Critical Safety Item Program (CSIP), August 29, 1990 describes a CSI and how it should be treated. Many programs will typically have a CSIP as part of the reliability or logistics process. U.S. Navy document Secretary of the Navy Instruction (SECNAVINST) 4140.2, Management of Aviation Critical Safety Items, January 25, 2006 is also a good resource for understanding and handling CSIs.

See *Safety-Critical Item (SCI)* for additional related information.

CULTURE

Culture is the knowledge and values shared by a society, the attitudes and behavior that are characteristic of a particular social group or organization. The safety culture of an organization is an important factor in implementing and maintaining safety in an operational or manufacturing program. It is also an important factor in an organization developing a safe system or product.

See *Safety Culture* for additional related information.

CUT SET (CS)

CS is a term used in FTA that identifies a unique set of events that together can cause the top UE of the FT to occur. The elements or events comprising a CS can be failures, human errors, software anomalies, environment conditions, or normal system actions. A CS can have any number of events in it; for example, a CS could be composed of one event or 10 events. Multiple events in a CS indicate that the events are ANDed together. A large FT can have well over 100,000 CSs. A minimal cut set (MCS) is a CS where none of the set elements can be removed from the set and still cause the top event to occur. Figure 2.13 shows an example FT and the CSs derived from it; note that a 1 order CS is an SPF.

84   SYSTEM SAFETY TERMS AND CONCEPTS

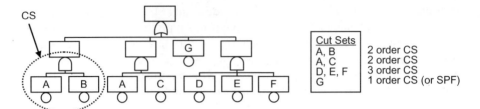

Figure 2.13   FT with cut sets shown.

CUT SET (CS) ORDER

CS order refers to the quantity of elements within a CS. Usually, the higher the CS order, the lower the probability of the CS, since CSs of order 2 or greater represent AND gate conditions, and the input probabilities are multiplied together.

CUT SET (CS) TRUNCATION

CS truncation refers to truncating or eliminating certain CSs from the overall FT probability calculation in order to make the calculation less complex and faster. Truncation is typically done by affecting the probability or CS order. For example, all CSs with a probability of 10^{-9} or smaller, or of order 6 or greater, are dropped from the overall calculation (because their contribution to the top probability would most likely be negligible). Truncation parameters should be carefully selected and the final calculation results carefully checked, as truncation can sometimes introduce errors or misunderstandings.

DAMAGE

Damage is the loss of value or utility of an item resulting from an event or condition. Damage is the undesired effects or consequences of an event, mishap, or enemy action. Damage is an outcome with undesirable effects, such as physical material loss or destruction.

In system safety, when an HA is performed, the outcome of an identified hazard is the expected damage that would result when the hazard transforms into a mishap. Damage, or damage effects, is typically measured in terms of death, severity of injury, functional and physical loss of a system or product, and/or the monetary losses incurred. MIL-STD-882 provides a method for categorizing the amount of damage by severity categories or levels.

See *Damage Effects* and *Hazard Severity Levels* for additional related information.

DAMAGE EFFECTS

Damage effects are the results or consequences of a mishap or UE. Damage, or damage effects, is typically measured in terms of death, severity of injury, functional and physical loss of a system or product, and/or the monetary losses incurred. MIL-STD-882 provides a method for categorizing the amount of damage by severity categories or levels. Damage effects are sometimes classified as primary damage effects and secondary damage effects. Primary damage effects are the direct results or consequences of a mishap, primarily affecting the system and/or system operators. Secondary damage effects are the indirect results or consequences of a mishap. Secondary damage would be considered collateral damage.

DAMAGE MODE AND EFFECTS ANALYSIS (DMEA)

DMEA is the analysis of a system or equipment conducted to determine the extent of damage sustained from given levels of hostile weapon damage mechanisms and the effects of such damage modes on the continued controlled operation and mission completion capabilities of the system or equipment. The DMEA is not the same as an HA, which would assess the damage effects resulting from potential system mishaps resulting from actuated hazards.

DANGER

Danger is the condition of being susceptible to harm or injury. A hazard presents danger; danger characterizes a relative exposure to a hazard, and therefore to a potential mishap. Danger also expresses a level of risk. A hazard may be present, but there may be little danger because of the precautions or hazard mitigations that are in place. It is recommended that this term be used carefully and sparingly as it can easily lead to confusion when attempting to describe a safety situation.

See *Hazard* and *Risk* for additional related information.

DANGER ZONE

A danger zone is a physical area or location within which a danger exists. For example, the work space surrounding a robotic manufacturing device is typically a danger zone.

DATA

Data are a collection of facts or information from which conclusions or decisions may be drawn. Data are recorded information, regardless of form or characteristics. Data may include numbers, words, pictures, and so on. Data can cover all types of information, including administrative, managerial, financial, scientific, technical, engineering, and logistics data. Data include written and electronically stored documents including enclosures. Data can consist of raw or refined information. Erroneous and/or misleading data can be a safety concern and an element of a hazard.

DATABASE

A set of related data or information, usually organized in such a manner as to permit efficient retrieval of specified subsets. Typically, significant operational data is stored in an electronic database, which the system utilizes for correct system decisions and operation. In training, simulation databases are often used for environment models especially for visual and radar landmass simulation. In safety-critical application, databases can be of system safety concern. Many different hazards can be spawned from erroneous or corrupted safety-critical data in an electronic database.

DATA ELEMENT LIST (DEL)

The DEL is a document containing the typical certification source data identified and listed in NAVAIRINST 13034.1C, Flight Clearance Polices for Air Vehicles and Aircraft Systems, September 28, 2004.

See *Engineering/Data Requirements Agreement Plan (E/DRAP or EDRAP)* for additional related information.

DATA FLOW DIAGRAM

Data flow diagrams model system behaviors that a system must perform, along with the interconnections between the behaviors, including all inputs and outputs along with the data stores that each behavior path must access. These diagrams are useful for system design, as well as for system safety HA.

DATA ITEM DESCRIPTION (DID)

DIDs are documents describing the content and format for specific types of data required by contract in the CDRL. The contractor prepares the required data in a format specified in a DID. Preexisting DIDs are indexed in the Acquisition Management System Data List (AMSDL). The following preexisting DIDs are associated with system safety:

1. DI-SAFT-81626, System Safety Program Plan.
2. DI-SAFT-80100A, System Safety Program Plan.
3. DI-SAFT-80101B, System Safety Hazard Analysis Report.
4. DI-SAFT-80102B, Safety Assessment Report.
5. DI-SAFT-80103A, Engineering Change Proposal System Safety Report.
6. DI-SAFT-80104A, Waiver or Deviation System Safety Report.
7. DI-SAFT-80105A, System Safety Program Progress Report.
8. DI-SAFT-80106B, Health Hazard Assessment Report.
9. DI SAFT-80931, Explosive Ordnance Disposal Data.
10. DI-SAFT-81299, Explosive Hazard Classification Data.
11. DI-SAFT-81300A, Mishap Risk Assessment Report.
12. DI-MISC-80370, Safety Engineering Analysis Report.
13. DI-SAFT-80183A, Flight Termination System Report.
14. DI-SAFT-80184, Radiation Hazard Control Procedures.
15. DI-SAFT-81065, Safety Studies Report.
16. DI-SAFT-81066, Safety Studies Plan.
17. DI-R-7085A, Failure Mode, Effects Criticality Analysis Report.

DATA LINK

A data link is the telecommunication means of connecting one location to another for the purpose of transmitting and receiving digital information. It can also refer to a set of electronics assemblies, consisting of a transmitter and a receiver, and the interconnecting data telecommunication circuit. These are governed by a link protocol enabling digital data to be transferred from a data source to a data receiver.

There are three basic types of data link configurations:

- Simplex communications, meaning all communications in one direction only.
- Half-duplex communications, meaning communications in both directions, but not both ways simultaneously.
- Duplex communications, meaning communications in both directions simultaneously.

In civil aviation, a data link system (known as Controller Pilot Data Link Communications) is used to send information between aircraft and air traffic controllers (ATCs) when an aircraft is too far from the ATC to make voice radio communication and radar observations possible. Such systems are used for aircraft crossing the Atlantic and Pacific oceans. In military aviation, a data link carries weapons targeting information, and it can also carry information to help warplanes land on aircraft carriers. In UA, land vehicles, boats, and spacecraft, a two-way (full-duplex or half-duplex) data link is used to send control signals and to receive telemetry. Examples of technologies used as data links are RF, fiber optics, and laser.

DATA PACKAGE

A data package is packet of technical and/or management material that is provided to a review board that is evaluating the safety of a product or system. The review board uses the information and data contained in the data package to draw conclusions regarding the safety of a product or system. The data package must contain correct, current, and appropriate information in order for the review board to make a favorable determination.

See *Technical Data Package (TDP)* for additional related information.

DEACTIVATED CODE

Deactivated code is computer code that resides in an executable program, but is not used by the program. It is a segment of unreachable code left in a computer program. It is unreachable because the code was deactivated, typically through the use of comments. Deactivated code raises the following software safety (SwS) questions:

- Why was the code deactivated; is there any safety significance?
- If it was deactivated for testing purposes, was it reactivated?
- Should deactivated code be considered dead code and thus be removed?

See *Dead Code* for additional related information.

DEAD CODE

Dead code is computer code that resides in an executable program but is not used by the program. Dead code is a segment of unreachable code left in a computer program. It is unreachable because the code logic does not call for it, either intentionally or unintentionally. The primary causes for unreachable or dead code include:

1. The code is intentionally commented-out; this can include both comments and code left in the module but intended to be nonfunctional.
2. The code segment is erroneously commented-out; it is intended functional code.

3. An error in the program decision logic such that there is no path (or call) to the code segment as intended by the designer; it is intended functional code.
4. Nonfunctional code intentionally left in the software module by the designer; however, there is intentionally no call to the specific code.
5. Code segments left in the operational code for testing purposes, but commented-out for operational use.
6. Code segments in the operational code that is for different versions or models of the system. The code recognizes the particular system and uses the appropriate code.
7. Code segments left in the operational because of a design modification eliminating a function.

In SwS there is a concern that dead or deactivated code may be accidentally reached and used if there is a fault whereby a jump is inadvertently made into the code segment. The outcome in such a situation would be unpredictable in most cases. However, it should be noted that if the code is commented-out, there is no chance of this happening; commented code is totally un-executable. It should also be noted that an inadvertent jump to dead code constitutes a failure that itself may be more significant than the operation of the dead code.

DEBUGGING

Debugging is a process to detect and remedy inadequacies in an item. Debugging should not to be confused with burn-in, fault-isolation, or screening. Locating errors in software code is referred to as debugging.

DEDUCTIVE REASONING

In deductive reasoning, the conclusion does not exceed or imply more than the premises or data it is based upon. Deductive reasoning is a logical process in which a conclusion is drawn from a set of premises and contains no more information than the premises taken collectively. For example, all dogs are animals; this is a dog; therefore, this is an animal. The truth of the conclusion is dependent upon the premises; the conclusion cannot be false if the premises on which it is based are true. For example, all men are apes; this is a man; therefore, this is an ape. The conclusion seems logically true; however, it is false because the premise is false.

See *Deductive Safety Analysis* and *Inductive Reasoning* for additional related information.

DEDUCTIVE SAFETY ANALYSIS

A deductive analysis is one that would conclude no more than the data provides. The specific causal factors supporting the conclusion must be identified and established, and then the conclusion will be true. It seems like reverse logic; however, the hazard can be validly deduced only from the specific detailed causal factors. Deductive and inductive qualities have become intangible attributes of HA techniques. An inductive analysis would be used to broadly identify hazards without proven assurance of the causal factors, while a deductive analysis would attempt to find the specific causal factors for identified hazards. An inductive analysis can be thought of as a "what-if" analysis. The analyst asks, "what if" this part failed, what are the consequences. A deductive analysis can be thought of as a "how-can" analysis. The analyst asks, if this event was to occur, "how can" it happen or what are the causal factors?

Deductive analysis tends to be a top-down approach (going from the general to the specific) and an example of this is the FTA methodology. An FMEA is just the reverse, it goes from the specific to the general and is known as an inductive analysis.

See *Deductive Reasoning*, *Inductive Reasoning*, and *Inductive Safety Analysis* for additional related information.

DEFECT

A defect is any nonconformance of an item, product, or activity with specified requirements. A defect is also any deviation or nonfulfillment of an intended requirement or reasonable expectation for use. A defect is an error or omission which, if left uncorrected, will result in an item which does not fulfill internal or customer requirements. Product improvements and changes in scope are not defects.

Defects are typically grouped into one of the following categories:

1. Critical defect—A defect that constitutes a hazardous or unsafe condition, or as determined by experience and judgment could conceivably become so, thus making the aircraft, system, or equipment unsafe for flight or endangering operating personnel.
2. Major defect—A defect, other than critical, that could result in failure or materially reduce the usability of the unit or part for its intended purpose.
3. Minor defect—A defect that does not materially reduce the usability of the unit or part for its intended purpose or is a departure from standards but which has no significant bearing on the effective use or operation of the unit or part.

Defects weaken the properties of an item; for example, a material defect weakens the structural integrity of an item, a software defect weakens the ability of a computer program to always work correctly, and a requirements defect weakens the design of a product or system. Defects can be causal factors in hazards, but not all defects are necessarily hazards or part of HCFs. Critical and major defects are typically of concern to system safety.

DEFENSE-IN-DEPTH

Defense-in-depth is the provision for multiple defenses (or safety features) as mitigations against a hazard or a top-level mishap (TLM). This approach is similar to the layers of protection (LOP) concept. Defense-in-depth is a design approach taken to preclude single failures from causing a particular UE through the incorporation of various LOP. Protection methods may include any one or more of the following:

- Multiple functional and/or engineered barriers
- Use of large design safety margins where possible
- High-quality in design and manufacture
- Operation within design limits
- Testing and inspection to maintain design limits
- Fault tolerance design
- Fail-safe design

See *Layers of Protection (LOP)* for additional related information.

DEFLAGRATION

With regard to explosives, deflagration is the ignition and burning of confined energetic materials leading to nonviolent pressure releases as a result of low strength containers or venting through container closures. The container might rupture but does not fragment; closures might be thrown about and fire may spread. Propulsion might launch an unsecured item. No blast or significant fragmentation damage to the surroundings; only heat and smoke damage from burning energetic material. Deflagration involves a chemical reaction proceeding at subsonic velocity along the surface of and/or through an explosive, producing hot gases at high pressure. With regard to explosives, this is the fourth most violent type of explosive event.

See *Explosive Event* for additional related information.

DEGRADATION

Degradation is deterioration of the physical properties or state of an item. System degradation is generally a reduced system state with reduced operational capabilities.

DEMILITARIZATION

Demilitarization (or demil) is the act of destroying the military advantages inherent in certain types of equipment or material. The term includes mutilation, dumping at sea, scrapping, melting, burning, or alteration designed to prevent the further use of this equipment and material for its originally intended military or lethal purpose and applies equally to material in unserviceable or serviceable condition. Demilitarization is also the steps taken in the disposal of military equipment. System safety is concerned with demil and disposal of equipment due to the hazards involved in the process and the potential hazardous and toxic materials involved.

DEPARTMENT OF DEFENSE EXPLOSIVES SAFETY BOARD (DDESB)

The DDESB is chartered in Title 10 of the U.S. Code. The DDESB authors DOD 6055.9-STD, Ammunition and Explosives Safety Standards. It also evaluates scientific data which may adjust those standards, reviews and approves all explosives site plans for new construction, and conducts worldwide visits to locations containing U.S. title munitions.

DEPENDABILITY

Dependability is a measure of the degree to which an item is operable and capable of performing its required function at any (random) time during a specified mission profile, given that the item is available at mission start. Examples of dependability measures are availability, interoperability, compatibility, reliability, repeatability, usage rates, vulnerability, survivability, penetrability, durability, mobility, flexibility, and reparability.

DEPENDENCE (IN DESIGN)

Design dependence is a design whereby the failure of one item directly causes, or leads to, the failure of another item. This happens when the functional status of one component is affected by the functional status of another component. CCF dependencies normally stem from the way the system is designed to perform its intended function. Dependent failures are those failures that defeat redundancy or diversity, where redundancy is intentionally employed to improve safety and/or reliability.

In some system designs, dependency relationships can be very subtle, such as in the following cases:

1. Standby redundancy—When an operating component fails, a standby component is put into operation, and the system continues to function.

Failure of an operating component causes a standby component to be more susceptible to failure because it is now under load.
2. Common loads—When failure of one component increases the load carried by other components. Since the other components are now more likely to fail, we cannot assume statistical independence.
3. Mutually exclusive events—When the occurrence of one event precludes the occurrence of another event.

Dependence is a design concept (and implementation) in which a system element or component has a link, or potential link, of some type with other system elements or components. Dependency is a unique relationship between two elements, that is, where something is contingent on something else. In a dependency situation, the functional status of one component is affected by the functional status of another component. The type and strength of the dependency link determines if the dependency is harmless or potentially detrimental. For system safety, this design dependency link is significant when it provides for the possibility of the failure of one item to cause the failure of another item. To some degree, all system aspects may have some form of dependency. For example, a component may have no dependencies on any other components in the subsystem it resides in; however, it may be susceptible to extreme heat from an external source, which is a provisional dependency. Dependency safety is contingent on the dependency type, form, amount, and safety-criticality involved. Dependence and independence have a profound effect on system success, safety, reliability, and probability calculations.

Dependence and independence have a profound effect on probability calculations. Dependence is a design concept (and implementation) which does not guarantee that the failure of one item does not cause a failure of another item.

The various types of dependencies can be described as follows:

- Functional link—A dependency established by one function that must occur before another, or two functions that share a common element
- Timing link—A dependency established by a set of functions or tasks that must occur in a specific timed order
- Schedule link—A timing dependency that exists between project events (shown by a Gantt chart)
- Organizational link—Dependencies within an organization, such as children are dependent on their parents, managers are dependent on their workers
- Event link—When the occurrence of one event causes or influences the occurrence of another event
- Statistical link—When the occurrence of one event makes it more probable that another event will occur

- Failure link—The failure of one component causes or influences the failure of another component
- Zonal link—Design is susceptible to common faults within a system zone because of activities within that zone
- Environmental link—Design is susceptible to environment faults or attacks (temperature, water, fire, and so on)
- Manufacturing link—Design is susceptible to common manufacturing defects

Dependencies can be classified as either intrinsic or extrinsic to the system as follows:

1. Intrinsic dependency refers to dependencies where the functional status of one component is affected by the functional status of another. These dependencies normally stem from the way the system is designed to perform its intended function. This type of dependency is based on the type of influence that components have on each other.
2. Extrinsic dependency refers to dependencies where the couplings are not inherent and are not intended in the designed functional characteristics of the system. Such dependencies are often physically external to the system, such as vibration, heat, RF, environment, and mission changes beyond original design limits.

See *Dependent Event* and *Independence (In Design)* for additional related information.

DEPENDENT EVENT

Events are dependent when the outcome of one event directly affects or influences the outcome of a second event (probability theory). To find the probability of two dependent events both occurring, multiply the probability of A and the probability of B after A occurs; $P(A \text{ and } B) = P(A) \cdot P(B \text{ given } A)$ or $P(A \text{ and } B) = P(A) \cdot P(B|A)$. This is known as conditional probability. For example, assume a box contains a nickel, a penny, and a dime; find the probability of choosing first a dime and then, without replacing the dime, choosing a penny. These events are dependent. The first probability of choosing a dime is $P(A) = 1/3$. The probability of choosing a penny is $P(B|A) = 1/2$ since there are now only two coins left. The probability of both is $1/3 \cdot 1/2 = 1/6$. Keywords such as "not put back" and "not replace" suggest that events are dependent. Two failure events A and B are said to be dependent if $P(A \text{ and } B) \neq P(A)P(B)$. In the presence of dependencies, often, but not always, $P(A \text{ and } B) > P(A)P(B)$. This increased probability caused by two (or more) dependent events is the reason dependent events are of safety concern.

DEPENDENT FAILURE

A component failure is a dependent failure when a conditional relationship exists between two components, whereby the failure of one is conditional upon failure of the other; the second failure depends on the first failure occurring. In probability theory, events are dependent when the outcome of one event directly affects or influences the outcome of a second event. To find the probability of two dependent events both occurring, multiply the probability of A and the probability of B after A occurs; $P(A \text{ and } B) = P(A) \cdot P(B \text{ given } A) = P(A) \cdot P(B|A)$. This is known as conditional probability. Two failure events A and B are said to be dependent if $P(A \text{ and } B) \neq P(A)P(B)$. In the presence of dependencies, often, but not always, $P(A \text{ and } B) > P(A)P(B)$. This increased probability of two (or more) events is the reason dependent failures are of safety concern. It should also be noted that when redundant dependent components are used in a system, they are both susceptible to the same dependent failure, which creates a CCF dependency.

It should be noted that in system safety and FTA, a *secondary* failure is typically a dependent failure mode. A secondary failure occurs when the failure of an item is caused by an external source or force, such as radio frequency (RF) energy and heat. For example, when excessive heat on a transistor from an external source causes the transistor to fail, this is referred to as a secondary type failure of the transistor (vs. a primary inherent failure mode). Conditional probability should be used on secondary failures because it is dependency situation; however, in FTA, the conditional aspect is often ignored and the failure event is simply treated as an independent failure and is assigned an appropriate failure that considers both the component failure and the external event failure rate. The mathematical error produced by this approach is typically minimal.

See *Dependent Event* for additional related information.

DEPENDENT VARIABLE

A dependent variable is the factor or element in an experiment that is expected to be affected by the manipulation or control of the independent variable.

DESIGN LOAD

The design load is the load a device, component, or structure is designed to handle, usually specified by a design requirement. For example, a resistor design load might be 12 amps or a bridge design load might be 50 tons. This is the load the device is expected to handle under normal operation. Quite often an SF is applied to the design load to cover unexpected load conditions.

See *Safety Factor (SF)* for additional related information.

DETERMINISTIC PROCESS

A deterministic process is a repeatable process that always results in the same outcome given the same set of input. The process is predictable and generally understood in how it works. A deterministic process is sure or certain, and is the opposite of a random process.

DERATING

Derating is the reduction of the applied load (or rating) of a device to improve reliability or to permit operation at high ambient temperatures. Derating involves using an item in such a way that applied stresses are below rated values, or the lowering of the rating of an item in one stress field to allow an increase in another stress field. Derating is the operation of an electrical or electronic device at less than its rated maximum power dissipation, taking into consideration the case or body temperature, ambient temperature, and the type of cooling mechanism used. When a component or circuit is operated at a lower design load (derated) than the component is rated for, the component or circuit can be assigned a higher reliability rating. This is an intentional design method for improved reliability.

DERIVED REQUIREMENTS

Derived requirements establish those characteristics typically identified during synthesis of preliminary product or process solutions and during related trade studies and verifications. They generally do not have a parent function and/or performance requirement but are necessary to have generated system elements accomplish their intended function. They were not identified during establishment of the original design requirements.

DESIGN

A "design" is a product consisting of those characteristics of a system that are selected by the developer in response to the development requirements. Some designs will match the requirements, while others will be elaborations of requirements, such as the definition and wording of error messages. Some designs will be implementation related, such as decisions, about what software units and logic to use to satisfy the requirements. A design is somewhat akin to an architecture. "To design" is the process of defining, selecting, and describing solutions to requirements in terms of products and processes. A design is the product of the process of designing that describes the solution (conceptual, preliminary, or detailed) of the system, system elements, or system end-items.

See *Architecture* for additional related information.

DESIGN ASSURANCE LEVEL (DAL)

The term DAL comes from Radio Technical Commission for Aeronautics (RTCA)/DO-254, Design Assurance Guidance for Airborne Electronic Hardware, 2000. In this document hardware is classified into five levels based on a set of criteria for each level. The derived software level (SL) is based on the contribution of the software to potential failure conditions as determined by the system safety assessment (SSA) process.

The DAL is an index number ranking the safety-criticality of the system functions. This ranking implies that in order to make the system safe, greater development rigor must be applied to each successively critical level. Table 2.3 correlates the hardware DALs to the five classes of failure conditions and provides definitions of hardware failure conditions and their respective DALs. Initially, the hardware DAL for each hardware function is determined by the SSA process using a functional hazard analysis (FHA) to identify potential hazards and then the preliminary system safety assessment (PSSA) process allocates the safety requirements and associated failure conditions to the function implemented in the hardware.

Given the safety, functional, and performance requirements allocated to the hardware by the system process, the hardware safety assessment determines the hardware DAL for each function and contributes to determining the appropriate design assurance strategies to be used.

See *Development Assurance Level (DAL)* and *Software Level (SL)* for additional related information.

DESIGNATED ENGINEERING REPRESENTATIVE (DER)

A DER is an engineer who is appointed by the FAA to act on behalf of a company or as an individual consultant to represent the FAA at the company's facility. Company DERs act on behalf of their employer and may only approve, or recommend approval, of technical data to the FAA for this company. Consultant DERs are appointed to act as an independent DER to approve or recommend approval of technical data to the FAA.

See *Designated Representative (DR)* for additional related information.

DESIGNATED REPRESENTATIVE (DR)

A DR is an individual designated and authorized to perform a specific function for a contractor or government agency. Responsibilities may include evaluation, assessment, design review, participation, and review/approval of certain documents or actions. For example, National Aeronautics and Space Administration (NASA), the FAA, or the DoD may designate a representative at a contractor's facility to facilitate a particular contractual effort. A contractor may do the same at a subcontractor's facility.

TABLE 2.3 Hardware Design Assurance Level Definitions and Their Relationships to Systems Development Assurance Level

Systems Development Assurance Level	Failure Conditions Classifications	Failure Conditions Description	Hardware Design Assurance Level Definitions
Level A	Catastrophic	Failure conditions that would prevent continued safe flight and landing.	A: Hardware functions whose failure or anomalous behavior, as shown by the hardware safety assessment, would cause a failure of system function resulting in a catastrophic failure condition for the aircraft.
Level B	Hazardous/ Severe-Major	Failure conditions that would reduce the capability of the aircraft or the ability of the flight crew to cope with adverse operating conditions to the extent there would be a large reduction in safety margins or functional capabilities, physical distress or higher workload such that the flight crew could not be relied upon to perform their tasks accurately or completely, or adverse effects on occupants, including serious or potentially fatal injuries to a small number of those occupants.	B: Hardware functions whose failure or anomalous behavior, as shown by the hardware safety assessment, would cause a failure of system function resulting in a hazardous/ severe-major failure condition for the aircraft.
Level C	Major	Failure conditions that would reduce the capability of the aircraft or the ability of the flight crew to cope with adverse operating conditions to the extent there would be a significant reduction in safety margins or functional capabilities, a significant increase in flight crew workload or in conditions impairing flight crew efficiency, or discomfort to occupants, possibly including injuries.	C: Hardware functions whose failure or anomalous behavior, as shown by the hardware safety assessment, would cause a failure of system function resulting in a major failure condition for the aircraft.

TABLE 2.3 *Continued*

Systems Development Assurance Level	Failure Conditions Classifications	Failure Conditions Description	Hardware Design Assurance Level Definitions
Level D	Minor	Failure conditions that would not significantly reduce aircraft safety, and which would involve flight crew actions that are well within their capabilities, minor failure conditions may include a slight reduction in safety margins or functional capabilities, a slight increase in flight crew workload, such as routine flight plan changes, or some inconvenience to occupants.	D: Hardware functions whose failure or anomalous behavior, as shown by the hardware safety assessment, would cause a failure of system function resulting in a minor failure condition for the aircraft.
Level E	No effect	Failure conditions that do not meet the operational capability of the aircraft or increase flight crew workload.	D: Hardware functions whose failure or anomalous behavior, as shown by the hardware safety assessment, would cause a failure of system function with no effect on aircraft operational capability or flight crew workload. For a function determined to be level E, no further guidance of this document need apply; however, it may be used for reference.

See *Designated Engineering Representative (DER)* for additional related information.

DESIGN DIVERSITY

Design diversity is a special design safety method whereby redundant components or elements are developed diversely and independently. Design diversity means that redundant components are developed using different design setups. This is a design strategy employed to avoid CCF problems to decrease the probability that redundant components might fail both at the same time under identical conditions. The strategy can be used on both hardware and software. For example, an aircraft with two redundant flight control computers would apply design diversity in the hardware by utilizing a different type of CPU, made by different manufacturers, in each computer. And, the software developed for each computer could be developed in a different programming language, as well as being developed by separate and independent programming teams.

See *Common Cause Failure (CCF)* and *Common Mode Failure (CMF)* for additional related information.

DESIGN-FOR-RELIABILITY (DFR)

DFR is the process of applying reliability engineering techniques to the development of a system or product in order to intentionally build reliability into the system or product from the outset, as opposed to trying to add it in at a later time. DFR applies the established reliability process.

DESIGN-FOR-SAFETY (DFS)

DFS is the process of applying system safety discipline and engineering techniques to the development of a system or product in order to intentionally build safety into the system or product from the outset, as opposed to trying to add it in at a later time. DFS applies the established system safety process.

DESIGN QUALIFICATION TEST

A test intended to demonstrate that the test item will function within performance specifications. The purpose is to uncover deficiencies in design and the method of manufacture. Sometimes the test is under simulated conditions more severe than those expected from handling, storage, and operations. The test is intended to exceed design safety margins or to introduce unrealistic modes of failure.

DESIGN REQUIREMENT

A design requirement is a written statement describing a design characteristic for the design of a product or system. Design requirements are a set of those requirements specifying how to develop and produce products and processes.
See *Requirements* for additional related information.

DESIGN SAFETY FEATURE (DSF)

A DSF is a special and intentional feature in the design of a system or product employed specifically for the purpose of eliminating or mitigating the risk presented by an identified hazard. A DSF may not be necessary for system function, but it is necessary for superior safety (i.e., risk reduction). A DSF can be any device, technique, method, or procedure incorporated into the design to specifically eliminate or reduce the risk factors comprising a hazard. A feature is a prominent attribute or aspect of something; a DSF is a special attribute intentionally placed in the system design for the purpose of mitigating a hazard and its associated risk. DSFs can take many different forms and methods such as:

- Redundancy design
- Fail-safe design
- Fault-tolerant design
- Single-point failure design
- Design diversity
- Special procedures
- Special training
- Warnings and cautions
- Safety margins
- Barriers
- Partitions
- Interlock design
- Fail-operational design
- Fail-nonoperational design
- X-ray analysis
- Stress testing
- Fire detection and suppression
- Personal protective equipment (PPE)

In order to effectively eliminate or mitigate a hazard, it is necessary to understand the risk drivers involved in order to select a DSF that appropriately addresses the risk. This involves understanding the HCFs and the risk they present, and then applying DSFs that can counter these causal factors and reduce the risk. DSFs for each hazard should be selected based on effectiveness, cost, and feasibility. Feasibility includes consideration of both means and schedule for accomplishment. When developing DSF options, it is recommended that the SOOP established by MIL-STD-882 be followed. The SOOP provides a preferred order for implementing different DSFs, each of which affects the level of risk differently; each SOOP level provides a higher level of safety in a step-like manner. During product and system development design reviews, the DSFs should be identified to show that an effective SSP is in effect. This is also a time to identify and take credit for safety features in the product/system design.

It should be noted that the terms safety feature, safety measure, safety mechanism, DSF, design safety measure, design safety mechanism, and hazard countermeasure are synonymous and can be used interchangeably; however, most system safety practitioners tend to use the terms safety feature or safety measure most often.

See *Safety Order of Preference (SOOP)* for additional related information.

DESIGN SAFETY MEASURE

A design safety measure is a special and intentional feature in the design of a system or product employed specifically for the purpose of eliminating or mitigating the risk presented by an identified hazard. A design safety measure may not be necessary for system function, but it is necessary for safety (i.e., risk reduction). A design safety measure can be any device, technique, method, or procedure incorporated into the design to specifically eliminate or reduce the risk factors comprising a hazard.

See *Design Safety Feature (DSF)* and *Safety Order of Preference (SOOP)* for additional related information.

DESIGN SAFETY MECHANISM

A design safety mechanism is a special and intentional feature in the design of a system or product employed specifically for the purpose of eliminating or mitigating the risk presented by an identified hazard. A design safety mechanism may not be necessary for system function, but it is necessary for safety (i.e., risk reduction). A design safety mechanism can be any device, technique, method, or procedure incorporated into the design to specifically eliminate or reduce the risk factors comprising a hazard.

See *Design Safety Feature (DSF)* and *Safety Order of Preference (SOOP)* for additional related information.

DESIGN SAFETY PRECEPT (DSP)

A DSP is a safety precept that is directed specifically at system design. These precepts are general design objectives intended to guide and facilitate the design of more detailed solutions, without dictating the specifics within the precept. General design direction allows for the selection of specific solutions that are focused on the particular application, along with the technologies available. An example DSP might be "The unmanned system shall be designed to only perform valid commands issued from a valid authorized source."

See *Safety Precept*s for additional related information.

DESIGN SPECIFICATION

A design specification is a document that describes the functional and physical prerequisites for an article in the form of design requirements. The specification and requirements can be at the system, subsystem, or component level. In its initial form, the design specification is a statement of functional requirements with only general coverage of physical and test requirements. The design specification evolves through the project life cycle to reflect progressive refinements in performance, design, configuration, and test requirements. Design specifications provide the basis for technical and engineering management control. A specification is a document prepared to support system acquisition and life-cycle management, which clearly and accurately describes essential technical requirements and verification procedures for items, materials, and services. When invoked by a contract, it is legally enforceable and its requirements are contractually binding.

See *Requirements* and *Specification* for additional related information.

DESTRUCTIVE PHYSICAL ANALYSIS (DPA)

An internal destructive examination of a finished part or device to assess design, workmanship, assembly, and any other processing associated with fabrication of the part. The part is typically damaged by the testing performed on it to determine if any flaws exist. Nondestructive test methods are preferred in order to save the item for usage; however, some types of items cannot be evaluated except by a destructive test. For example, a new wing design may be mechanically flexed many millions of times until it breaks, to determine if the design achieved the predicted goals.

DETECTABLE FAILURE

A detectable failure is a failure at the component, equipment, subsystem, or system (product) level that can be identified through periodic testing or revealed by an alarm or an indication of an anomaly during system operation.

DETERMINISTIC (ANALYSIS)

A deterministic analysis is an analysis where the given situation is deterministic and the results are an inevitable consequence of the antecedent causes. Given the same input, the results will always produce the same answer; they are repeatable. For example, 2 + 2 is a deterministic situation where the result will always equal 4. In a nondeterministic situation, the results will not always be the same. A nondeterministic situation occurs in parallel programming, where depending on timing between branches, the results may not always be identical or the steps achieving the results may not be identical.

DETONATION

With regard to explosives, detonation is the most violent type of explosive event. A supersonic decomposition reaction propagates through the energetic material producing an intense shock in the surrounding medium (e.g., air or water) with very rapid plastic deformation of metallic cases followed by extensive fragmentation. All the energetic material will be consumed. The effects include ground craters for items on or close to the ground, holing/plastic flow damage/fragmentation of adjacent metal, and blast overpressure to nearby structures.

See *Explosive Event* for additional related information.

DETONATION VELOCITY

Detonation velocity is the rate at which a self-sustaining supersonic shock wave (detonation wave) travels through an explosive material.

DEVELOPMENT ASSURANCE LEVEL (DAL)

The term Development Assurance Level (DAL) comes from Society of Automotive Engineers/Aerospace Recommended Practice (SAE/ARP)-4754, Certification Considerations for Highly-Integrated or Complex Aircraft Systems, Aerospace Recommended Practice, 1996. This document establishes system DALs and the process of assurance rigor required for each level. The DAL also determines the necessary software and hardware design levels of DO-178B and DO-254. The DAL is an index number ranking the safety-criticality of the system or software. This ranking implies that in order to make the

TABLE 2.4 System DAL Assignment

Failure Condition Classification	System Development Assurance Level
Catastrophic	A
Hazardous/severe major	B
Major	C
Minor	D
No safety effect	E

system safe, greater development rigor must be applied to each successively critical level.

The system development assurance level is assigned based on the most severe failure condition classification associated with the applicable aircraft-level function(s), shown in Table 2.4. This table departs slightly from Advisory Circular (AC) 25.1309-1A and Advisory Material Joint (AMJ) 25.1309 by establishing level E as "no safety effect."

System architectural features, such as redundancy, monitoring, or partitioning, may be used to eliminate or contain the degree to which an item contributes to a specific failure condition. System architecture may reduce the complexity of the various items and their interfaces and thereby allow simplification or reduction of the necessary assurance activity. If architectural means are employed in a manner that permits a lower assurance level for an item within the architecture, substantiation of that architecture design should be carried out at the assurance level appropriate to the top-level hazard (TLH). This does not preclude assignment of item levels that are lower than the level associated with the TLH, but assurance that the item level assignments and their independence are acceptable should be validated at the higher level and verified by the SSA. Detailed guidelines and methods for conducting the various assessments are described in SAE/ARP-4761.

The Development Assurance Level of an aircraft function depends on the severity of the effects of failures or development errors of that function on the aircraft, crew, or occupants. The Development Assurance Level of each item depends on both the system architecture and the resulting failure effects of the item on the functions performed by the system. DO-178 procedures should be used to verify that the software implementation meets the required DALs. The hardware DALs are verified via procedures that are to be defined by RTCA DO-254.

The DAL of an item may be reduced if the system architecture:

- Provides multiple independent implementations of a function (redundancy)
- Isolates potential faults in part of the system (partitioning)
- Provides for active (automated) monitoring of the item
- Provides for human recognition or mitigation of failure conditions

See *Design Assurance Level (DAL)* and *Software Level (SL)* for additional related information.

DEVELOPER

A developer is the individual or organization assigned a responsibility for a development effort, which typically designs and produces a product or system. Developers can be either internal to the government or contractors.

See *Contractor* for additional information.

DEVIATION

A deviation is written authorization, granted after contract award and prior to the manufacture of an item, to depart from a particular performance or design requirement of a contract, specification, or referenced document, for a specific number of units or a specified period of time. Deviations are intended only as one-time departures from an established configuration for specified items or lots and are not intended to be repeatedly used in place of formal engineering changes.

Deviations can be classified into categories by the following criteria:

- Critical deviation—A deviation is designated as critical when the deviation consists of a departure involving safety or when the configuration documentation defining the requirements for the item classifies defects in requirements and the deviations consist of a departure from a requirement classified as critical.
- Major deviation—A deviation is designated as major when the deviation consists of a departure involving health, performance, interchangeability, reliability, survivability, maintainability, or durability of the item or its repair parts; effective use or operation; weight; or appearance (when a factor) or when the configuration documentation defining the requirements for the item classifies defects in requirements and deviations consist of a departure from a requirement classified as major.
- Minor deviation—A deviation is designated as minor when it consists of a departure that does not qualify as critical or major or when the configuration documentation defining the requirements for the item classifies defects in requirements and the deviations consist of a departure from a requirement classified as minor.

DIAGNOSTICS

Diagnostics refers to the hardware, software, or other documented means used to determine that a malfunction has occurred and to isolate the cause of the malfunction. It refers to the action of detecting and isolating failures or faults.

DISEASE

A disease or illness is nontraumatic physiological harm or loss of capacity produced by systemic, continued, or repeated stress or strain; exposure to toxins, poisons, fumes, and so on; or other continued and repeated exposures to conditions of the environment over a long period of time. For practical purposes, an occupational illness or disease is any reported condition that does not meet the definition of injury (from a mishap). Illness includes both acute and chronic illnesses, such as, but not limited to, a skin disease, respiratory disorder, or poisoning.

See *Illness* for additional related information.

DISCREPANCY

A discrepancy is a departure from the development requirements specifications, resulting in a product or system characteristic that is in nonconformance. A discrepancy usually results in the failure of an item to perform its required function as intended. An item may be a system, subsystem, product, software, human, or component.

See *Deviation*, *Failure*, and *Nonconformance* for additional related information.

DISSIMILAR DESIGN

Dissimilar design is a special design safety method whereby redundant components or elements are developed diversely and independently; in addition, they are developed using different approaches and methods. In order to make design redundancy safer, two redundant components are designed and/or manufactured differently, but to the same requirements. Both components perform the same functions, but possibly in slightly different ways; that is, design diversity is applied. The objective is to ensure that both redundant components in a system do not fail simultaneously due to CCFs. Since they have been developed differently, they should not have the same inherent failure susceptibilities. For example, a system using two redundant MOVs in a process system would apply dissimilar design by using MOVs from two different manufacturers. In a design such as this, it is important to ensure the operating and maintenance manuals contain a warning note that the valves cannot be from the same manufacturer.

See *Design Diversity* for additional related information.

DISSIMILAR SOFTWARE

Dissimilar multiple-version software is a system design technique that involves producing two or more software components that provide the same function, but in different ways. This approach is intended to avoid sources of common errors between redundant components. Multiple-version dissimilar software is also referred to as multi-version software, dissimilar software, N-version programming, or software diversity. The degree of dissimilarity and hence the degree of safety protection is not usually measurable with dissimilar software. In this approach, software modules for redundant subsystems are developed to the same design requirements using design diversity. The probability of loss of system function will decrease if the module results are compared during operation and adjustments are made when error differences are discovered. Software verification is the first line of defense against errors, while dissimilar software provides a second line of defense. For example, an aircraft flight control system that has two redundant computers might develop the software for one computer in the C language and the software for the second computer in the Ada language.

See *Design Diversity* and *Dissimilar Design* for additional related information.

DIVERSITY

In systems theory, diversity refers to diversity in design, whereby redundant components or elements are developed in diverse (i.e., different) methods. Diversity is a design strategy employed to avoid CCF problems to decrease the probability that redundant components might fail both at the same time under identical conditions. The strategy can be used on both hardware and software. For example, an aircraft with two redundant flight control computers would apply design diversity in the hardware by utilizing a different type of CPU, made by different manufacturers, in each computer. And, the software developed for each computer could be developed in a different programming language, as well as being developed by separate and independent programming teams.

See *Design Diversity* for additional related information.

DORMANT CODE

Dormant code is software code that is included in the executable software, but is not meant to be used. Dormant code is similar to dead code. Dormant code is usually the result of COTS or reused software that includes extra functionality over what is required.

See *Dead Code* for additional related information.

DORMANT FAILURE (OR DORMANCY)

A dormant failure refers to a component that is not checked for operability before the start of a mission; thus, it could unknowingly be failed when required for use. When failure of the component occurs, it is not detected or annunciated. Certain dormant faults can be of system safety concern because they are involved in system designs where operation is critical. Dormant failures are also referred to as latent failures.

See *Latent Failure* for additional related information.

DOWNTIME

Downtime is a period of time when workers are unable to perform their tasks, or a system is unable to perform its tasks. Downtime can be due to many different reasons, such as failures, maintenance schedules, setup time, and worker absenteeism.

DUD

A dud is an explosive munition which has not been armed as intended or which has failed to explode after being armed.

ELECTRICAL/ELECTRONIC/PROGRAMMABLE ELECTRONIC SYSTEM (E/E/PES)

The term E/E/PES typically represents a system for control, protection, or monitoring based on one or more E/E/PES devices. This includes all elements of the system such as power supplies, sensors and other input devices, data and communication paths, actuators, and other output devices. A programmable electronic system (PES) is an electronic computer system that does not utilize mechanical parts for decision making. It involves a computer or a programmable logic device (PLD) that can be programmed to perform the desired system functions. For example, a PES may be a controller unit for a numerical control milling machine. Quite often, a PES is used as a safety system in a larger system, such as a safety monitoring device for a chemical process plant. E/E/PES devices are typically certified to Functional Safety standards, such as:

- International Electrotechnical Commission (IEC) 61508, Functional Safety of Electrical/Electronic/Programmable Electronic Safety-Related Systems, 1999.
- IEC 61511, Functional Safety—Safety Instrumented Systems for the Process Industry Sector.

See *Programmable Electronic System (PES)* for additional related information.

ELECTRICALLY ERASABLE PROGRAMMABLE READ-ONLY MEMORY (EEPROM)

An EEPROM is a memory device composed of arrays of floating-gate transistors. It is a type of nonvolatile memory used in computers and other electronic devices to store small amounts of data that must be saved when power is removed, for example, calibration tables or device configuration. EEPROMs can be programmed and erased electrically using field electron emission. EEPROMs have two limitations: endurance and data retention. During rewrites, the gate oxide in the floating-gate transistors gradually accumulates trapped electrons. The electric field of the trapped electrons adds to the electrons in the floating gate, lowering the window between threshold voltages for zeros versus ones. After sufficient number of rewrite cycles, the difference becomes too small to be recognizable, the cell is stuck in programmed state, and endurance failure occurs. Manufacturers typically specify the maximum number of rewrites being 10^6 or more. During storage, the electrons injected into the floating gate may drift through the insulator, especially at increased temperature, and cause charge loss, thus reverting the cell into an erased state. Manufacturers typically guarantee data retention of 10 years or more.

ELECTRIC SHOCK

Electric shock is the sudden pain or convulsion which results from the passage of an electric current through the body. Minor electrical shocks may cause mishaps due to involuntary reactions. Major electrical shocks may cause death due to burns or paralysis of the heart or lungs. An electric shock results from the passage of direct or alternating electrical current through the body or a body part. Electrical current/voltage in a system is a basic hazard source for many types of electrical related hazards, all of which should be identified in a system safety HA.

MIL-HDBK-454, General Guidelines for Electronic Equipment, Guideline 1, Safety Design Criteria—Personnel Hazards, provides useful safety design information. It states the following regarding shock hazards:

> 5.2.1 Shock Hazards
> Current rather than voltage is the most important variable in establishing the criterion for shock intensity. Three factors that determine the severity of electrical shock are: (1) quantity of current flowing through the body; (2) path of current through the body; and (3) duration of time that the current flows through the body. The voltage necessary to produce the fatal current is dependent upon the resistance of the body, contact conditions, and the path through the body. See Table 1.1. Sufficient current passing through any part of the body will cause severe burns and hemorrhages. However, relatively small current can be lethal if the path includes a vital part of the body, such as the heart or lungs. Electrical

TABLE 2.5 Probable Effects of Shock (MIL-HDBK-454 Table 1.1)

Current values (milli amperes)		
AC 25 Hz–400 Hz	DC	Effects
0–1	0–4	Perception
1–4	4–15	Surprise
4–21	15–80	Reflex action
21–40	80–160	Muscular inhibition
40–100	160–300	Respiratory block
Over 100	Over 300	Usually fatal

burns are usually of two types, those produced by heat of the arc which occurs when the body touches a high-voltage circuit, and those caused by passage of electrical current through the skin and tissue. While current is the primary factor which determines shock severity, protection guidelines are based upon the voltage involved to simplify their application. In cases where the maximum current which can flow from a point is less than the values shown in Table 1.1 (Table 2.5 here) for reflex action, protection guidelines may be relaxed.

MIL-HDBK-454 identifies suitable protection measures against accidental contact with voltage sources in Table 1.11 (Table 2.6 here). It should be noted that this table is for reference only and the applicable paragraphs of MIL-HDBK-454 should be read for clarification and guidance.

ELECTROCUTION

Electrocution is death caused by electrical shock.

ELECTRO-EXPLOSIVE DEVICE (EED)

An EED is any explosive device such as a blasting cap, squib, explosive switch, or igniter which is designed to be initiated by an electric current.

ELECTROMAGNETIC (EM)

EM refers to energy fields that contain physical properties that are both electric and magnetic in nature. These fields are produced by electronic equipment and contain EM energy and radiation.

See *Electromagnetic Radiation (EMR)* for additional related information.

TABLE 2.6 Suitable Protection Measures

Voltage Range	Type of Protection						
	None	Guards and Barriers	Enclosures	Marking		Interlocks		Discharge Devices	
				Caution	Danger	Bypassable	Non-Bypassable	Automatic	Shorting Rods
0–30	X								
>30–70	X								
>70–500		X		X		X		X	
>500			X		X		X	X	X

Source: MIL-HDBK-454 Table 1.11.

ELECTROMAGNETIC COMPATIBILITY (EMC)

EMC is the ability of systems, equipment, and devices that utilize the EM spectrum to operate in their intended operational environments without suffering unacceptable degradation or causing unintentional degradation because of EMR or response. It involves the application of sound EM spectrum management; system, equipment, and device design configuration that ensures interference-free operation; and clear concepts and doctrines that maximize operational effectiveness. EMC is the condition that prevails when various electronic devices in a system are performing their functions according to design in a common electromagnetic environment (EME). EMC is a system safety concern because many safety-critical electronic systems must work correctly in an EME. EMC safety also applies to the potential ignition of explosive or ordnance devices.

See *Electromagnetic Radiation (EMR)* for additional related information.

ELECTROMAGNETIC ENVIRONMENT (EME)

An EME is the resulting product of the power and time distribution, in various frequency ranges, of the radiated or conducted EM emission levels that may be encountered by a military force, system, or platform when performing its assigned mission in its intended operational environment. It is the sum of EMI; electromagnetic pulse (EMP); hazards of EMR to personnel, ordnance, and volatile materials; and natural phenomena effects of lightning and precipitation static.

See *Electromagnetic Radiation (EMR)* for additional related information.

ELECTROMAGNETIC FIELD (EMF)

An EMF is the field of EMR produced by an EMR source, such as a microwave, cell phone, radio, and radar.

See *Electromagnetic Radiation (EMR)* for additional related information.

ELECTROMAGNETIC INTERFERENCE (EMI)

EMI refers to EM energy that interrupts, obstructs, or otherwise degrades or limits the effective performance of electrical equipment. It is any EM disturbance that interrupts, obstructs, or otherwise degrades or limits the effective performance of electronics and electrical equipment. It can be induced intentionally, as in some forms of electronic warfare, or unintentionally, as a result of spurious emissions and responses, intermodulation products, and the like. EMI is a system safety concern because many safety-critical electronic systems must work correctly in an EME. EMI safety applies to the potential failure or malfunction of electronic controls due to EMI.

See *Electromagnetic Radiation (EMR)* for additional related information.

ELECTROMAGNETIC PULSE (EMP)

An EMP is a burst of EMR that results from an explosion (especially a nuclear explosion) or a suddenly fluctuating magnetic field. The resulting electric and magnetic fields may couple with electrical/electronic systems to produce damaging current and voltage surges. EMP is a broadband, high-intensity, short-duration burst of EM energy.

EMP is a potential system safety concern because electronic equipment can undergo upsets and failures due to EMP. Although the strongest part of the pulse lasts for only a fraction of a second, any unprotected electrical equipment and anything connected to electrical cables, which act as giant lightning rods or antennas, will be affected by the pulse. Vacuum tube-based equipment is much less vulnerable to EMP than newer solid-state equipment. Although vacuum tubes are far more resistant to EMP than solid-state devices, other components in vacuum tube circuitry can be damaged by EMP.

See *Electromagnetic Radiation (EMR)* for additional related information.

ELECTROMAGNETIC RADIATION (EMR)

EMR is the energy radiated from an EMF, consisting of alternating electric and magnetic fields that travel through space at the velocity of light. Radiation includes gamma radiation, X-rays, ultraviolet (UV) light, visible light, infrared radiation (IR), and radar and radio waves. An EMR source can be natural, such as from sunlight or lightning. Most EMR sources of concern are man-made and propagate from electronic equipment, such as radar, microwaves, and computers.

The thermal effects of RF energy are of safety concern and the source for many different types of hazards. Both RF and 60 Hz fields are classified as "nonionizing radiation" because the frequency is too low to produce sufficient photon energy to ionize atoms. Still, at sufficiently high power densities, EMR poses certain health hazards. It has been known since the early days of radio that RF energy can cause injuries by heating body tissue. In extreme cases, RF-induced heating can cause blindness, sterility, and other serious health problems. These heat-related health hazards are called "thermal effects" of EMR. These effects depend on the frequency of the energy, the power density of the field, and even such factors as the polarization of the wave. At frequencies near the body's natural resonant frequency, RF energy is absorbed more efficiently, and maximum heating occurs. In adults, this frequency usually is about 35 MHz (35,000 Hz) if the person is grounded and about 70 MHz if the

ELECTROMAGNETIC RADIATION (EMR) 115

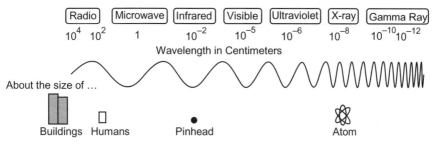

Figure 2.14 Electromagnetic field wavelengths.

person's body is insulated from the ground. Also body parts may be resonant; the adult head for instance is resonant at around 400 MHz, while a baby's smaller head resonates near 700 MHz. As the frequency is increased above resonance, less RF heating generally occurs. However, additional longitudinal resonance occurs at about 1 GHz (1,000,000 Hz) near the body surface.

Radio waves, television waves, and microwaves are all types of EM waves, which only differ from each other in wavelength. Wavelength is the distance between one wave crest to the next. As depicted in Figure 2.14, waves in the EM spectrum vary in size from very long radio waves the size of buildings, to very short gamma-rays smaller than the size of the nucleus of an atom.

Electric or magnetic fields can move through empty space and various other media, including air. These fields are generated whenever electricity flows, whether the electricity is manmade or produced in nature. The energy level of an EMF decreases rapidly with distance from the source, and natural and man-made objects in its path can further decrease the level.

All electrical devices, industrial, commercial or residential, generate EM fields when they operate. These fields can be detected and measured with the proper scientific equipment. Some devices, such as radio and television transmitters, are designed specifically to produce EMFs. Others, such as a video display terminal (VDT), produce small fields only incidentally near the device. While some of these fields are confined and can only be detected close to the source, others are detectable at great distances. These propagating fields travel through the atmosphere or space as EM waves. These waves, just like ocean waves, are characterized by length from crest to crest (wavelength), the rate at which they are produced (frequency), and their energy level (amplitude).

EM waves are not only described by their wavelength, but also by their energy and frequency. All three of these aspects are related to each other mathematically. EM waves carry photon energy; the higher the frequency, the higher the energy. Figure 2.15 shows the EM spectrum in terms of wavelength, frequency, and energy.

EM waves carry energy, and the higher the frequency, the greater the energy. EM waves from the upper end of the spectrum can be powerful enough to break the internal bonds of atoms and molecules, creating charged particles called ions. The powerful radiation of the upper end is therefore called *ionizing*

116 SYSTEM SAFETY TERMS AND CONCEPTS

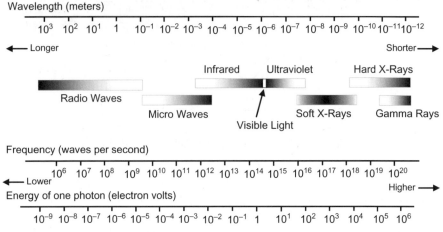

Figure 2.15 The electromagnetic spectrum.

radiation and, because the weaker waves of the lower end can not alter the bonds of molecules, they are called *nonionizing*.

Certain bands of the EMR spectrum have been allocated specific applications. The various radio frequency bands and their frequency ranges are shown in Table 2.7.

EMR is a hazardous energy source that is present in many system environments, from both external sources and from system self-generating sources (internal). This hazardous energy source can be the root cause of many different system safety hazards. For example, EMR can cause ignition of ordnance devices, and EMR can cause erroneous action of aircraft flight control electronics. For this reason, it is imperative that the system safety engineer be cognizant of EMR characteristics, hazards, and hazard mitigating design methods. Table 2.8 lists some of the basic hazard types associated with EMR.

Hazards are caused by transmitter or antenna installations that generate electromagnetic radiation (EMR) in the vicinity of ordnance, personnel, or fueling operations in excess of established safe levels or that increase the existing levels to a hazardous level. System safety typically establishes safe distances from EMR antennas to prevent personnel injury, fuel ignition, and explosives initiation.

Hazards of EMR to flammable items refer to the ignition capacity of EMR on flammable system components. The thermal effect of EMR can provide sufficient energy to ignite liquids and flammable vapors. The thermal effect can also vaporize certain materials, thus causing a fire in the process. An EMR field can also cause arcing and sparking of certain system components, which in turn can cause ignition of liquids and flammable vapors. Analysis, test, and design measures are necessary to ensure that the flammable limits of vulnerable components are not reached during system operation.

TABLE 2.7 Radio Frequency Bands

VLF	LF	MF	HF	VHF	UHF	SHF
Very Low Frequency	Low Frequency	Medium Frequency	High Frequency	Very High Frequency	Ultra High Frequency	Super High Frequency
3–30 KHz	30–300 KHz	300–3000 KHz	3–30 MHz	30–300 MHz	300–3000 MHz	3–30 GHz
Naval Telecomm	LORAN Navigation	AM Radio, Ham Radio	Ham Radio, International Radio	FM Radio, Cordless Phones	Cell Phones, Microwave Ovens	Microwave Relay, Police Radar

TABLE 2.8 EMR-Related Hazards

Hazard	Causal Factors
Eye damage	Eye exposure to EMR energy exceeding eye damage limits.
Skin burn	Skin exposure to EMR energy exceeding burn limits.
Organ damage	Organ exposure to EMR energy exceeding damage limits.
Fuel ignition source	Fuel vapor exposure to EMR energy exceeding ignition limits.
Material ignition source	Physical material exposure to EMR energy exceeding ignition limits.
Explosives ignition source	Exposure of ordnance to EMR energy exceeding limits.
Electronics disruption	Exposure of system electronics to disrupting levels of EMI and HIRF. Safety-critical components are of safety concern. Also a source of common cause failure.
Software disruption	Exposure of computer electronics to disrupting levels of EMI can cause memory and register bit errors, thus modifying the software.

ELECTROMAGNETIC (EM) SUSCEPTIBILITY

EM susceptibility refers to the susceptibility of components, subsystems, or systems to conducted or radiated EM emissions. EM susceptibility refers to EMI.

See *Electromagnetic Interference (EMI)* and *Electromagnetic Radiation (EMR)* for additional related information.

ELECTRONIC SAFETY AND ARMING DEVICE (ESAD)

An ESAD is a safe and arm device that operates by electronic methods, as opposed to mechanical methods.

ELECTROSTATIC DISCHARGE (ESD)

ESD is a charge of static electricity inadvertently discharged to the ground—after it has built up on a system, component, or human being. Personnel generate personnel electrostatic discharge (PESD); helicopters generate helicopter electrostatic discharge (HESD). ESD can cause damage to electronic components.

ELECTROSTATIC DISCHARGE (ESD) SAFETY

ESD safety covers those safety aspects involved with the hazardous effects of electrostatic discharge. ESD is a charge of static electricity inadvertently discharged to the ground—after it has built up on a system, component, or human being. Personnel generate PESD; helicopters generate HESD. ESD damage is similar to damage associated with HF radar pulses.

Due to static effects, the human body can be loaded from zero to several thousand volts. This electrostatic voltage can destroy susceptible electronics, usually without a visible spark, and voltages as low as 10–50 volts. People working in an ESD environment should use conductive shoes and a grounded band on their wrist to keep themselves grounded. This arrangement is very effective for avoiding electrostatic voltage buildup, but it also makes the working environment potentially dangerous to the people who are working with the electricity. If someone is well grounded and touches something with some high voltage on it, they can easily receive a very large and dangerous electrical shock. To avoid this danger, people in this kind of environment are not be directly grounded, they should be grounded through a resistor which limits the current flow through the people to a safe value in case they touch something with high voltage. ESD safety testing should be performed on EEDs and ESADs to ensure sufficient safety margins over the max no-fire energy (MNFE) levels.

EMBEDDED SYSTEM

An embedded system is a computer system designed to perform one or a few dedicated functions, usually with real-time computing constraints. It is embedded as part of a complete device often including hardware and mechanical parts. By contrast, a general-purpose computer is designed to be flexible and to meet a wide range of end-user needs. Embedded systems are controlled by one or more main processing cores that are typically either a microcontroller or a digital signal processor. The key characteristic is however being dedicated to handle a particular task, which may require very powerful processors. For example, air traffic control systems may usefully be viewed as embedded, even though they involve mainframe computers and dedicated regional and national networks between airports and radar sites. Since the embedded system is dedicated to specific tasks, design engineers can optimize it, reducing the size and cost of the product and increasing the reliability and performance. Some embedded systems are mass-produced, benefiting from economies of scale. Embedded systems range from portable devices such as digital watches and MP3 players, to large stationary installations like traffic lights, factory controllers, aircraft flight control systems, or the systems controlling nuclear power plants.

EMERGENCE (EMERGENT)

Emergence is the appearance of novel system characteristics exhibited on the level of the whole system, but not by the components in isolation. The term is often used when recognizing an *emergent* system property. System safety and reliability are referred to as emergent system properties.

EMERGENCY SHUTDOWN (ESD) SYSTEM

An ESD system is a system whose sole function is to safely shut down a process or a system when anomalies occur or process parameter limits are exceeded. It is typically implemented as a safety instrumented system (SIS). The SIS performs specified functions to achieve or maintain a safe state of the process when dangerous process conditions occur. SISs are separate and independent from regular process control systems.

See *Safety Instrumented System (SIS)* for additional related information.

EMERGENCY STOP

An emergency stop is the capability to immediately override all controls to a device and bring it to an immediate stop state. It may be a button or switch that is used to cut power to a piece of machinery or to stop movement of an unmanned ground vehicle (UGV) system. An emergency stop is a DSF.

EMERGENT PROPERTY

An emergent property is a novel system characteristics exhibited on the level of the whole system, but not by the components in isolation. It is a special attribute or characteristic, such as system safety and reliability, which result as a unique property of the final system design and manufacture.

EMPTY AMMUNITION

Empty ammunition is an ammunition item or component whose explosive material has been completely removed and not replaced by other materials, or which was not loaded at the time of manufacture.

ENABLING

The act of removing or activating one or more safety features designed to prevent arming, thus permitting arming to occur.

END-TO-END TESTS

End-to-end tests are the suite of test performed to ensure that a product or system fulfills its requirements and is ready for manufacture and operation.

For example, in aircraft development is involved the tests performed on the integrated ground and flight system, including all elements of the payload, its control, stimulation, communications, and data processing to demonstrate that the entire system is operating in a manner to fulfill all mission requirements and objectives.

END USER

End users are the intended operators of the system; they may also include personnel that will maintain or install system equipment.

ENERGETICS

The term energetics is tied to the branch of physics that deals with energy, which can include lasers, explosives, fuels, and so on. For the explosives community, the term is more often used in the context of materials or items, which by the nature of their design, expend or liberate energy in a short period of time and can be used in weapon systems. Probably the most important measurable safety characteristic of energetics is the ability of the material to initiate and/or propagate to a fire or detonation. Initiation is caused by the addition of energy to a chemical system which, already loaded with energy, cannot remain at equilibrium and decomposes in an uncontrolled manner, such as a fire or detonation. The forces and energies involved in initiation of the energetic is a guide for the evaluation of hazardous conditions in the processing, handling, storage, shipping, use, and disposal of energetics.

Since energetics are a primary hazard source, their use in a system will generate many different types of hazards. System safety should be involved in the design and development of energetic systems to ensure that the risk is mitigated to an acceptable level. The following properties of energetics may be used as aids in characterizing energetics-related hazards:

1. Initiation and propagation
2. Stability
3. Compatibility
4. Ignitability
5. Toxicity
6. Heat of explosion and combustion
7. Thermal conductivity
8. Specific heat
9. Thermal diffusivity
10. Erosive burning

11. Burning rate
12. Resonant burning
13. Dielectric constant
14. Electrical conductivity
15. Vapor pressure
16. Flash point

See *Explosives* for additional related information.

ENERGETIC MATERIALS

Energetic materials are substances or mixtures of substances that are capable of undergoing a chemical reaction, releasing a large amount of energy. This term applies to high explosives, propellants (fuels), and pyrotechnics that detonate, deflagrate, burn vigorously, generate heat, light, smoke, or make a sound.

ENERGY

Energy is a scalar physical quantity that describes the amount of work that can be performed by a force; it is an attribute of objects and systems that is subject to a conservation law. Different forms of energy include kinetic, potential, thermal, gravitational, sound, light, elastic, fluid, electrical and EM energy. The forms of energy are often named after a related force. Any form of energy in a system provides a hazard source—the basic component of a hazard. For this reason, it is important for system safety to identify and evaluate all system energy sources when performing an HA. Any single energy source can spawn many different hazards. For example, fuel spawns hazards associated with fire, explosion, toxicity, component damage, and engine failure due to lack of fuel.

See *Hazard* for additional related information.

ENERGY BARRIER

An energy barrier is any design or administrative method that prevents a hazardous energy source from reaching a potential target in sufficient magnitude to cause damage or injury. Barriers separate the target from the source by various means involving time or space. Barriers can take many forms, such as physical barriers, distance barriers, timing barriers, procedural barriers, and the like. In system safety, this is a term often used when conducting a BA or HA. Understanding the energy path and placing a barrier of some type in this path is a method for mitigating these types of hazards.

See *Barrier Analysis (BA)* for additional related information.

ENERGY PATH

An energy path is the path of energy flow from energy source to target. In system safety, this is a term often used when conducting a BA or HA. Understanding the energy path and placing a barrier of some type in this path is a method for mitigating these types of hazards. For example, an electrical wire typically provides the energy path for electrical energy.

See *Barrier Analysis (BA)* for additional related information.

ENERGY SOURCE

An energy source is any material, mechanism, or process that contains potential energy that can be released for operational aspects of a system. Energy sources include kinetic, potential, thermal, gravitational, sound, light, elastic, fluid, hydraulic, electrical, and EM. Energy sources are a major concern to system safety because released energy can cause harm to a potential target. For example, gasoline and fuels are an energy source for fire and explosion hazards. Electrical power is an energy source for personnel electrocution or burn injuries, as well as ignition source for fires. Energy sources provide the source component for hazards and therefore are a prime element in HA. Energy is a prime factor when conducting a BA or HA.

See *Barrier Analysis (BA)*, *Energy*, and *Hazard* for additional related information.

ENGINEERING CHANGE

An engineering change is a modification to the current approved configuration documentation of an item at any point in the life cycle of the item. Since engineering changes modify the existing design baseline, it is necessary for system safety to evaluate the change and ensure the level of safety mishap risk is not degraded. System safety must ensure that DSFs are not eliminated or reduced in effectiveness, and that no new hazards are introduced by the engineering change.

ENGINEERING CHANGE PROPOSAL (ECP)

ECPs are proposed changes to the already solidified design. An ECP is the documentation by which a proposed engineering change is described, justified, and submitted to the responsible design control authority for approval or disapproval and to the procuring activity for approval or disapproval of implementing the design change in assets to be delivered or retrofitted into assets already delivered. ECPs are part of the formal engineering CCB process.

System safety should be actively involved as a team member in the ECP process. All ECPs must be evaluated by an experienced system safety analyst to determine if any hazards are associated with the proposed change, the associated risk, and the safety impact of the ECP on the existing system. The PM should be notified when an ECP will decrease the level of safety of the existing system. It is important that only qualified safety engineers make the safety determination on proposed ECPs because they have the most experience and knowledge in regard to system hazards.

See *Change Control Board (CCB)* and *Configuration Management (CM)* for additional information.

ENGINEERING CRITICAL

Engineering critical is a term used to describe a part so crucial that independent malfunction or failure could be catastrophic and result in personal injury or loss of life, jeopardize a military mission, or loss of military weapons system or equipment. Engineering critical parts require special documentation, controls, and testing beyond normal requirements. Engineering critical items are typically CSIs.

ENGINEERING/DATA REQUIREMENTS AGREEMENT PLAN (E/DRAP OR EDRAP)

The EDRAP supports the flight clearance process for NAVAIR; it is a U.S. Navy term for a document containing criteria for flight clearance certification. It represents the negotiated written agreement established during the flight clearance planning process. The plan contains a detailed description of the engineering data that the engineering competencies require to establish the system airworthiness with confidence. The EDRAP requires that a DEL be established for each technical specialty or engineering discipline.

The EDRAP documents the sources of evidence required by MIL-HDBK-516B for airworthiness certification. The EDRAP is a requirements and evidence capture process. The EDRAP is essentially a plan for airworthiness certification, which is continually updated until a flight clearance is granted. NAVAIRINST 13034.1C, Flight Clearance Polices for Air Vehicles and Aircraft Systems, September 28, 2004, documents the Navy flight clearance process and requires the EDRAP. The EDRAP includes reference to the configuration, usage limits, and flight envelope to which the clearance applies. It also identifies the validation data (or evidence) that will be used to evaluate airworthiness. The EDRAP consists of a list of required reports, data, engineering analyses, and/or product descriptions required in order to provide sufficient evidence to clear an aircraft for a specific usage and envelope.

The intent of the E/DRAP and its associated process is to insure that the airworthiness process holders, the subject matter experts (SMEs), and program management remain coordinated throughout the program and that all constraints are fairly represented. The primary reason for the EDRAP is to bound data requests and submissions so that the process is finite and predictable. It allows the contractor to plan for the effort required for a flight clearance and to give the contractor a good idea of what to expect during the review process. An EDRAP does not imply any sort of review deadline or clearance timetable, though there are guidelines in the NAVAIR flight clearance instruction, NAVAIRINST 13034.1C. An EDRAP does not specify any exit or success criteria; it does identify the participants in the process and an overall time frame for the process in general and for the submittal of the requested data. The EDRAP includes a section on system safety, an SSP, and a SwSP.

See *Airworthiness* and *Flight Clearance* for additional information.

ENGINEERING DEVELOPMENT MODEL

System development is the process of designing, developing, and testing a system design until the final product meets all requirements and fulfills all objectives. There are several different models by which a system can be developed. Each of these models has advantages and disadvantages, but they all achieve the same end—development of a system using a formal process. Figure 2.16 shows the five-system life-cycle phases that play into the development processes.

The various development models include the following:

Sequential Engineering Development Model

This is the standard traditional approach that has been in use for many years. This method performs the system life-cycle phases sequentially. The development and test phase is subdivided into preliminary design, final design, and test for more refinement. Under this model, each phase must be complete and successful before the next phase is entered. This method normally takes the longest length of time because the system is developed in sequential stages. It can also present the greatest risk in cost and schedule because unknown factors are usually discovered late in the sequential process. Three major design reviews are conducted for exit from one phase and entry into the next. These are the system design review (SDR), PDR, and CDR. These design reviews are an important aspect of the HA. This is the traditional development process.

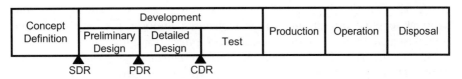

Figure 2.16 Engineering development life cycle.

Concurrent Engineering

This method performs several of the standard development tasks concurrently, in an attempt to save development time. For example, production may begin before final design and test are complete. This method has a higher probability for technical risk problems since some items are in preproduction before full development and testing.

Spiral Development

In the spiral development process, a desired capability is identified, but the end-state requirements are not known at program initiation. Requirements are refined through experimentation, demonstration, risk management, and continuous user feedback. Development progresses in incremental spirals of preliminary design, detailed design, and test, but the requirements for future increments depend on user feedback and technology maturation.

In this methodology, the user is provided the best possible capability within each increment, and continuous user feedback is important. The requirements for future increments are dependent on the feedback from users and technology maturation. It is an iterative process designed to assess the viability of technologies while simultaneously refining user requirements. Spiral development complements an evolutionary approach by continuing in parallel with the acquisition process to speed the identification and development of the technologies necessary for follow-on increments. Each incremental spiral provides the best possible capability. Spiral development is a form of evolutionary acquisition.

Incremental Development

In the incremental development process, a desired capability is identified, an end-state requirement is known, and that requirement is met over time by developing several increments, each dependent on available mature technology. This method breaks the development process into incremental stages, in order to reduce development risk. Basic designs, technologies, and methods are developed and proven before more detailed designs are developed. It is basically a form of progressive prototyping. Incremental development is a form of evolutionary acquisition.

It is during the development stages that system safety is "designed into" the product for safe operational usage. Concept definition, preliminary design, final design, and test are the most significant phases for applying system safety.

See *Concurrent Development Model*, *Incremental Development Model*, *Spiral Development Model*, and *System Life-Cycle Model* for additional information.

ENTROPY

Entropy is a measure of disorder; the more ordered something is, the lower its entropy. Temperature is related to entropy; cold bodies have low entropy because their atoms are less ordered than hot bodies where the atoms are moving around more. In thermodynamics, entropy is a measure of the amount of energy in a system that is no longer available for doing useful work. In communication theory, entropy is a measure of the uncertainty of an outcome.

ENVIRONMENT

The term environment is an all-encompassing term. It typically refers to the air, water, land, plants, animals, and weather. It also refers to conditions surrounding an employee's workplace, such as noise, pollution, and temperature. It also refers to the environment which might affect a system, such as vibration, EMR, heat, dust, and sand. An environment consists of the totality of surrounding conditions. In systems design and engineering, the environment is the context within which a system exists. It is composed of all things that are external to the system, and it includes everything that may affect the system, and may be affected by it at any given time. For example, typical environmental conditions a system must operate in include temperature (hot and cold), rain, ice, snow, sand, humidity, salt, RF radiation, and EMR.

Environment is the natural environment (weather, climate, ocean conditions, terrain, vegetation, space conditions); combat environment (dust, fog, nuclear-chemical-biological); threat environment (effects of existing and potential threat systems to include electronic warfare and communications interception); operations environment (thermal, shock, vibration, power variations); transportation and storage environment; maintenance environment; test environments; manufacturing environments (critical process conditions, clean room, stress); and other environments (e.g., software engineering environment, EM) related to system utilization.

ENVIRONMENTAL IMPACT STATEMENT (EIS)

An EIS is an analysis of the environmental impact for a proposed action, such as construction, modification, test, or operation of a system. The EIS is submitted to the Environmental Protection Agency (EPA) and the public for review and comment. The EIS process is formal and rigorous and imposes an exacting scrutiny from an environmental standpoint. Depending on the complexity of the proposed action, the time required to complete and process an EIS can range from 12 to 24 months or longer. An EIS is required by the National Environmental Policy Act (NEPA) of 1969.

ENVIRONMENTAL REQUIREMENTS

Environmental requirements are the requirements that characterize the impact of the environment on the system as well as the system impact on the natural environment. For example, systems are typically required to operate in specified levels of heat, rain, ice, salt, sand, and humidity. The specified levels are written as environmental design requirements.

ENVIRONMENTAL VALIDATION PROGRAM

Environmental validation programs are tests performed to verify adequate workmanship in the construction of a test item. It is often necessary to impose stresses beyond those predicted for the mission in order to uncover defects. Thus, random vibration tests are conducted specifically to detect bad solder joints, loose or missing fasteners, improperly mounted parts, and so on. Cycling between temperature extremes during thermal-vacuum testing and the presence of EMI during EMC testing can also reveal the lack of proper construction and adequate workmanship.

EQUIPMENT UNDER CONTROL (EUC)

EUC is a term that refers to equipment, machinery, apparatus, or plant used for manufacturing, process, transportation, medical, or other activities. This is a term often used in functional safety.

ERASABLE PROGRAMMABLE READ-ONLY MEMORY (EPROM)

An EPROM is a type of memory chip that retains its data when its power supply is switched off; it is nonvolatile. It is an array of floating-gate transistors individually programmed by an electronic device that supplies higher voltages than those normally used in digital circuits. Once programmed, an EPROM can be erased only by exposing it to strong UV light. EPROMs are easily recognizable by the transparent fused quartz window in the top of the package, through which the silicon chip is visible, and which permits exposure to UV light during erasing. Generally, EPROMs must be removed from the circuit board in order to be erased, since it is not usually practical to build in a UV lamp to erase parts in-circuit. Manufacturers typically guarantee data retention of 10 years or more, and the data can be read an unlimited number of times. The erasing window must be kept covered with an opaque label to prevent accidental erasure by sunlight.

EVALUATION ASSURANCE LEVEL (EAL)

The EAL is a numerical rating describing the depth and rigor of a software product security evaluation. Each EAL corresponds to a package of security assurance requirements (SARs) which covers the complete development of a product, with a given level of strictness. Common criteria (CC) lists seven levels, with EAL 1 being the most basic (and therefore cheapest to implement and evaluate) and EAL 7 being the most stringent (and most expensive). Higher EALs do not necessarily imply better security; they indicate that the claimed security assurance of the product has been more extensively validated.

The CC for Information Technology Security Evaluation is an International Standards Organization/International Electrotechnical Commission (ISO/IEC 15408) for computer security certification. CC is a framework in which computer system users can specify their security functional and assurance requirements, vendors can then implement and/or make claims about the security attributes of their products, and testing laboratories can evaluate the products to determine if they actually meet the claims. CC provides assurance that the process of specification, implementation, and evaluation of a computer security product has been conducted in a rigorous and standard manner. The EAL level does not measure the security of the system itself, it simply states at what level the system was tested to see if it met all the requirements.

The EAL definitions are as follows:

EAL 1—Functionally Tested

EAL1 is applicable where some confidence in correct operation is required, but the threats to security are not viewed as serious. It will be of value where independent assurance is required to support the contention that due care has been exercised with respect to the protection of personal or similar information. EAL1 provides an evaluation of the target of evaluation (TOE) as made available to the customer, including independent testing against a specification and an examination of the guidance documentation provided. It is intended that an EAL1 evaluation could be successfully conducted without assistance from the developer of the TOE, and for minimal cost. An evaluation at this level should provide evidence that the TOE functions in a manner consistent with its documentation and that it provides useful protection against identified threats.

EAL2—Structurally Tested

EAL2 requires the cooperation of the developer in terms of the delivery of design information and test results, but should not demand more effort on the part of the developer than is consistent with good commercial practice. As such it should not require a substantially increased investment of cost or time. EAL2 is therefore applicable in those circumstances where developers or users require a low to moderate level of independently assured security in the absence of ready availability of the complete development record. Such a situation may arise when securing legacy systems.

EAL3—Methodically Tested and Checked

EAL3 permits a conscientious developer to gain maximum assurance from positive security engineering at the design stage without substantial alteration of existing sound development practices. EAL3 is applicable in those circumstances where developers or users require a moderate level of independently assured security, and require a thorough investigation of the TOE and its development without substantial re-engineering.

EAL4—Methodically Designed, Tested, and Reviewed

EAL4 permits a developer to gain maximum assurance from positive security engineering based on good commercial development practices which, though rigorous, do not require substantial specialist knowledge, skills, and other resources. EAL4 is the highest level at which it is likely to be economically feasible to retrofit to an existing product line. EAL4 is therefore applicable in those circumstances where developers or users require a moderate to high level of independently assured security in conventional commodity TOEs and are prepared to incur additional security-specific engineering costs. Commercial operating systems that provide conventional, user-based security features are typically evaluated at EAL4. Operating systems that provide multilevel security are evaluated at a minimum of EAL4.

EAL5—Semiformally Designed and Tested

EAL5 permits a developer to gain maximum assurance from security engineering based upon rigorous commercial development practices supported by moderate application of specialist security engineering techniques. It is likely that the additional costs attributable to the EAL5 requirements, relative to rigorous development without the application of specialized techniques, will not be large. EAL5 is therefore applicable in those circumstances where developers or users require a high level of independently assured security in a planned development and require a rigorous development approach without incurring unreasonable costs attributable to specialist security engineering techniques. Numerous smart card devices have been evaluated at EAL5, as have multilevel secure devices.

EAL6—Semiformally Verified Design and Tested

EAL6 permits developers to gain high assurance from application of security engineering techniques to a rigorous development environment in order to produce a premium TOE for protecting high-value assets against significant risks. EAL6 is therefore applicable to the development of security TOEs for application in high-risk situations where the value of the protected assets justifies the additional costs. An example of an EAL6 certified system is the Green Hills Software INTEGRITY-178B operating system.

EAL7—Formally Verified Design and Tested

EAL7 is applicable to the development of security TOEs for application in extremely high-risk situations and/or where the high value of the assets justifies the higher costs. Practical application of EAL7 is currently limited to TOEs with tightly focused security functionality that is amenable to extensive formal analysis.

If a product is CC certified, it does not necessarily mean it is completely secure. This is possible because the process of obtaining a CC certification allows a vendor to restrict the analysis to certain security features and to make certain assumptions about the operating environment and the strength of threats, if any, faced by the product in that environment.

EVENT

An event is an undesirable change in the state of a system or product. It may involve structures, components, subsystems, humans, or organizations. An event is typically triggered by a failure, error, or combinations of failures and errors. Typically, an event results in a mishap.

EVENT SEQUENCE DIAGRAM (ESD)

An ESD is essentially a flowchart, with paths leading to different end states. Each path through the ESD flowchart is considered a scenario. Along each path, PEs are identified as either occurring or not occurring. An ESD can easily be mapped into an event tree (ET), which allows for the practical quantification of mishap scenarios; however, the ESD representation has the significant advantage over the ET of enhancing communication between risk engineers, designers, and operators. In situations that are well covered by operating procedures, the ESD flow can reflect these procedures, especially if the procedures branch according to the occurrence of PEs. Instrument readings that inform crew decisions can be indicated at the appropriate PE. At each PE along any given path, the events preceding that event are easily identified, so that their influence on the current PE can be modeled adequately. An ESD provides a very compact representation of system operation.

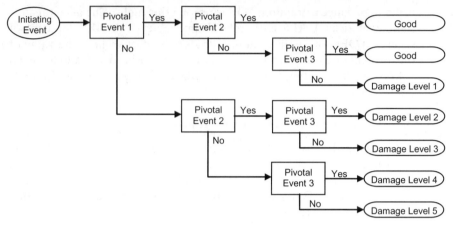

Figure 2.17 Event sequence diagram (ESD) concept.

In the scenario approach to mishap analysis, the scenario consists of a series of events that begins with the IE, which is then followed by one or more critical PEs until a final consequence has resulted. Typically, barriers are designed into the system to protect against IEs, and the PEs represent the failure or success of these barriers. Figure 2.17 shows the ESD concept. When all of the PE barriers work successfully, the system is usually operationally in a safe state. The ESD shows the various outcomes that are possible, given the different combinations of PE failure/success.

See *Event Tree Analysis (ETA)* and *Initiating Event Analysis (IEA)* for additional related information.

EVENT TREE (ET)

An ET is a graphical model of an accident scenario that yields multiple outcomes and outcome probabilities. An ET is constructed during ETA and is frequently used as one of the tools in a probabilistic risk assessment (PRA). ETs first made their appearance in risk assessment in the WASH-1400 nuclear power safety study, where they were used to generate, define, and classify scenarios specified at the PE level.

See *Event Tree Analysis (ETA)* and *WASH-1400* for additional related information.

EVENT TREE ANALYSIS (ETA)

ETA is a system safety analysis technique for identifying and evaluating the sequence of events in a potential accident scenario following the occurrence of a postulated IE. ETA utilizes a visual logic tree structure known as an ET.

The objective of ETA is to determine whether the IE will develop into a serious mishap, or if the event is sufficiently controlled by the safety systems and procedures implemented in the system design. An ETA can result in many different possible outcomes from a single IE, and it provides the capability to obtain a probability for each outcome. ETA combines a decision tree for the evaluation of multiple outcomes, with an FT to determine the cause and probability of certain failure events in the ET.

The ETA is a very powerful tool for identifying and evaluating all of the system consequence paths and outcomes that are possible after an IE occurs. The ETA model will show the probability of the system design resulting in a safe operation path, a degraded operation path, and an unsafe operation path. The intent of ETA is to evaluate all of the possible outcomes that can result from an IE. Generally, there are many different outcomes possible from an IE, depending upon whether design safety systems work properly or malfunction when needed. ETA provides a PRA of the risk associated with each potential outcome.

ETA can be used to model an entire system, with analysis coverage given to subsystems, assemblies, components, software, procedures, environment, and human error. ETA can be conducted at different abstraction levels, such as conceptual design, top-level design, and detailed component design. ETA has been successfully applied to a wide range of systems, such as: nuclear power plants, spacecraft, and chemical plants. The technique can be applied to a system very early in design development and thereby identify safety issues early in the design process. Early application helps system developers to design-in safety of a system during early development rather than having to take corrective action after a test failure or a mishap. Figure 2.18 is an example of an ETA model.

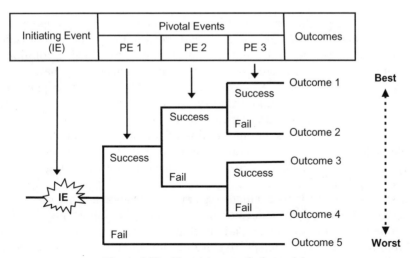

Figure 2.18 Event tree analysis model.

134 SYSTEM SAFETY TERMS AND CONCEPTS

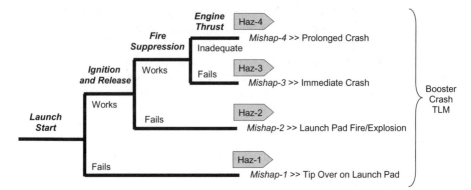

Figure 2.19 Event tree and hazard–mishap relationships.

Figure 2.19 uses ETA to show why hazards are unique and that generic TLMs cover families of hazards. In this ETA diagram of the TLM "Booster Crash" for a spacecraft rocket booster being launched from a launch pad.

Note that in this example, four discrete hazards are shown, with some common causal factors shared by the hazards. Also, note that each hazard has a slightly different outcome, each with a slightly different severity level. Booster crash is not a single hazard with multiple causal factors; it is a TLM category that encompasses four related hazards in a family of hazards.

ETA is an HA technique that is important and essential to system safety. For more detailed information on the ETA technique, see Clifton A. Ericson II, *Hazard Analysis Techniques for System Safety* (2005), chapter 12.

EVOLUTIONARY ACQUISITION

Evolutionary acquisition is an adaptive and incremental strategy applicable to high technology and software intensive systems when requirements beyond a core capability can generally, but not specifically, be defined. Evolutionary acquisition is DoD's preferred strategy for rapid acquisition of mature technology for the warfighter. An evolutionary approach delivers capability in increments, with the recognition that future improvements in capability will be needed. It is substantially dependent on the consistent and continuous definition of requirements and maturation of technologies that lead to disciplined development and production of systems that provide increasing capability toward a materiel concept. Evolutionary acquisition programs that develop successive increments of mission capability require corresponding successive test periods to evaluate system performance against operational requirements. Evolutionary development provides the opportunity for interaction between the user, tester, and developer.

From a system safety perspective, it does not matter which development model is used because they each ultimately undergo the design review process.

Each method must utilize design and development reviews in order to verify and certify the implementation. Even prototype increments must be developed with design reviews to manage the process. Therefore, the standard SSP and tasks apply to each development approach; it is just a matter of level, detail, and timing.

See *Engineering Development Model* and *System Life-Cycle Model* for additional related information.

EXCLUSIVE OR GATE

An exclusive OR gate is a logic gate used in FTA where the gate logic states that gate output occurs only if at least one of the inputs occurs, but not both. See Figure 2.27 below for an example of an exclusive OR gate.

See *Fault Tree Symbols* for additional related information.

EXEMPTED LASERS

Exempted lasers are military lasers exempted from 21 Code of Federal Regulations (CFR) 1040 when compliance would hinder mission fulfillment during actual combat or combat training operations or when the exemption is necessary in the interest of national security. These lasers must comply with MIL-STD-1425.

EXIGENT CIRCUMSTANCES

Exigent circumstances refer to circumstances demanding attention. Exigent circumstances typically involve an event where it is clearly evident that rapid intervention is required to prevent the immediate loss of life or property.

EXIT CRITERIA

Exit criteria are the specific accomplishments or conditions that must be satisfactorily demonstrated before an effort can progress further in the current acquisition phase or transition to the next acquisition phase. Technical exit criteria are used for acquisition phase milestone reviews.

EXPERIMENTAL DESIGN

An experimental design is a procedure or test whereby the independent variable is manipulated to determine the effect upon the dependent variable.

EXPLOSION

An explosion is a violent release of energy caused by a chemical or nuclear reaction; a terminal force that is built up and then quickly released, as in the act of exploding or bursting. It is a chemical reaction of any chemical compound or mechanical mixture that, when initiated, undergoes a very rapid combustion or decomposition releasing large volumes of highly heated gases that exert pressure on the surrounding medium. Depending on the rate of energy release, an explosion can be categorized as a deflagration, a detonation, or pressure rupture. An explosion involves the ignition and rapid burning of a confined energetic material causing high local pressures leading to violent pressure rupturing of the confining structure. Metal cases are fragmented into large pieces that can be thrown long distances. Unreacted and /or burning energetic material is thrown also. Fire and smoke hazards will exist and air shocks produced cause damage to nearby structures. Blast overpressures are lower than a detonation reaction. When an explosion occurs, the pressure release produces a large volume of gas. The effective volume of gas is increased by the effect of liberated heat.

See *Explosive Event* for additional related information.

EXPLOSION PROOF

Explosion proof refers to an item enclosed in a case that is capable of withstanding an explosion of a specified gas or vapor that may occur within it and of preventing the ignition of a specified gas or vapor surrounding the enclosure. It prevents ignition of a surrounding flammable atmosphere.

EXPLOSIVE DEVICE

An explosive device is any item that contains explosive material(s) and is configured to provide quantities of gas, heat, or light by a rapid chemical reaction initiated by an energy source usually electrical or mechanical in nature.

EXPLOSIVE EVENT

An explosive event is any event involving conventional ordnance, ammunition, explosives, explosive systems, and devices that results in an unintentional detonation, firing, deflagration, burning, launching of ordnance material, leaking or spilled propellant fuels and oxidizers, or chemical agent release. An explosive event is essentially an explosion event.

Explosive events of energetics can be categorized into the following types of reactions:

Detonation

This is the most violent type of explosive event. A supersonic decomposition reaction propagates through the energetic material to produce an intense shock in the surrounding medium, air, or water. It produces a very rapid plastic deformation of metallic cases followed by extensive fragmentation. All energetic material will be consumed. The effects will include large ground craters for munitions on or close to the ground, holing/plastic flow damage/fragmentation of adjacent metal plates, and blast overpressure to nearby structures.

Partial Detonation

This is the second most violent type of explosive event. In a partial detonation, some, but not all, of the energetic material reacts as in detonation. An intense shock wave is formed; some of the case is broken into small fragments; a ground crater can be produced, adjacent metal plates can be damaged as in a detonation, and there will be blast overpressure damage to nearby structures. A partial detonation can also produce large case fragments as in a violent pressure rupture (brittle fracture). The amount of damage, relative to a full detonation, depends on the portion of material that detonates.

Explosion

This is the third most violent type of explosive event. Ignition and rapid burning of the confined energetic material builds up high local pressures leading to violent pressure rupturing of the confining case or structure. Metal cases are fragmented (brittle fracture) into large pieces that are often thrown for long distances. Unreacted and/or burning energetic material is also thrown about. Fire and smoke hazards will exist. Air shocks are produced that can cause damage to nearby structures. The blast and high velocity fragments can cause minor ground craters and damage (breaking, tearing, gouging) to adjacent metal plates. Blast pressures are lower than for a detonation.

Deflagration

This is the fourth most violent type of explosive event. Deflagration involves a chemical reaction proceeding at subsonic velocity along the surface of and/or through an explosives, producing hot gases at high pressure. Ignition and burning of the confined energetic material leads to nonviolent pressure release, as a result of low case strength or venting through the case. The case might rupture but does not fragment; closure covers might be expelled, and unburned or burning energetic material might be thrown about and spread the fire. Propulsion might launch an unsecured test item, causing an additional hazard. No blast or significant fragmentation damage to the surroundings; only heat and smoke damage from the burning energetic material.

Burning

This is the least violent type of explosive event. The energetic material ignites and burns, non-propulsively. The case may open, melt, or weaken sufficiently to rupture nonviolently, allowing mild release of combustion gases. Debris stays mainly within the area of the fire. The debris is not expected to cause fatal wounds to personnel or be a hazardous fragment beyond 15 m (49 ft).

EXPLOSIVE MATERIAL

An explosive material is a chemical, or a mixture of chemicals, which undergoes a rapid chemical change (with or without an outside supply of oxygen), liberating large quantities of energy in the form of blast, light, or hot gases. Incendiary materials and certain fuels and oxidizers made to undergo a similar chemical change are also explosive materials.

Examples of explosive materials include:

1. Explosives—Trinitrotoluene (TNT), pentaerythritol tetranitrate (PETN), polymer-bonded explosive (PBXN), Research Department Explosive (RDX) compositions, explosive D, tetryl, fulminate of mercury, black powder, smokeless powder, flashless powder, and rocket and missile propellants.
2. Fuels and oxidizers—OTTO Fuel II, mixed amine fuel, inhibited red fuming nitric acid, and ethylene oxide.
3. Incendiaries—Napalm, magnesium, thermite, and pyrotechnics.

EXPLOSIVE ORDNANCE DISPOSAL (EOD)

EOD is the task of detection, identification, on-site evaluation, rendering safe, recovery, and final disposal of unexploded ordnance (UXO). EOD is normally an inherently hazardous task, but it is made even more so when the UXO has become more hazardous due to dropping, firing, launching, or deterioration.

EXPLOSIVES

The term explosive or explosives includes any chemical, compound, or mechanical mixture that, when subjected to heat, impact, friction, detonation, or other suitable initiation, undergoes a very rapid chemical change with the evolution of large volumes of highly heated gases which exert pressures in the surrounding medium. The term applies to high explosives, propellants, and pyrotechnics that detonate, deflagrate, burn vigorously, or generate heat, light, smoke, or sound.

EXPLOSIVES SAFETY

Explosives safety is the activity or process used to prevent premature, unintentional, or unauthorized initiation of explosives or explosive devices. Explosives safety also involves designing to minimize the deleterious effects of unintended initiations. Explosives safety includes all mechanical, chemical, biological, electrical, and environmental hazards associated with explosives or EM environmental effects. Equipment, systems, or procedures and processes whose malfunction would cause unacceptable mishap risk to manufacturing, handling, transportation, maintenance, storage, release, testing, delivery, firing, or disposal of explosives are also included. An explosives safety program follows a prevention focused process based on:

1. Reducing the probability of an explosives mishap from occurring
2. Reducing the consequences of an explosive mishap, should it occur
3. Continually informing and educating personnel on explosive mishap risks

There are many elements to an explosives safety program. Explosive safety is a joint effort involving many disciplines, such as weapon design, fuze design, explosives design, testing, insensitive munitions safety, environmental safety, and system safety. For this reason it is difficult to explicitly identify all tasks related to an explosive safety effort. Since a complete explosives safety program for A&E items requires an integrated effort involving several different disciplines, the Navy's explosives safety program is best managed and performed under the purview of the Weapon Systems Explosives Safety Review Board (WSESRB) and the Naval Ordnance Safety and Security Activity (NOSSA).

As a minimum, an explosives safety program should provide for the following:

1. Ensure compliance with all explosives policies, procedures, standards, regulations, and laws
2. Assessment of system designs incorporating explosives for hazards and mishap risk
3. The application of design mitigation measures to reduce mishap risk to an acceptable level
4. Acceptance of the design and design mishap risk by the appropriate review boards

5. The documentation and communication of system mishap risks
6. Establish Explosive Safety Quantity-Distance (ESQD) requirements for storage of A&E
7. Facility site approvals for storage of A&E
8. Explosives hazard classifications for transportation of A&E
9. A Hazards of Electromagnetic Radiation to Ordnance (HERO) program
10. An insensitive munitions assessment and test program
11. A fuze safety program to ensure compliance with fuze design guidelines and standards
12. An SSP for the identification and mitigation of system hazards

All acquisition programs that include or support A&E items must comply with DoD explosives safety requirements. The PM is responsible for implementing a safety program that covers all aspects of explosives safety and meets all Department of the Navy (DON) explosives safety policies and requirements, as well as Federal, state, and local regulations. The PM is responsible for design requirements, management, engineering, and hazard controls for conventional A&E, and conventional components of non-nuclear weapons systems, such as warheads, rocket motors, separation charges, igniters, and initiators. Consideration is applicable to A&E during their development manufacturing, testing, transportation, handling, storage, maintenance, and demilitarization (disposal).

When performing HA on explosive systems, the activities or situations that can adversely impact the safety of explosive items include the following:

1. Improper/inadequate design and testing
    - Failure to use IM
    - Failure to use adequate fuze systems
    - Failure to properly test all necessary parameters
    - Combining incompatible materials
    - Failure to design for EOD
2. Rough handling
    - Dropping of an item, during handling, beyond a specified threshold distance
    - Puncture of the protective case or the item by a forklift
    - Rough handling which results in dents, cuts, bends, or other deformity
    - Impact due to the fork of a fork lift
    - The impact of a bullet or piece of shrapnel
    - The impact from dropping the explosive item
    - The impact from the dropping of another item on the explosive item
    - Friction due to rubbing or dragging
    - Energy from the detonation of an adjacent explosive

3. Improper procedures
    - Electrical discharge generated by a moving vehicle
    - Electrical discharge due to an improper tool
    - Static electricity discharge due to improper grounding
    - Static electrical charge due to movement of explosive components
    - Static electrical charge due to dragging
    - Exposure to high powered RF fields
    - Heat and friction due to operations in areas containing explosive dust
    - Static electrical discharge from an explosive/ammunition worker in a plant
    - Contamination with chemicals, foreign materials, or other nonexplosive components
    - Unauthorized smoking by workers in A&E areas
    - Improper handling at very low temperature
    - Static charge generated by the motion of granular or powdered explosives
4. Excess exposure
    - Heat generated by a fire
    - Self-heating due to a combination of improper storage, poor ventilation, and high temperatures
    - Exposure to unauthorized "open flames"
    - Exposure of the items to conditions that cause corrosion and deterioration
    - Exposure to bullet strike, shrapnel, hot spalls, and so on
5. Natural events
    - Heat generated by a lightning strike
    - Prolonged storage at very high or very low temperatures
    - High solar radiation
    - Ageing
6. By-products or residue
    - Explosive dusts, vapors, or fumes
    - Solid and liquid propellants
    - Pyrotechnics
    - Leakage and exudates of explosive by-products

EXPLOSIVE SYSTEM

An explosive system is any system, weapon, device, or tool that uses or includes explosive material.

EXPLOSIVE TRAIN

An explosive train is the detonation or deflagration transfer mechanism (i.e., train) beginning with the first explosive element (e.g., primer, detonator) and terminating in the main charge.

EXPOSURE TIME

Exposure time refers to the period of time when an entity is vulnerable to some stated condition or event, such as a failure, hazard, or mishap. In a probability of failure calculation, exposure time is the operational time period during which the component is exposed to potential failure, for example, $P = 1 - e^{-\lambda T}$, where T is the exposure time.

FACTOR OF SAFETY

Factor of safety is a term describing the cushion between expected load and actual design strength in a component or product design. A factor of safety is typically the ratio of the actual safety design to the required design. The required design is what the item is required to be able to withstand for expected conditions, whereas the factor of safety includes a safety margin that provides a measure of how much more the actual design can tolerate. The terms SF, factor of safety, margin of safety, and safety margin are synonymous.

See *Safety Factor (SF)* for additional related information.

FAIL-SAFE

Fail-safe is a common industry term for a specific design safety method, whereby a system reverts to a safe mode or state when a failure occurs. Often a system design requirement will state that the system must be designed to be fail-safe. Although fail-safe design seems like an intuitively obvious method, and is used quite universally, it involves nuances that can allow it to easily be misused and misapplied. Many questions arise, such as does fail-safe cover all possible failures and does it cover only SPFs or multiple failures? Does fail-safe apply to components, black boxes, subsystems, functions, or combinations of these?

By the strictest definition, a fail-safe system is one that cannot cause harm when it fails. The term fail-safe is used to describe a device which, when it fails, fails in a way that will cause no harm or at least a minimum of harm to other devices or danger to personnel. Fail-safe is a system safety concept that, in theory, is intended to ensure a system remains safe, or in a safe state, in the event of a failure, thereby preventing a mishap while alternative action is being taken. It is a concept that must be intentionally and carefully implemented in order to achieve true fail-safe design that is cost-effective. A fail-safe design is a design that imposes a unique safeguard that functions in the event of a failure. This safeguard keeps the system in a safe operational mode, switches the system to a safe substate or mode, or safely shuts down the system, depending upon the design. Every fail-safe design is exclusive to the particular system in which it is implemented. A fail-safe design could also be classified as a fault-tolerant design and as an SPF-tolerant design.

Note that fail-safe is not intended to protect against all system failures; it is primarily intended to protect against catastrophic failures. Somewhat coincidentally, potential catastrophic failures typically lie within an SCF, which means that a fail-safe design usually protects an SCF. A fail-safe design must be carefully thought out and implemented. When developing a fail-safe design, there are several system unique questions that must be resolved before stabling a design. These questions include the following:

1. What is the failure, or failures, that the system must be protected from (i.e., tolerated)?
2. Is the failure of concern catastrophic in outcome?
3. What is being protected—a device, subsystem, system, or function?
4. What is the desired safe state and how safe is it (i.e., what is the risk)?
5. How is the safe state achieved?
6. Does the safe state involve continued operation, degraded operation, or shutdown entirely?
7. How many failures are to be tolerated in the fail-safe design?
8. Should the design be passive fail-safe or active fail-safe?
9. How reliable is the fail-safe design and implementation?

As an example, the fail-safe design of a fuzing system is to render the munition incapable of arming and functioning upon malfunction of safety feature(s), exposure to out of sequence arming stimuli, or incorrect operation of components. However, the fail-safe design of a spacecraft oxygen system would not likely be to shut down the oxygen system.

A fail-safe design must be thoroughly tested before the system is put into production and operation to verify that the actual implementation covers all possible potential failure conditions. It should be noted that fail-safe operation does not nominally apply to normal system operation, but rather only to abnormal system operation. The goal of a fail-safe design is to make the system as tolerant as possible to likely failures such that the system defaults to the safest state upon the occurrence of a failure.

Fail-safe design examples include:

- The safety glass used in modern automobile windows which is designed to shatter into very small pieces rather than in the long jagged fragments when common window glass breaks.

- Luggage carts in airports and lawn mowers in which the hand brake must be held down at all times. If it is released, the cart will stop (i.e., dead man's switch).
- Air brakes on railway trains, where the brakes are held in the off position by massive air pressure created in the brake system. Should a brake line split, or a carriage become de-coupled, the air pressure will be lost and the brakes applied. It is impossible for the train to be driven with a leak in the brake system
- Motorized gates where in case of a power outage, the gate can be pushed open by hand with no crank or key required.
- Avionics systems using redundant subsystems to perform the same computation with voting logic to determine the "safe" result when one subsystem fails.

Fail-safe is a system characteristic whereby any malfunction affecting safety will cause the system to revert to a state that is known to be within acceptable risk parameters. Fail-safe provides the ability to sustain a failure and retain safe control of the system or operation, or revert to a state which will not cause a mishap. A fail-safe design should be provided in those areas where failure can cause catastrophic damage to equipment, injury to personnel, or inadvertent operation of critical equipment (source: MIL-STD-1472E Human Engineering, Design Criteria Standard, 1998).

FAIL-SAFE INTERLOCK

A fail-safe interlock is an interlock where the failure of a single mechanical or electrical component of the interlock will cause the system to go into, or remain in, a safe state.

See *Interlock* for additional related information.

FAILURE

A failure is the inability of an item to perform its required or intended function. An item may be a system, subsystem, product, software, human, or component. A failure is departure from, or nonconformance with, requirements or specifications; the inability of an item to perform within previously prescribed limits. System safety is concerned with failures because a failure effectively causes a change in system state, which could have hazardous consequences. It should be noted that failures and hazards are not interchangeable, that is, not all failures will cause hazards.

See *Nonconformance* for additional information.

FAILURE CAUSE

Failure cause is the process or mechanism responsible for initiating the failure mode. The possible processes that can cause component failure include physical failure, design defects, manufacturing defects, environmental forces, and so on.

FAILURE EFFECT

Failure effect is the consequence or consequences a failure mode has on the operation, function, or status of an item and on the system.

FAILURE MODE

A failure mode is the manner or way in which the failure of an item occurs; the mode or state the item is in after it fails; the physical or functional manifestation of a failure. For example, two primary failure modes of a resistor are open and shorted.

FAILURE MODES AND EFFECT TESTING (FMET)

FMET is a form of off-nominal testing; system failures and anomalies are introduced into a test case to determine the effect. If the system does not respond as intended, it may indicate a hazard in the system design. FMET is a tool for identifying errors and deficiencies in the system design. FMET test cases are established from the FMEA and the HAs.

FAILURE MODES AND EFFECTS ANALYSIS (FMEA)

FMEA is an analysis tool for evaluating the effect(s) of potential failure modes of subsystems, assemblies, components, or functions. It is primarily a reliability tool to identify credible failure modes that would adversely affect overall system reliability. FMEA has the capability to include failure rates for each failure mode, in order to achieve a quantitative analysis. Additionally, the FMEA can be extended to evaluate failure modes that may result in an undesired system state, such as a system hazard, and thereby also be used for HA.

FMEA is a disciplined bottom-up evaluation technique that focuses on the design or function of products and processes, in order to prioritize actions to reduce the risk presented by product or process failures. In addition, the FMEA is a tool for documenting the analysis and capturing recommended design changes. Time and resources for a comprehensive FMEA must be

146 SYSTEM SAFETY TERMS AND CONCEPTS

allotted during design and process development, when recommended changes can most easily and inexpensively be implemented. The purpose of FMEA is to evaluate the effect of failure modes to determine if design changes are necessary due to unacceptable reliability or safety resulting from potential failure modes. When component failure rates are attached to the identified potential failure modes, a probability of subsystem or component failure can be derived. FMEA was originally developed to determine the reliability effect of failure modes, but it can also be used to identify mishap hazards resulting from potential failure modes.

FMEA is applicable to any system or equipment, at any desired level of design detail—subsystem, assembly, unit, or component. FMEA is generally performed at the assembly or unit level, because failure rates are more readily available for the individual embedded components. The FMEA can provide a quantitative reliability prediction for the assembly or unit that can be used in a quantitative safety analysis (e.g., FT). FMEA tends to be more hardware and process oriented but can be used for software analysis when evaluating the failure of software functions.

FMEA is a valuable reliability tool for analyzing potential failure modes and calculating subsystem, assembly, or unit failure rates. FMEA can be modified and extended to identify hazards resulting from potential failure modes and evaluating the resulting mishap risk. Note however, that an FMEA will likely not identify all system hazards, because it is only looking at single component failure modes, while hazards can be the result of multiple hazards and events other than failure modes. Also, an FMEA does not identify hazards arising from events other than failures (e.g., timing errors, radiation, and high voltage). For these reasons, FMEA is *not* recommended as the *sole tool* for hazard identification. FMEA should only be used for HA when done in conjunction with other HA techniques. Overall, an FMEA is a valuable support asset for an SSP. The FMEA can be used to identify hazards, and it can also provide failure rates for other quantitative analyses such as FTA.

Figure 2.20 depicts the FMEA concept. The subsystem being analyzed is divided into its relevant indenture levels, such as Unit 1, Unit 2, Unit 3. Each unit is then further subdivided into its basic items. Each item is listed down

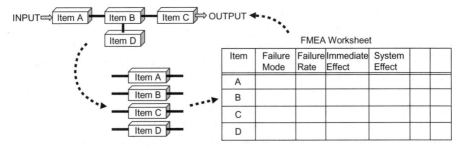

Figure 2.20 FMEA concept.

the left-hand column of the FMEA worksheet and individually evaluated. The idea is to analyze the effect of each individual failure mode. In effect, the subsystem is analyzed "top-down" when it is divided into indenture levels, and then it is analyzed "bottom-up" when each item is individually evaluated in the worksheet. An item can be a hardware part or component, or it can be a system, software, or process function. Each item is then singularly isolated, and all potential failure modes for this item are listed in the first column of the FMEA. The analyst must look back at the unit and system designs to determine failure effects.

In practice, there are three approaches to performing an FMEA:

1. **Structural approach**
   The structural FMEA is performed on hardware and focuses on potential hardware failure modes. The hardware can be at any hardware indenture level for the analysis; subsystem, unit, assembly, or part (component). The structural approach tends to be a detailed analysis at the component level.
2. **Functional approach**
   The functional FMEA is performed on functions. The functions can be at any functional indenture level for the analysis; system, subsystem, unit, or assembly. This approach focuses on ways in which functional objectives of a system go unsatisfied or are erroneous. The functional approach is also applicable to the evaluation of software through the evaluation of required software functions. The functional approach tends to be more of a system-level analysis.
3. **Hybrid approach**
   The hybrid FMEA is a combination of the structural and the functional approaches. The hybrid approach begins with the functional analysis of the system and then transitions to a focus on hardware, especially hardware that directly contributes to functional failures identified as safety-critical.

FMEA can also be applied to software. A software FMEA (SFMEA) normally involves performing an analysis of the software functions. An SFMEA would follow the same basic steps as a hardware FMEA; set up a starting point, understand the design, make a list of typical failure modes, and then perform the analysis. Software failure modes would be seen as types of erroneous behavior, and not typos in the code. Performing an FMEA on a mechanical or electrical system is generally more straightforward than performing an FMEA on software. Failure modes of components such as relays and resistors are generally well understood. Mechanical and electrical components fail due to aging, wear, or stress. For software, the situation is different because software modules do not fail per se, they only display incorrect behavior. A software-oriented FMEA can only address incorrect behavior of software (i.e., the software fails to perform or function as intended).

FMEA							
Component	Failure Mode	Failure Rate	Causal Factors	Immediate Effect	System Effect	Notes

Figure 2.21 Example FMEA worksheet.

Figure 2.21 shows an example FMEA form that is typically used. The form can be modified to suit the needs of a particular program.

The original standard for FMEA is MIL-STD-1629A, Procedures for Performing a Failure Mode, Effects and Criticality Analysis, 1980. A more recent standard is SAE/ARP-5580, Recommended FMEA Practices for Non-Automobile Applications, July 2001. Another more recent standard is SAE Standard J-1739, Potential Failure Mode and Effects Analysis in Design (Design FMEA) and Potential Failure Mode and Effects Analysis in Manufacturing and Assembly Processes (Process FMEA) and Effects Analysis for Machinery (Machinery FMEA), August 2002.

FMEA is an analysis technique that is important and essential to reliability engineering and system safety. For more detailed information on the FMEA technique, see Clifton A. Ericson II, *Hazard Analysis Techniques for System Safety* (2005), chapter 13.

See *Failure Modes and Effects and Criticality Analysis (FMECA)* for additional related information.

FAILURE MODES AND EFFECTS AND CRITICALITY ANALYSIS (FMECA)

FMECA is a more detailed version of the FMEA. FMECA requires that more information be obtained from the analysis, particularly information dealing with the detection methods for the potential failure modes and the reliability-oriented risk priority number (RPN), where RPN = Likelihood of Failure × Failure Effect Severity × Likelihood of Failure Detection.

Figure 2.22 shows an example FMECA worksheet form that is typically used. The form can be modified to suit the needs of a particular program.

See *Failure Modes and Effects Analysis (FMEA)* and *Risk Priority Number (RPN)* for additional related information.

FMECA									
Item	Failure Mode	Failure Rate	Causal Factors	Immediate Effect	System Effect	RPN	Method of Detection	Current Controls	Recommended Action

Figure 2.22 Example FMECA worksheet.

FAILURE REPORTING, ANALYSIS, AND CORRECTIVE ACTION SYSTEM (FRACAS)

A FRACAS is a process that provides for reporting, classifying, and analyzing failures, and planning corrective actions in response to those failures. FRACAS is typically used during system development to collect data, record, and analyze system failures, providing a history of failure and corrective actions. The FRACAS method was first introduced in the United States in the 1970s. The method calls for a systematic failure data collection, management, analysis, and corrective action implementation. The FRACAS process is a disciplined closed loop failure reporting, analysis, and corrective action system; it provides a useful tool in the achievement of product reliability and safety. A FRACAS is used to record all failures and problems related to a product or process and their associated root causes and failure analyses in order to assist in identifying and implementing corrective actions. Some organizations use simple in-house programs for failure data collection and management, while others may require the more powerful approach of web-based FRACAS software for cross-organizational data collection and analysis, allowing use by customers and developers.

FAMILY OF SYSTEMS (FoS)

An FoS is a set of systems that provide similar capabilities through different approaches to achieve similar or complementary effects. The mix of systems can be tailored to provide desired capabilities, dependent on the situation. For instance, the warfighter may need the capability to track moving targets. The FoS that provides this capability could include unmanned or manned aerial vehicles with appropriate sensors, a space-based sensor platform, or a special operations capability. Each can provide the ability to track moving targets, but with differing characteristics of persistence, accuracy, timeliness, and so on.

150 SYSTEM SAFETY TERMS AND CONCEPTS

Applying system safety to an FoS is a larger and more complex process than for a single system. Although the system safety process remains essentially the same, the necessary safety analyses and safety evidence on the various systems may not be done to the same level of detail or consider the FoS system interfaces. Integrating this safety knowledge into one super-system model may require a special FoS system safety effort.

See *System of Systems (SoS)* for additional related information.

FAULT

A fault is the occurrence of an undesired system state, which may be the result of a failure. A fault state can be reached by the occurrence of a failure or by a change in state caused by a command path set of cascading faults and/or failures. Typically, a fault is the correct operation of a function, component, or subsystem, except it occurs prematurely (i.e., it involves inadvertent operation). A fault state is the sequence of events that would normally happen, but unexpectedly due to initiation by a failure or error. For example, "light off" is an undesired *fault state* that may be due to light bulb failure, loss of power, or erroneous operator action. (Note: All failures are faults, but not all faults are failures).

Cascading failures can cause cascading fault states, whereby a sequence of normally "off" states can be initiated to the "on" fault states by a component failure. For example, the simple failure of a switch can result in prematurely (or inadvertently) applying power to a missile launch system. The switch failure results in providing the normally expected power to the launch system, except at the wrong time. This is considered a "command fault" state, which is used extensively as a fault guide in FTA.

Figure 2.23 contains an FT of an example circuit that provides light when the two switches are closed. This FT model shows the difference between a failure and a fault. In this example, the light is in the "off" fault state because

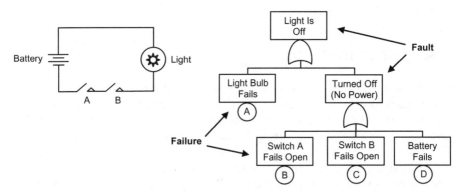

Figure 2.23 Fault versus failure.

the light bulb fails, or because it is (in a sense) commanded off due to lack of power or the switches fail to open, thereby removing power.

A failure is the occurrence of a basic component failure. A fault is the occurrence or existence of an undesired state for a component, subsystem, or system. In a fault, the component operates correctly, except at the wrong time, because it was commanded to do so (due to the system design). When a fault occurs, it is a preliminary indication that a failure may have occurred. A potential safety-critical fault state generally requires some sort of risk mitigation via design safety methods. In essence, a failure represents a component condition, whereas a fault represents a system condition (i.e., system state).

See *Cascading Failure* for additional related information.

FAULT HAZARD ANALYSIS (FHA)

Fault hazard analysis (FHA) is an analysis technique for identifying hazards arising from component failure modes. It is accomplished by examining the potential failure modes of subsystems, assemblies, or components, and determining which failure modes can form undesired states that could result in a mishap. Note that FHA deals with faults even though it looks at failure modes and is similar to an FMEA in structure. The technique was developed to allow the analyst to stop the analysis at a point where it becomes clear that a failure mode did not contribute to a hazard, whereas the FMEA requires complete evaluation of all failure modes.

FHA can be implemented on a subsystem, a system, or an integrated set of systems; it can be performed at any level from the component level through the system level. It is hardware oriented and not well suited for software analysis. The FHA is a thorough technique for evaluating potential failure modes. However, it has the same limitations as the FMEA. It looks at single failures and not combinations of failures. FHAs generally overlook hazards that do not result entirely from failure modes, such as poor design and timing errors.

In general, FHAs are more easily and quickly performed than other HA techniques. FHAs can be performed with minimal training. FHAs are inexpensive to perform in terms of time and manpower.

FHAs force the analyst to focus on system elements and hazards. The disadvantage of the FHA is that it focuses on single failure modes and not combinations of failure modes, which allows for the overlooking of other types of hazards (e.g., human errors, radiation hazards). FHA is sometimes used to meet the needs for performing an SSHA-type HA. In this situation the FHA evaluates component failure modes in a subsystem.

FHA is a formal and detailed HA utilizing structure and rigor through the use of a standardized worksheet. An example FHA worksheet is shown in Figure 2.24.

Fault Hazard Analysis									
Subsystem_____				Assembly/Unit_____			Analyst_____		
Component	Failure Mode	Failure Rate	System Mode	Effect on Subsystem	Secondary Causes	Upstream Command Causes	MRI	Effect on System	Remarks

Figure 2.24 Example fault hazard analysis worksheet.

FHA is an HA technique that is important and essential to system safety. For more detailed information on the FHA technique, see Clifton A. Ericson II, *Hazard Analysis Techniques for System Safety* (2005), chapter 14.

See *Mishap Risk Index (MRI)* and *Safety Order of Precedence (SOOP)* for additional related information.

FAULT INJECTION

Fault injection is the process of deliberately inserting faults into a system (by manual or automatic methods) to test the ability of the system to safely handle the fault or to fail to a safe state. Usually, fault injection criteria is defined by system safety and is implemented by the software test engineering group to measure the system's ability to mitigate potential mishaps to an acceptable level of risk.

FAULT ISOLATION (FI)

FI is the process of determining the location of a fault to the extent necessary to affect repair.

FAULT TREE ANALYSIS (FTA)

FTA is a safety analysis technique that develops an FT diagram that logically models and graphically represents the various combinations of possible system events that can lead to a UE, such as a mishap. The analysis is deductive in nature, in that it transverses from the general problem to the specific causes. The FT develops the logical fault paths from the UE at the top, to all of the

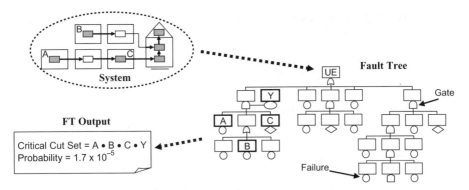

Figure 2.25 Fault tree analysis example.

possible root causes at the bottom. The strength of FTA is that it is easy to perform, easy to understand, provides useful system insight, and shows all of the possible causes for a problem under investigation. The default standard for FTA is documented in N. Roberts, W. Vesely, D. Haasl, and F. Goldberg, Nuclear Regulatory (NUREG)-0492, *Fault Tree Handbook* (1981).

Figure 2.25 shows an overview of an FT structure for an example missile system.

FTA can be used to model an entire system, with analysis coverage given to subsystems, assemblies, components, software, procedures, environment, and human error. FTA can be conducted at different abstraction levels, such as conceptual design, top-level design, and detailed component design. FTA has been successfully applied to a wide range of systems, such as missiles, ships, spacecraft, trains, nuclear power plants, aircraft, torpedoes, medical equipment, and chemical plants. FTs are graphical models using defined logic gates and events to model the cause–effect relationships involved in causing the UE. The graphical model can be translated into a mathematical model to compute failure probabilities and system importance measures. FT development is an iterative process, where the initial structure is continually updated to coincide with design development.

In the analysis of systems, there are two types of FTA application. The most commonly used application is the proactive FTA, performed during system development to influence design by predicting and preventing future problems. The other application is the reactive FTA, performed after an accident or mishap has occurred. The FT methodology used for both types of FTA applications is identical except the reactive FTA may include the use of mishap evidence and the evidence event gate. FTA is one of the few tools that is useful in performing a PRA because it can be quantified. There are many benefits to FTA, and it is a highly respected and used methodology for the evaluation of both simple and complex system designs.

FTA is an analysis technique that is important and essential to system safety and reliability engineering. For more detailed information on the FTA technique, see Clifton A. Ericson II, *Hazard Analysis Techniques for System Safety* (2005), chapter 11.

FAULT TREE (FT) SYMBOLS

FTA is a graphical safety analysis technique that develops an FT diagram that utilizes special symbols. Basic event FT symbols and their descriptions are shown in Figure 2.26.

Basic FT gate symbols are shown in Figure 2.27, along with their definition and mathematical formula.

Figure 2.28 shows some alternative symbols that are in use throughout industry.

FAULT-TOLERANT

Fault-tolerant is the capability to provide continued correct system operation in the presence of a defined set of hardware and/or software faults.

FEEDBACK

Feedback is a process by which information concerning the adequacy of the system, its operation, and its outputs are introduced back into the system.

Symbol	Use	Description
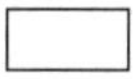 ▭	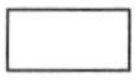 ▭	**Node Text Box** Contains the text for all FT nodes. Text goes in the box, and the node symbol goes below the box.
○	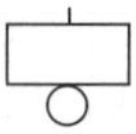 ▭ / ○	**Primary Failure** A basic component failure; the primary, inherent, failure mode of a component; a random failure event.
 ◇	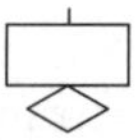 ▭ / ◇	**Secondary Failure** An externally induced failure or a failure mode that could be developed in more detail if desired.
 ⌂	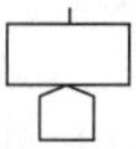 ▭ / ⌂	**Normal Event** An event that is expected to occur as part of normal system operation.
 ⬯	 ▭ / ⬯	**Condition Event** A conditional restriction or probability.
 △	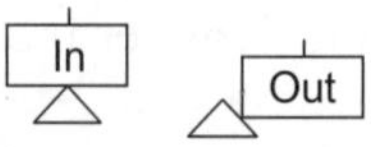 In △ / Out △	**Transfer** Indicates where a branch or sub-tree is marked for the same usage elsewhere in the tree. Transfer IN and transfer OUT.

Figure 2.26 FT symbols for basic events, conditions, and transfers.

Symbol	Gate Type	Description
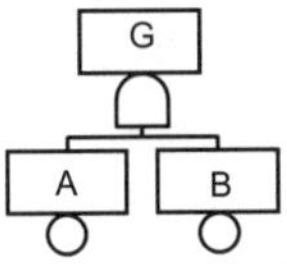 G / A B (AND shape)	AND Gate	The output occurs only if all of the inputs occur together. $P = P_A \bullet P_B = P_A P_B$ (2 inputs) $P = P_A \bullet P_B \bullet P_C = P_A P_B P_C$ (3 inputs)
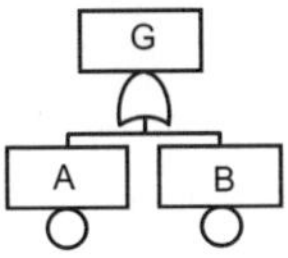 G / A B (OR shape)	OR Gate	The output occurs only if at least one of the inputs occurs. $P = P_A + P_B - P_A P_B$ (2 inputs) $P = (P_A + P_B + P_C) - (P_{AB} + P_{AC} + P_{BC}) + (P_{ABC})$ (3 inputs)
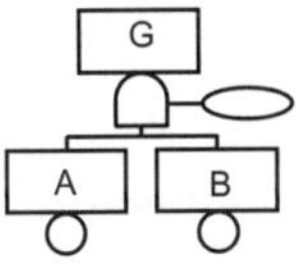 G / A B with condition	Priority AND Gate	The output occurs only if all of the inputs occur together, and A must occur before B. The priority statement is contained in the Condition symbol. $P = (P_A P_B) / N!$ Given $\lambda_A \approx \lambda_B$ and N = number of inputs to gate
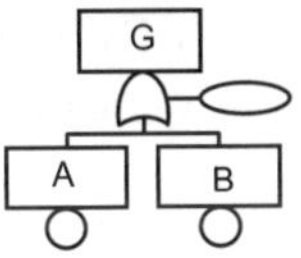 G / A B with condition	Exclusive OR Gate	The output occurs if either of the inputs occurs, but not both. The exclusivity statement is contained in the Condition symbol. $P = P_A + P_B - 2(P_A P_B)$
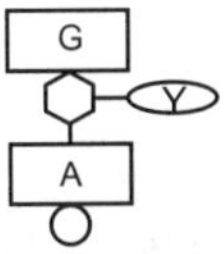 G / A with Y condition	Inhibit Gate	The output occurs only if the input event occurs and the attached condition is satisfied. $P = P_A \bullet P_Y = P_A P_Y$

Figure 2.27 FT gate symbols.

Typical Symbol	Action	Description	Alternate Symbol
	Exclusive OR Gate	Only one of the inputs can occur, not both. Disjoint events.	
	Priority AND Gate	All inputs must occur, but in given order, from left to right.	
M/N	M of N Gate	M of N combinations of inputs causes output to occur.	 m/n

Figure 2.28 Alternative FT symbols.

Feedback indicates that there is a discrepancy between what the system is producing and what it should be producing. Negative feedback means the output is too low, while positive feedback means the output is too high. Feedback is a functional monitoring signal obtained from a given dynamic and continuous system. A feedback function only makes sense if this monitoring signal is looped back into an eventual control structure within a system. This monitoring shall be compared with a known desirable state. The difference between the feedback monitoring signal and the desirable state of the system gives the notion of error. The amount of error can guide corrective actions to the system in order to generate trends to bring the system gradually back to the desirable state.

FEEDFORWARD

Feedforward is a process, akin to feedback, that informs current operations with future ideals, and adjusts the output model accordingly.

FIELD-LOADABLE SOFTWARE (FLS)

FLS is software that can be loaded without removal of the equipment from the operational installation. FLS can be either executable code or data. FLS might also include software loaded into a line replaceable unit (LRU) at a repair station or shop. FLS should be under tight configuration control for safety purposes.

FIELD-PROGRAMMABLE GATE ARRAY (FPGA)

An FPGA is one type of PLD. FPGAs use a grid of logic gates, similar to that of an ordinary gate array, but the programming is done by the customer, not by the manufacturer. The term "field-programmable" means the array is done outside the factory, or "in the field." FPGAs are usually programmed after being soldered down to the circuit board. In larger FPGAs, the configuration is volatile and must be reloaded into the device whenever power is applied or different functionality is required. Configuration is typically stored in a configuration programmable read-only memory (PROM) or EEPROM. The more modern FPGAs have the ability to be reprogrammed at "run time," which leads to the idea of reconfigurable computing or reconfigurable systems, that is, CPUs that reconfigure themselves to suit the task at hand. FPGAs are of concern to system safety because of their nondeterministic nature in implementing the program logic, which in turn affects potential failure modes and hazards.

FINAL (TYPE) QUALIFIED (FTQ) EXPLOSIVE

An FTQ explosive is a material in a specific application or weapon system that has been formally approved for service usage. The approval is based, in part, on an assessment of the explosive material as part of the design of the item or configuration in which it will be used, and for which it has been adjudged to be safe and suitable for use. This approval step is required and the data has to be presented to the appropriate review board before a weapon or other device is approved for limited or full production. All explosives that are FTQ are also, by inference "qualified" when being considered for the same type of role, for example, a main-charge application; however, this is not so if an explosive is being considered for a different role, for example, a booster explosive being considered as a main-charge explosive. FTQ explosive is the NATO-accepted term that is equivalent to the term "final qualified" or "service approved" explosive used by the United States.

FINDING

A "finding" is a product of an audit; it is the identification of a failure to show compliance to one or more of the objectives or requirements. A finding may involve an error, deficiency, or other inadequacy. A finding might also be the identification of the nonperformance of a required task or activity. Typical categories of items resulting from an audit include: compliance, finding, observation, issue, and action.

See *Audit* and *Safety Audit* for additional related information.

FIREBRAND

A firebrand is a projectile of burning or hot fragment(s) whose thermal energy may be transferred to a receptor that is hit by the projectile.

FIRE CLASSES

Fire classes refer to the classification of fires are according to the type of combustible involved:

- Class A—Ordinary combustibles such as wood, cloths, paper, rubber, and certain plastics.
- Class B—Flammable or combustible liquids, gases, greases, and similar materials.
- Class C—Energized electrical equipment.
- Class D—Combustible metals such as magnesium, titanium, zirconium, sodium, or potassium.

FIRE POINT

Fire point is the lowest temperature at which a liquid gives off sufficient flammable vapor to produce sustained combustion after removal of the ignition source.

FIRE DOOR

A fire door is tested and rated for resistance to varying degrees of fire exposure; it is used to prevent the spread of fire through horizontal and vertical openings. Fire doors must normally remain closed or close automatically in the presence of a fire.

FIRE RESISTIVE

Fire resistive is a term applied to materials or structures that are capable of resisting the effects of fire. Fire-resistive materials or structures are noncombustible.

FIRE RETARDANT

Fire retardant is a treatment applied to the surfaces of combustible materials, or their coverings, to prevent or retard ignition or the spread of fire.

FIRE WALL

A fire wall is a fire-resistant wall designed to prevent the horizontal spread of fire to adjacent areas.

FIRMWARE

Firmware is the combination of a hardware device and computer instructions, or computer data that resides as read-only software in the hardware device. The computer instructions are embedded in the firmware device. The software cannot be readily modified without reprogramming or replacing the memory in the firmware device. Firmware is subject to the same development, analysis, and test rigor as standard software for purposes of SwS.

FISH BONE DIAGRAM

A fish bone diagram is an Ishikawa diagram, which is used to analyze problems and show the various causes of an event. It is sometimes used in accident/mishap investigations.

See *Ishikawa Diagram* for additional related information.

FLAMMABLE

Flammable refers to any substance or material that is easily ignited, burns intensely, or has a rapid rate of flame spread. Flammable and inflammable are identical in meaning.

FLASH ARRESTOR

A flame arrestor, or flash arrestor, is a device used on vents for flammable liquid or gas tanks, storage containers, and so on, which prevent flashback when a flammable or explosive mixture is ignited.

FLASH MEMORY

A type of nonvolatile memory which is capable of being erased electrically and reprogrammed, but only in blocks, as opposed to one byte increments.

FLASH POINT

The flash point of a volatile liquid is the lowest temperature at which it can vaporize to form an ignitable mixture in air. Measuring a liquid's flash point requires an ignition source. At the flash point, the vapor may cease to burn when the source of ignition is removed. Flash point should not to be confused with autoignition temperature, which does not require an ignition source. Fire point is a slightly higher temperature and is defined as the temperature at which the vapor continues to burn after being ignited. Neither the flash point nor the fire point is related to the temperature of the ignition source or of the burning liquid, which are much higher. Flash point refers to both flammable liquids and combustible liquids. There are various international standards for defining each, but most agree that liquids with a flash point less than 43°C are flammable, while those having a flash point above this temperature are combustible.

The automobile provides a good flash point example, where gasoline is used in an engine which is driven by a spark as an ignition source. The fuel is premixed with air within its flammable limits and heated above its flash point, then ignited by the park plug. The fuel should not pre-ignite in the hot engine. Therefore, gasoline is required to have a low flash point and a high autoignition temperature.

Table 2.9 shows flash point and autoignition temperatures for several fuels.

TABLE 2.9 Example Flash Point and Autoignition Temperatures

Fuel	Flash Point	Autoignition Temperature
Ethanol	12.8°C (55°F)	365°C (689°F)
Gasoline	<−40°C (−40°F)	246°C (475°F)
Diesel	>62°C (143°F)	210°C (410°F)
Jet Fuel	>60°C (140°F)	210°C (410°F)
Kerosene	>38–72°C (100–162°F)	220°C (428°F)

FLASHOVER (ELECTRICAL)

An electrical flashover is an electric arc between two components or a component and ground.

FLASHOVER (FIRE)

A fire flashover is an event that sometimes occurs during a fire in an enclosed area where the thermal radiation feedback from walls and ceilings reach a temperature that causes autoignition of materials in the room.

FLIGHT ACCEPTANCE

Flight acceptance is the concurrence that a product or system is ready for flight following the appropriate acceptance testing. An acceptance test is the validation process that demonstrates that hardware and/or software is acceptable for use. It provides the basis for delivery of an item under terms of a contract, where the contract should specify acceptance criteria. It also serves as a quality control screen to detect deficiencies. A product can be accepted or rejected based on how well the product's performance compares with preestablished acceptance criteria.

System safety is generally involved in the acceptance test process in some manner. Safety may review test results to ensure safety concerns are adequately resolved and that design safety requirements are met. Safety may also review acceptance test procedures to ensure that no test hazards exist and that the potential mishap risk of a test is acceptable. It may be necessary for safety to be present as a witness to the conduct of certain acceptance tests, generally on safety-critical components.

FLIGHT CERTIFICATION

Flight certification is a term used to describe the process by which a system is deemed airworthy. Flight certification is analogous to airworthiness certification.

See *Airworthiness Certification* for additional related information.

FLIGHT CLEARANCE

The flight clearance is evidence that an independent engineering assessment of airworthiness has been performed and the assessment indicates the aircraft system can be operated with an acceptable level of technical risk. There are two categories of flight clearances: interim and permanent. An interim flight clearance is temporary approval for flight of an aircraft system in a nonstandard configuration or operation outside the envelopes defined in Naval Air Training and Operating Procedures Standardization (NATOPS). The permanent flight clearance is an aircraft system's NATOPS.

The flight clearance defines the configuration and an operating envelope that is deemed safe and appropriate for that configuration. Any information with respect to operating envelope, usage, or flight configuration listed in the EDRAP is advisory in nature and is not a flight clearance precursor, though the rough draft of the flight clearance distributed to the SMEs for review and comment may be drawn from the EDRAP.

See *Engineering/Data Requirements Agreement Plan (E/DRAP or EDRAP)* for additional information.

FLIGHT CRITICAL

Flight critical is a designation given to the minimum subset of equipment and assets necessary for safe flight and landing of an aircraft or air vehicle. This includes flight and landing under both normal and abnormal conditions. The failure or malfunction of this equipment would preclude safe flight and landing. The equipment item can be any system level of hardware or software and should be designated as safety-critical. Flight-critical equipment should be identified early in the development program and receive the highest level of safety integrity during design, development, and manufacture. As a general rule, should a contingency necessitate shutting down certain equipment, the flight critical must not be shut down.

See also *Flight Essential* and *Mission Critical* for additional related information.

FLIGHT ESSENTIAL

Flight essential refers to the subset of equipment and assets necessary for continued flight and landing of an aircraft or air vehicle. Typically, this subset of equipment applies to normal operation (i.e., no failures) and is therefore larger than the subset of equipment designated as flight critical.

See *Flight Critical* for additional related information.

FLIGHT OPERATING LIMITATION (FOL)

An FOL is a temporary hazard mitigating control (flight operation restriction) that is established for a flight test to prevent the occurrence of an identified hazard or safety concern that cannot be mitigated through design prior to the flight test.

FLIGHT OPERATING LIMITATIONS DOCUMENT (FOLD)

A FOLD is a document containing temporary flight limitations placed on flight testing due to potential safety concerns with unresolved variables. During the design and development of new air vehicles, safety and hazard concerns may arise on certain functions, subsystems, components, or tasks. If a hazard cannot be completely resolved prior to flight testing, an FOL is documented to establish a temporary hazard mitigating control that is in effect until a detailed investigation of the safety concern has been completed and, if found to be valid, any required permanent changes to the air vehicle have been incorporated. FOLs are typically managed by system safety and are collected within a single document: the FOLD. Quite often, a procedural or training change can be incorporated into the flight test program in a relatively short period of time compared to the incorporation and evaluation of any changes to the hardware or software. Once permanent design changes are made to mitigate the identified hazard, the FOLD can be removed and closed. FOLDs can be subdivided into categories of flight, ground, and carrier operations.

FLIGHT READINESS REVIEW (FRR)

The FRR examines tests, demonstrations, analyses, and audits that determine the system's readiness for a safe and successful flight or launch and for subsequent flight operations. It also ensures that all flight and ground hardware, software, personnel, and procedures are operationally ready.

See *Critical Design Review (CDR)*, *Preliminary Design Review (PDR)*, and *Systems Engineering Technical Reviews (SETR)* for additional information.

FLIGHT SAFETY-CRITICAL AIRCRAFT PART (FSCAP)

FSCAP is any aircraft part, assembly, or installation containing a critical characteristic whose failure, malfunction, or absence may cause a catastrophic failure resulting in loss or serious damage to the aircraft or an uncommanded engine shutdown resulting in an unsafe condition. For the most purposes, "CSI," "FSCAP," "flight safety part," and "flight safety-critical part" are synonymous. The term CSI shall be the primary term used.

FOREIGN OBJECT DAMAGE (FOD)

FOD is damage caused by foreign object debris (also abbreviated FO). FOD is any damage attributed to a foreign object (i.e., any object that is not part of the vehicle) that can be expressed in physical or economic terms and may or may not degrade the product's required safety or performance characteristics. FOD is an abbreviation often used in aviation to describe both the damage done to aircraft by foreign objects, and the foreign objects themselves.

Internal FOD is used to refer to damage or hazards caused by foreign objects inside the aircraft. For example, "cockpit FOD" might be used to describe a situation where an item gets loose in the cockpit and jams or restricts the operation of the controls. "Tool FOD" is a serious hazard caused by tools left inside the aircraft after manufacturing or servicing. Tools or other items can get tangled in control cables, jam moving parts, short out electrical connections, or otherwise interfere with safe flight. Aircraft maintenance teams usually have strict tool control procedures, including toolbox inventories to make sure all tools have been removed from an aircraft before it is released for flight.

FOREIGN OBJECT DEBRIS (FOD)

Foreign object debris (FOD) is a substance, debris, or article alien to a vehicle or system that has potential to cause damage to the vehicle or system. FOD can be either internal or external to the system. This term is typically used in regard to aircraft, as FOD is a serious problem that can cause aircraft damage or the crash of an aircraft. For example, an aircraft mechanic could leave a wrench (FOD) in the mechanical linkage of an aircraft's flight controls, resulting in jamming.

FORMAL QUALIFICATION TESTING (FQT)

FQT is a process to determine whether an item complies with the allocated requirements for that item.

FORMAL METHODS

Formal methods are mathematically based techniques for the specification, development, and verification of hardware and software systems. The use of formal methods is motivated by the expectation that design requirements can be proven correct mathematically. However, the high cost of using formal methods means that they are usually only used in the development of high-integrity systems where safety is of utmost importance.

FORMAL SPECIFICATION

A formal specification is a specification using mathematical formalism, where a well-defined, logically sound mathematical notation expresses the requirements of a system. This is opposed to the typical informal specification whereby requirements are expressed in a natural language.

FORMULA TRANSLATION (FORTRAN)

Fortran (or FORTRAN) is a computer programming language developed for mathematical and scientific operations. It is known as a HOL. Fortran stands for FORmula TRANslation.
See *High-Order Language (HOL)* for additional related information.

FRACTURE CONTROL PROGRAM

A systematic project activity to ensure that a space payload intended for flight has sufficient structural integrity as to present no critical or catastrophic hazard. This activity also ensures quality of performance in the structural area for any payload. Central to the program is fracture control analysis, which includes the concepts of fail-safe and safe-life, defined as follows:

1. Fail-safe—Ensures that a structural element, because of structural redundancy, will not cause collapse of the remaining structure or have any detrimental effects on mission performance.
2. Safe-life—Ensures that the largest flaw that could remain undetected after nondestructive examination would not grow to failure during the mission.

FRATRICIDE

Fratricide is defined as an unintentional attack on friendly forces by other friendly forces. In the Navy, fratricide is used in the context of ownship fired or launched ordnance intersecting with other ownship ordnance or intersecting with the ship's structure. These collisions would result in an energetic reaction leading to significant damage and/or death/injury.

FUEL

A compound or mixture that is capable of reacting with an oxidizer. In the reaction, the fuel is said to be oxidized. In a system design, fuel is typically used in an engine to achieve a desired objective, such as in an automobile, motorcycle, boat, aircraft, and spacecraft.

FUNCTION

A system function is a task, action, or activity that must be performed in order to achieve a desired outcome. It is an action required to be carried out in order to meet systems requirements and attain the purpose(s) of the system. For example, in a weapon system, some typical functions include "weapon arm," "weapon safe" and "weapon fire." Functions are identifiable entities, which may exist in a hierarchical set of functions. A function may be a concrete and physical entity, such as the "left and right movement of an aircraft aileron." A function may also be an abstract entity, such as a mathematical function that associates an input to a corresponding output according to some rule, or a software function that is a portion of code that performs a specific task. At the top level, most functions are defined by a system requirement, while lower level functions are typically developed via derived requirements.

Since a function performs a desired task, the erroneous performance of a function, or the failure to perform a function when needed, may result in serious safety consequences. Erroneous function and failure to function can result from many different or combined causal factors, such as hardware failures, hardware tolerance errors, system timing errors, software errors, human errors, sneak circuits, and environmental factors. The criticality of the function typically determines the safety vulnerability involved. Functions can be real or abstract entities in a system, and they should be recognized in an overall system functional hierarchy.

FUNCTIONAL BLOCK DIAGRAM (FBD)

An FBD is a systems engineering tool for showing how a system operates, by depicting components, functions, and interfaces in a single diagram. The FBD is also known as the single line diagram (SLD) or the functional flow diagram (FFD). The FBD provides a means whereby individuals not familiar with the detailed design can quickly and easily grasp how the system operates. For example, an FBD could be used to translate a complicated electric circuit diagram into a simplified series of functions.

The concept of FBDs can also show hierarchy, in addition to system function. The top-level functions can be decomposed into lower-level functions. And, each of the lower-level functions can be decomposed into lower levels, and so on, until the bottom level is reached. Figure 2.29 shows the FBD concept.

Figure 2.30 shows system components in the form of an FBD for the SCF of "missile fire." This type of system diagram helps visualize system operation for a particular function and is useful in HAs. Each of the blocks in this diagram could be broken down into its own lower level of detail FBD.

The safety analyst typically utilizes the FBD to understand how the system operates and to identify hazards by analysis of the tasks and functions.

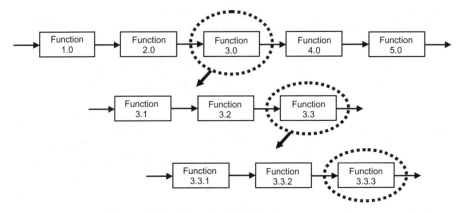

Figure 2.29 Functional block diagram concept.

Figure 2.30 Functional block diagram (FBD) of safety-critical function.

FUNCTIONAL CONFIGURATION AUDIT (FCA)

An FCA is a formal review to verify that the system and all subsystems can perform all of their required design functions in accordance with their functional and allocated configuration baselines. FCA is a system review used to verify that actual performance of the CI meets specification requirements. The FCA is a formal examination of the functional characteristics of a CI as demonstrated by test data to verify that the item has achieved the performance specified in its functional or allocated configuration prior to acceptance. The FCA verifies that all requirements established in the specifications, associated test plans, and related documents have been tested and that the item has passed the tests or corrective action has been initiated. As part of an FCA, system safety typically makes a presentation summarizing the safety of the functional configuration. This may include the functional DSFs in the system design, verification of the functional safety design requirements, and the current level of mishap risk which the design presents.

FUNCTIONAL HAZARD ANALYSIS (FHA)

FHA is a system safety analysis tool for identifying hazards through the rigorous evaluation of system and/or subsystem functions, including software functions. Systems are designed to perform a series of functions, which can be broken into sub-functions, sub-sub-functions, and so on. Functional objectives are usually well understood even when design details are not available or understood. FHA is an inductive HA approach (inductively determines effect of potential fault events) that evaluates the functional failure, malfunction, and corruption of functions.

The purpose of FHA is to identify system hazards by the analysis of system functions. Functions are the means by which a system operates to accomplish its mission or goals. System hazards are identified by evaluating the safety impact of a function failing to operate, operating incorrectly, or operating at the wrong time. When a function's failure can be determined hazardous, the casual factors of the malfunction should be investigated in greater detail. Also, this technique identifies function that can be classified as SCFs.

FHA is applicable to the analysis of all types of systems, equipment, and software. FHA can be implemented on a single subsystem, a complete functional system, or an integrated set of systems. The level of analysis detail can vary, depending upon the level of functions being analyzed. For example, analysis of high-level system functions will result in a high-level HA, whereas analysis of low-level (detailed design) subsystem functions will yield a more detailed functional analysis. Through logical analysis of the way a system is functionally intended to operate, the FHA provides for the identification of hazards early in the design process. A basic understanding of system safety concepts and experience with the particular type of system is essential to create a correct list of potential hazards.

FHA is a powerful, efficient, and comprehensive system safety analysis technique for the discovery of hazards. It is especially powerful for the safety assessment of software. Since software does not have discrete failure modes as hardware does, the best way to identify software-related hazards is by evaluating the effect of potential software functions failing. Software is built upon performing functions; therefore, FHA is a very natural and vital tool. After a functional hazard is identified, further analysis of that hazard may be required to determine if the causal factors of the functional failure are possible. Since the FHA focuses on functions, it might overlook other types of hazards, such as those dealing with hazardous energy sources, sneak circuit paths, and hazardous material (HAZMAT). For this reason, the FHA should not be the sole HA performed, but should be done in support of other types of HA, such as PHA and SSHA.

When performing an FHA, some of the functional safety considerations to evaluate include:

- Fails to function/operate
- Functions incorrectly/erroneously
- Functions inadvertently
- Functions at wrong time (early, late)
- Unable to stop functioning
- Function receives erroneous data

- Function sends erroneous data
- Function has conflicting data or information

An FHA is typically performed utilizing a worksheet, which helps to add structure and rigor to the analysis, record the process and data, and help support justification for the identified hazards. The following basic information should be obtained from the FHA analysis worksheet:

- Hazards
- Hazard effects (mishaps)
- HCFs (to subsystem identification)
- Safety-critical factors or parameters
- Risk assessment (before and after DSFs are implemented)
- Derived safety requirements for eliminating or controlling the hazards

An example FHA analysis worksheet is shown in Figure 2.31. This particular FHA analysis worksheet utilizes a columnar-type format. Other worksheet formats may exist because different organizations often tailor their FHA analysis worksheet to fit their particular needs. The specific worksheet to be used may be determined by the SSP, system safety working group (SSWG), or the FHA customer.

Note that in this analysis methodology, *every system function* is listed and analyzed. For this reason, not every entry in the FHA form will constitute a hazard, since not every function is hazardous. The analysis documents, however, that all functions were considered by the FHA. Note also that the analysis becomes a traceability matrix, tracing each function, its safety impact, and resulting safety requirements. The FHA can become somewhat of a living

System: Subsystem:				Functional Hazard Analysis			Analyst: Date:		
Function	Hazard No.	Hazard	Effect	Causal Factors	IMRI	Recommended Action	FMRI	Comments	Status

Figure 2.31 Example FHA worksheet.

document that is continually being updated as new information becomes available.

Among the advantages of the FHA technique are that it is easily and quickly performed, it provides rigor for focusing on hazards associated with system functions, it is a valuable tool for SwS analysis, and it helps to identify SCFs in the system. Since the technique focuses on functions, it might overlook other types of hazards, such as those dealing with hazardous energy sources or sneak circuit paths. This is why other analysis tools are also required.

After a functional hazard is identified, further analysis may be required to identify the specific causal factors.

SAE/ARP-4761, Guidelines and Methods for Conducting the Safety Assessment Process on Civil Airborne Systems and Equipment, Appendix A—Functional Hazard Assessment, 1996, is a useful reference on FHA that is recommended by the FAA for use on civil aircraft systems.

FHA is an HA technique that is important and essential to system safety. For more detailed information on the FHA technique, see Clifton A. Ericson II, *Hazard Analysis Techniques for System Safety* (2005), chapter 15.

See *Mishap Risk Index (MRI)* and *Safety Order of Precedence (SOOP)* for additional related information.

FUNCTIONAL HIERARCHY

A functional hierarchy is a hierarchical list of functions, similar to the system hardware hierarchy. The objective is to identify and list system functions. The identification of SCFs is a primary goal of system safety.

See *System Hierarchy* for additional related information.

FUNCTIONAL LOGIC DIAGRAM (FLD)

An FLD is a graphical representation of the functions in a system, showing logic gates, timers, and logic signal interfaces.

FUNCTIONAL REQUIREMENT

A functional requirement defines what the user needs for the system to operate in a desired manner; it delineate system functions and features. This should be contrasted with functional requirements that define specific behavior or functions. A nonfunctional requirement is a requirement that specifies criteria that can be used to judge the operation of a system, rather than specific behaviors. In general, functional requirements define what a system is supposed to do whereas nonfunctional requirements define how a system is supposed to be. Functional requirements specify particular results of a system whereas nonfunctional requirements specify overall characteristics such as cost and reliability. Functional requirements drive the application architecture of a system, while nonfunctional requirements drive the technical architecture of a system.

In system safety, one of the first HAs to be performed would be an FHA based on the functional requirements and the functional architecture. It is very valuable and useful for system safety to identify those functional requirements that are safety-related and safety-critical. Once SCFs are identified, components within that function can be considered safety-critical also.

See *Requirement* and *Specification* for additional related information.

FUNCTIONAL TEST

Functional tests involve the operation of a unit in accordance with a defined operational procedure to determine whether functional performance is within the specified requirements.

FUSE

A fuse is a component used in an electrical circuit to open the circuit when a certain level of current is reached. For example, household circuits typically use 15, 20, and 30 amp fuses. Fuses are actually safety devices used to prevent over temperature and fires if a fault in the circuit causes excessive current. A fuse is also a simple initiation mechanism for simple explosives, such as a burning fuse used to ignite firecrackers; as opposed to a fuze which is used in complex explosive systems.

FUSIBLE LINK

A fusible link is a device consisting of two strips of metal soldered together with a fusible alloy that is designed to melt at a specific temperature. This allows the two pieces to intentionally separate at a given temperature. Fusible links are produced in a variety of designs and different temperature ratings. Fusible links are utilized as the triggering device in fire sprinkler systems and mechanical automatic door release mechanisms that close fire doors. Some high-security safes also utilize fusible link-based relockers as a defense against torches and heat-producing tools.

A fusible link is a design safety mechanism used for hazard risk mitigation.

FUZE (FUZING SYSTEM)

A fuze, or fuzing system, is a physical device in a system intended for the purpose of detonating an explosive device or munition when intended, and for preventing device initiation when not intended. It is typically designed to sense a target or respond to one or more prescribed conditions, such as elapsed time, pressure, or command, which initiates a train of fire or detonation in a munition. Safety and arming (S&A) are primary roles performed by a fuze to preclude inadvertent ignition of a munition and intended ignition before the desired position or time. It should be noted that the term *fuse* is typically used to represent a simple burning fuse, for explosives such as firecrackers, while the term *fuze* is used for more sophisticated munitions containing mechanical and/or electronic components. In military munitions, a fuze is the part of the device that initiates function, and it typically includes some type of S&A device.

A fuzing system has two primary objectives: (1) to reliably initiate an explosive system, and (2) to reliably prevent initiation until the intended safety conditions have been met. MIL-STD-1316, DoD Design Criteria Standard, Safety Criteria for Fuze Design, provides requirements for the design of a safe fuzing system.

A fuze is a device used in munitions which is designed to detonate, or to set forces into action to ignite, detonate, or deflagrate the charge (or primer) under specified conditions. In contrast to a simple pyrotechnic fuse, a munitions fuze always has some form of safety/arming mechanism, designed to protect the user from premature or accidental detonation. A munition fuze assembly may contain only the electronic or mechanical elements necessary to signal or actuate the detonator, but some fuzes contain a small amount of primary explosive to initiate the detonation. Fuze assemblies for large explosive charges may include an explosive booster.

Fuzes typically employ one or more of the following activation mechanisms:

- Time fuzes
- Impact fuzes
- Proximity fuzes
- Remote detonator fuzes
- Barometric or altitude fuzes

A fuze must be designed to function appropriately considering relative movement of the munition with respect to its target. The target may move past stationary munitions like land mines or naval mines; or the target may be approached by a rocket, torpedo, artillery shell, or air-dropped bomb. Timing of fuze function may be described as optimum if detonation occurs when target damage will be maximized, early if detonation occurs prior to optimum, late if detonation occurs past optimum, or dud if the munition fails to detonate. Any given batch of a specific design should be tested to determine the expected percentage of early, optimum, late, and dud initiations.

172 SYSTEM SAFETY TERMS AND CONCEPTS

Explosives are used in many different types of devices for many different applications. In order to start these devices in a controlled manner, an effective initiation system is necessary. The initiation system should be reliable so that it performs when intended, and it should be safe in order that it does not perform or function when not intended. An explosives initiation system is called the fuzing system or fuze. The purpose of the fuze is to reliably initiate the device, but only after all safety criteria have been met. The basic concept of a fuzing system is shown in Figure 2.32 using a warhead as an example system.

A fuze is composed of several different components or subsystems which achieve the primary functions of (a) arming/safing and (b) initiating. The arming/safing function physically allows or prevents the initiating process by acting as a physical barrier in the initiation train, which is typically achieved with an S&A device. The initiating function is the physical process of detonating the weapon explosives.

The initiating function utilizes an explosive train that detonates the main explosive charge through a series of incremental steps. The main charge is typically a large quantity of an insensitive munition, which requires a high level of energy for detonation (i.e., safety). Thus, smaller amounts of more sensitive explosives are used in the explosive train to incrementally produce a higher energy level, which eventually is sufficient to cause detonation of the main charge. Safety is achieved by only allowing very small amounts of the more sensitive explosives components, and highly protecting them from inadvertent ignition sources. It is customary to use more than one type of energetic material in the explosives train. The sequence of energetic materials begins with a

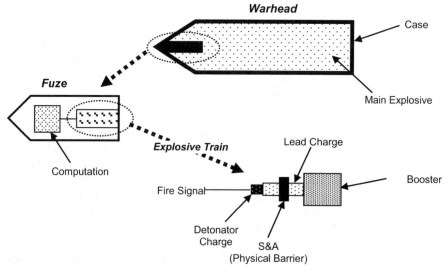

Figure 2.32 Fuze system concept.

small amount of a relatively sensitive material in the initiation system, and proceeds through a series of explosives of increasing insensitivity and increasing quantity. The output from one component in the sequence of explosives (the explosives train) is used to initiate the reaction in the next member of the sequence. The amount of explosive present in the detonator and lead is quite small, while the booster, which increases the output from the relatively low level produced by the detonator, leads to a higher level that is sufficient to initiate the main charge. The requirements for explosives that are used in the various parts of the explosive train reflect the quantities used and the sensitivities of the different types of explosives.

As seen in Figure 2.32, the explosive train (or initiator train) is an integral part of the fuze. The detonator and lead are shown with an interrupter interposed between them, to physically prevent the initiation of the main charge if the most sensitive component should accidentally be initiated. The interrupter part of the fuze is the S&A device. The computation part of the fuze determines when the detonator should be fired. The computation part can be electrical, mechanical, or a combination. Depending upon the particular design, the S&A device will close to allow the lead to initiate the booster, based on its own built-in decision criteria, decision criteria from the computation device, or a combination of both. The S&A is generally a mechanical device which requires a minimum of two different sensed physical environments to cause arming, such as velocity (setback) and rotation (spin).

The primary safety aspects of a fuze involve arming and initiation. The fuze design should include positive design measures that prevent inadvertent and premature arming, within an acceptable level of risk. This is primarily achieved through the design of the S&A device. Another aspect of fuze safety is safe separation. The fuze should not initiate before a safe distance is reached where personnel and/or equipment at the firing point will not be impacted by exploding fragments. While Figure 2.32 displayed the general concept for a fuze, Figure 2.33 shows fuze S&A design concept. The detail to note from this

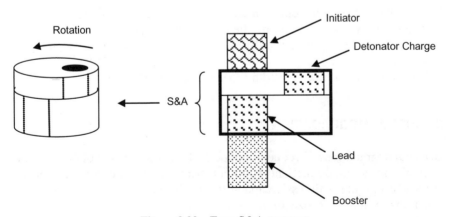

Figure 2.33 Fuze S&A concept.

diagram is that the detonator charge is physically isolated within the S&A from the lead charge. The top portion of the S&A must rotate around until the detonator and lead are in alignment before an initiation train can be executed.

Explosives and fuze safety is the process used to prevent premature, unintentional, or unauthorized initiation of explosives and devices containing explosives and to minimize the effects of explosions, combustion, toxicity, and any other deleterious effects. Explosives safety includes all mechanical, chemical, biological, electrical, and environmental hazards associated with explosives or EM environmental effects. Equipment, systems, or procedures and processes whose malfunction would cause unacceptable mishap risk to manufacturing, handling, transportation, maintenance, storage, release, testing, delivery, firing, or disposal of explosives are also included.

GOVERNMENT FURNISHED EQUIPMENT (GFE)

GFE is property in the possession of, or acquired directly by, the government, and subsequently delivered to or otherwise made available to the contractor. For example, when starting development on a new fighter aircraft, the government may require the use of the radar already in inventory from an existing fighter aircraft. In this case, the radar is already in use, and was designed and built to the specifications and requirements of a different aircraft system.

GFE can present many problems for safety, particular when used in SCFs or applications. The use of GFE in a system should not preclude a complete SHA which includes GFE items; GFE is not exempt from system safety when included in a new development system.

See *Commercial Off-the-Shelf (COTS)* for additional information.

GOVERNMENT-OFF-THE-SHELF (GOTS)

GOTS refers to government created software, usually from another project. The software was not created by the current developers. Source code, documentation, analysis, and test and results are generally included with GOTS software.

See *Reusable Software* for additional related information.

GRACEFUL DEGRADATION

Graceful degradation is a reduction of performance as a result of failure, while maintaining essential function(s) and performance. It is the capability of continuing to operate with lesser capabilities in the face of faults or failures, or when the number or size of tasks to be done exceeds the capability to complete.

GRAPHICAL USER INTERFACE (GUI)

The term *user interface (UI)* is used to describe the controls and displays that interface a human operator with a system. In today's environment, many UIs involve computer controlled electronic devices. A basic goal of human systems integration (HSI) is to improve the interactions between users and computer-controlled systems by making the interface more usable and receptive to the user's needs, as well as making it less error prone. HSI applies the principles of human perception and information processing to create an effective UI. A few of the many potential benefits that can be achieved through utilization of HSI principles include a reduction in errors, a reduction in required training time, an increase in efficiency, and an increase in user satisfaction. Usability is the degree to which the design of a particular UI takes into account the human psychology and physiology of the users, and makes the process of using the system effective, efficient, satisfying, and safe.

A GUI (pronounced "gooey") is a type of UI that allows humans to interact with computers and electronic devices using graphical images on a computer screen rather than text-based commands or mechanical devices. A GUI offers graphical icons, and visual indicators, as opposed to straight text-based interfaces. The GUI consists of graphical elements such as windows, menus, radio buttons, check boxes, and icons. The GUI may employ a pointing device, touch screen, and/or a keyboard. These aspects can be emphasized by using the alternative acronym WIMP, which stands for Windows, Icons, Menus, and Pointing device. GUIs (or WIMPs) present the user with numerous widgets (e.g., buttons, dials, and check boxes) that represent and can trigger some of the system's available commands. Digital screens using GUIs provide savings in cost, weight, and space over the mechanical devices they replace. GUIs, however, require considerable CPU power and a high-quality display, whereas text-based systems are not as demanding. Just as mechanical UI devices can present safety problems, GUIs are not immune to this potential and can also present safety issues, only in different ways.

On the surface it might seem that UIs and GUI designs are relatively benign and would not pose any safety concerns. However, clearly the opposite is true, as demonstrated by past mishaps in a variety of different systems. Just as a mechanical switch can fail *on* or *off* and affect an SCF, so too can a GUI switch; however, the likelihood is apparently much smaller for a GUI. The primary safety concerns with GUIs tend to deal with user confusion and user overload (too many GUIs and GUI options) which could cause a safety error to be committed by the user.

Some generic GUI safety concerns include the following:

- A GUI switch failing on or off when not commanded to do so
- Operator erroneously initiates a command by touching a GUI control
- Operator error due to GUI mode confusion
- Operator error due to GUI data confusion

- Operator GUI error due to overload and stressful scenario
- Erroneous command due to someone wiping the GUI screen

GUARD (SAFETY GUARD)

A safety guard is a device, barrier, or enclosure that prevents accidental contact with an item, such as a control switch. The guard typically prevents accidental operation of machinery, weapons, and so on.

HANG FIRE

A hang fire is a malfunction that causes an undesired delay in the functioning of a firing system, typically a weapon firing system. On an aircraft, a hang fire weapon will often stay with the aircraft, in an unknown safety state.

HARDWARE

In a system, items that have physical being are refereed to as hardware. For example, hardware refers to items such as motors, pumps, circuit cards, and power supplies. Hardware is a system element, as are software and humans.

HARDWARE DESCRIPTION LANGUAGE (HDL)

In electronics design, HDL is any programming language providing formal description of electronic circuits, especially digital logic. HDL can describe a circuit's operation, design, organization, and tests to verify its operation by means of simulation. HDLs are standard text-based expressions of the spatial and temporal structure and behavior of electronic systems. HDL syntax and semantics includes explicit notations for expressing concurrency; HDLs also include an explicit notion of time, which is a primary attribute of hardware. HDLs are used to write executable specifications of a piece of hardware; simulation program allows the hardware designer with the ability to model a piece of hardware before it is created physically. HDLs are processed by a compiler, usually called a synthesizer, where synthesis is a process of transforming the HDL code listing into a physically realizable gate netlist. This is software that should be considered in a SwSP.

HARM

Harm involves physical injury or damage to health, property, or the environment, or some combination of these. It is any physical damage to the body caused by violence, accident, mishap, fracture, and so on. Harm can be expressed as the potential expected outcome from a potential mishap or the actual harm resulting from an actual event or mishap that has occurred. In system safety, harm is typically the expected or actual outcome resulting from a mishap. This outcome is usually stated in terms of death, injury, system damage, and/or environmental damage.

HAZARD

The term "hazard" is a common, well-used term that most people feel they understand completely and that its definition is intuitively obvious. In its colloquial usage, the definition is fairly simple, straightforward, and broad. However, in the field of system safety, the technical definition of a hazard is a little more complex, narrower in scope, and possibly less well understood.

In the common or colloquial sense, a hazard is thought of as a source of danger; the possibility of incurring loss; a risk. A typical industry definition provided by MIL-STD-882D states, "A hazard is any real or potential condition that can cause injury, illness, or death to personnel; damage to or loss of a system, equipment, or property; or damage to the environment." This implies that a hazard is a condition that can cause a mishap. These are good generic definitions; however, the "potential condition," "source of danger," and "how the mishap is caused" are not sufficiently defined by these definitions for engineering purposes.

For technical purposes, a hazard can be defined as the existence of a specific set of system conditions that form a unique potential mishap event. In other words, a hazard is an existing system state that is dormant, but which has the potential to result in a mishap when the inactive hazard state components are actualized. A hazard is a potential mishap, and a mishap is an UE that has occurred as a result of an actualized hazard. This more technical definition is necessary because in order to mitigate the risk presented by a hazard, all of the components and parameters comprising the hazard must be identified and understood.

If a hazard is a potential mishap, then the hazard introduces a risk of loss that is presented by the mishap and its potential outcome. Hazard–mishap risk can be changed (mitigated) but only when the hazard components are known and understood. Therefore, it is necessary to identify and understand the composition of all hazards within a system in order to understand and mitigate the risk before a mishap actually occurs. A hazard is an entity that contains only the specific elements *necessary and sufficient* to result in a mishap. The components of a hazard define the necessary conditions for a mishap and the end outcome or effect of the mishap. A mishap is the result of an actuated hazard, and a hazard is a unique system entity. This entity is a set of prearranged or predesigned hazardous conditions that are inadvertently built into a system through human design. In the safety sense, mishaps are formulated events in that they are actually created through the involvement of hazardous components, poor design, and/or inadequate design foresight.

In order for a hazard to exist, three hazard components must be present to form the hazard: (1) the hazard source (HS) which provides the basic source of danger, (2) the potential IMs that will transition the hazard from an inactive state to a mishap event, and (3) the target-threat outcome (TTO) that will be negatively impacted by the mishap event outcome. The existence of a hazard requires the existence of these three components as a prerequisite. These components are necessary to assess the risk and to know where and how to mitigate the hazard.

Another way to technically define a hazard is: "A hazard is a set of inactive conditions, which consist of a Hazardous Source, an Initiating Mechanism and a Target-Threat Outcome, which leads to a mishap when the Initiating Mechanism is actualized." A hazard is a physical entity that characterizes a potential mishap. A hazard is a condition that is prerequisite to a mishap; that is, it is a blueprint for a mishap. When a hazard exists, a *hazard triangle* is created (see discussion below).

If a *mishap* is an *actual event* that has occurred and resulted in death, injury, and/or loss, and a *hazard* is a *potential condition* that can potentially result in death, injury, and/or loss, then a hazard and a mishap must be linked by a transition mode. These definitions lead to the principle that a hazard is the precursor to a mishap; a hazard *defines* a potential event (i.e., mishap), while a mishap is the occurred event. This means that there is a direct relationship between a hazard and a mishap, as depicted in Figure 2.34.

The concept conveyed by Figure 2.34 is that a hazard and a mishap are two separate states of the same phenomenon, linked by a state transition. You can think of these states as the before and after states. A hazard is a "potential event" at one end of the spectrum, that may be transformed into an "actual event" (the mishap) at the other end of the spectrum, based upon occurrence of the state transition.

Mishaps are the immediate result of actualized hazards. The state transition from a hazard to a mishap is based on two factors: (1) the unique set of *hazard components* involved and (2) the *mishap risk* presented by the hazard components. The hazard components are the items comprising a hazard, and the mishap risk is the probability of the mishap occurring and the severity of the resulting mishap loss.

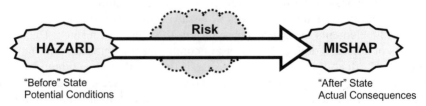

Figure 2.34 Hazard/mishap relationship.

Hazard/mishap risk is a fairly straightforward concept, where risk is defined as:

$$\text{Risk} = \text{Likelihood} \times \text{Severity}$$

The mishap likelihood factor is the probability of the hazard components occurring and transforming into the mishap. The mishap severity factor is the overall consequence of the mishap, usually in terms of loss resulting from the mishap (i.e., the undesired outcome). Both probability and severity can be defined and assessed in either qualitative terms or quantitative terms. Time is factored into the risk concept through the probability calculation of a fault event, for example, $P_{FAILURE} = 1.0 - e^{-\lambda T}$, where T = exposure time and λ = failure rate.

Hazards primarily exist in most systems because of the need for HSs in the system, coupled with the fact that eventually everything fails, and these failures can unleash the undesired effects of the hazard sources. Hazards also exist due to the need for safety-critical system functions, coupled with the potential for failures and human error. Hazards typically exist in a system for the following basic reasons, singularly or in combinations:

- The use of hazardous system elements (e.g., fuel, explosives, electricity, velocity, stored energy)
- Operation in hazardous environments (e.g., flood zones, ice, humidity, heat)
- The need for hazardous functions (e.g., aircraft fueling, welding)
- The use of SCFs (e.g., flight control, weapon arming)
- The inclusion of unknown design flaws and errors
- The potential for hardware wear, aging, and failure
- Inadequacy in designing for tolerating critical failures

These three required components of a hazard form what is known as the *hazard triangle*, which is illustrated in Figure 2.35. The hazard triangle conveys by the idea that a hazard consists of *three necessary* and *coupled* components, each of which forms the side of a triangle. All three sides of the triangle are

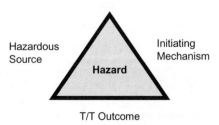

Figure 2.35 Hazard triangle.

essential and are required in order for a hazard to exist (i.e., HS, IM, and TTO). Remove any one of the triangle sides, and the hazard is eliminated because it is no longer able to produce a mishap (i.e., the triangle is incomplete). For example, remove the human operator from an aircraft, and the hazard "aircraft crashes resulting in pilot death" is eliminated because the TTO is removed. Reduce the probability of the IM side and the mishap probability is reduced. Reduce an element in the HS or the TTO side of the triangle and the mishap severity is reduced.

A hazard can be eliminated by eliminating any one of the three basic components of a hazard (HS, IM, or TTO). Practically speaking, however, hazards are primarily eradicated by eliminating the HS component. Hazards are predominantly mitigated in risk by reducing the probability of the IM. Hazard risk can also be mitigated by reducing the effective danger of the HS, by protecting the target, or by reducing the amount of the threat. It should be noted however, that most hazards are mitigated by reducing the actuating probabilities of the IM. Changes in the probability and/or severity of a hazard modify the risk of the hazard. The hazard triangle concept is useful in determining if a conceptualized hazard meets the necessary criteria and when determining where to mitigate a hazard. It also demonstrates that when a hazard is mitigated, it is not eliminated (a common error of safety beginners) because all three sides are still present. It is interesting to note that these mitigation truisms correlate appropriately with the SOOP.

Basically, a hazard exists as a result of a system liability component being present in the system (i.e., the HS), combined with a system design that poorly tolerates various mechanisms that can impinge on the liability component. The actual amount of risk presented by the hazard is a function of the system failure influence on the liability component from factors such as hardware failures, human errors, sneak electrical paths, software errors, and incorrect interfaces. In order to mitigate a hazard, the hazard must be recognized and understood, and then the influence factors appropriately modified. Table 2.10 provides some example items and conditions for each of the three hazard components.

TABLE 2.10 Example Hazard Components

Hazardous Element	Initiating Mechanism	Target: Threat
Ordnance	Inadvertent signal; RF energy	Personnel: Explosion, death/injury
High pressure tank	Tank rupture	Personnel: Explosion, death/injury
Fuel	Fuel leak and ignition source	Personnel: Fire, loss of system, death/injury
High voltage	Touching an exposed contact	Personnel: Electrocution, death/injury

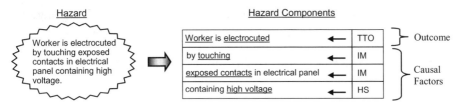

Figure 2.36 Example of hazard components.

To demonstrate the hazard component concept, consider a detailed breakdown of the following example hazard: "Worker is electrocuted by touching exposed contacts in electrical panel containing high voltage." Figure 2.36 shows how this hazard is divided into the three necessary hazard components to validate the hazard. Note in this example that all three hazard components are present and can be clearly identified. In this particular example, there are actually two IMs involved. The TTO defines the mishap outcome, while the combined HS and TTO define the mishap severity. The HS and IM are the HCFs that are used to determine the mishap probability. If the high-voltage component can be removed from the system, the hazard is eliminated. If the voltage can be reduced to a lower, less harmful level, then the mishap severity is reduced, and the hazard is mitigated to a lesser level of risk.

A hazard is an existing potential condition (inactive) that will result in a mishap when actualized. The hazard condition is a potential state formed by the HS (e.g., energy) and the potential IMs (e.g., failures) that will transform the hazard into a mishap and the hazard outcome. The hazard outcome is predicted in the expected mishap target (e.g., personnel) and the expected mishap threat (e.g., death or injury).

HAZARD ACTION RECORD (HAR)

An HAR is a record that contains all information relevant to the identification, assessment, mitigation, and closing of a hazard. A HAR is a key instrument or tool employed for tracking hazards. Note: It is referred to as a record, rather than a report, because database elements are known as records and HARs are generally maintained in a database (as a record). Many different types of reports can be derived from the information in an HAR. HARs have also been referred to as Safety Action Records (SARs); however, this term is less desirable because it conflicts with the more standard acronym for the term Safety Assessment Report (SAR).

HARs are generally recorded on a formal HAR form, and it is important that the HAR form have sufficient fields for recording all the necessary and relevant information. Needed information may include identification date, initial and final risk assessments, mitigation methods employed, verification test data that the safety requirements were successfully implemented, status (open, closed) of the hazard, and any other relevant information generated from the analyses. The HAR form should be formalized and described in the SSPP.

See *Hazard Tracking System (HTS)* for additional related information.

HAZARD ANALYSIS (HA)

HA is the act of performing a special analysis for the identification and evaluation of hazards. An HA typically tries to answer three questions:

1. What can go wrong that will lead to a mishap (how will it happen)?
2. What are the consequences?
3. What is the likelihood of occurrence?

The primary purpose of HA is to identify hazards and to obtain sufficient hazard data for risk assessments. HA is applied to hardware, software, functions, procedures, and human tasks. HA can be applied at all stages of the system life cycle; the HA process becomes more detailed and accurate as more information about the system becomes available. Different HA techniques and approaches to hazard identification may be required at different stages of the system life cycle to ensure all types of hazards are identified. HA involves:

- Acquiring system data, knowledge, and understanding.
- Identifying the hazards that exist within a system.
- Determining the chain of events that could potentially lead to each hazard.
- Determining the consequences resulting from an occurrence of the hazard.
- Investigating any safeguards already in place to address the hazards.
- Assessing the risk presented by the identified hazards.
- Establishing design requirements to mitigate hazard risk.

HA is the systematic examination of a system, item, or product within its life cycle, to identify hazardous conditions including those associated with human, product, and environmental interfaces, and to assess their consequences to the functional and safety characteristics of the system or product. In order to design-in safety, hazards must be designed-out (eliminated) or mitigated (reduced in risk), which can only be accomplished through HA. Hazard identification is a critical system safety function and is one of the basic required elements of an SSP.

HA provides the basic foundation for system safety. HA is performed to identify hazards, hazard effects, and HCFs. HA is used to determine system risk, to determine the significance of hazards, and to establish design measures that will eliminate or mitigate the identified hazards. HA is used to systematically examine systems, subsystems, facilities, components, software, personnel, and their interrelationships, with consideration given to logistics, training, maintenance, test, modification, and operational environments. In order to effectively perform HAs, it is necessary to understand what comprises a hazard, how to recognize a hazard, and how to define a hazard. To develop the skills needed to identify hazards and HCFs, it is necessary to understand the nature of hazards, their relationship to mishaps, and their effect upon system design.

Potential mishaps exist as hazards, and hazards exist in system designs. Hazards are actually designed into the systems we design, build, and operate. Sometimes this happens intentionally and most often unintentionally. In order to perform HA, the analyst must first understand the nature of hazards. Hazards are predictable, and what can be predicted can also be eliminated or controlled.

There are seven basic HAs types recommended by MIL-STD-882 for application on all major and/or complex systems. These types are:

- Preliminary hazard list (PHL)
- PHA
- SSHA
- SHA
- Operating and support hazard analysis (O&SHA)
- Health hazard assessment (HHA)
- Safety requirements/criteria analysis (SRCA)

Within the system safety discipline, over 100 different HA techniques have been developed, some of which are very unique and some of which are variants of others. Just to name a few, techniques include analyses such as:

- FTA
- BPA
- Threat hazard analysis (THA)
- Sneak circuit analysis (SCA)
- Code safety analysis
- BA
- Change analysis
- Interlock analysis
- Task analysis
- ETA

See *Hazard*, *Safety Analysis Techniques*, and *Safety Analysis Types* for additional related information.

HAZARD ANALYSIS AND CRITICAL CONTROL POINT (HACCP)

HACCP is a systematic approach to the identification, evaluation, and control of food safety hazards. This methodology, which in effect seeks to plan out unsafe practices, differs from traditional "produce and test" quality assurance methods which are less successful and inappropriate for highly perishable foods. HACCP is used in the food industry to identify potential food safety hazards, so that key actions, known as critical control points (CCPs) can be taken to reduce or eliminate the risk of the hazards being realized. HACCP is a systematic preventative approach to food safety that addresses physical, chemical, and biological hazards as a means of prevention rather than finished product inspection. The system is used at all stages of food production and preparation processes. HACCP is also being applied to industries other than food, such as cosmetics and pharmaceuticals.

HACCP is based around seven established principles:

- Principle 1: Conduct an HA.

    Plants determine the food safety hazards and identify the preventive measures the plant can apply to control these hazards. A food safety hazard is any biological, chemical, or physical property that may cause a food to be unsafe for human consumption.
- Principle 2: Identify CCPs.

    A CCP is a point, step, or procedure in a food process at which control can be applied and, as a result, a food safety hazard can be prevented, eliminated, or reduced to an acceptable level.
- Principle 3: Establish critical limits for each CCP.

    A critical limit is the maximum or minimum value to which a physical, biological, or chemical hazard must be controlled at a CCP to prevent, eliminate, or reduce to an acceptable level.
- Principle 4: Establish CCP monitoring requirements.

    Monitoring activities are necessary to ensure that the process is under control at each CCP. The United States Department of Agriculture (USDA) Food Safety Inspection Services (FSIS) requires that each monitoring procedure and its frequency be listed in the HACCP plan.
- Principle 5: Establish corrective actions.

    These are actions to be taken when monitoring indicates a deviation from an established critical limit. The final rule requires a plant's HACCP plan to identify the corrective actions to be taken if a critical limit is not met. Corrective actions are intended to ensure that no product injurious to health or otherwise adulterated as a result of the deviation enters commerce.
- Principle 6: Establish record keeping procedures.

    HACCP regulation requires that all plants maintain certain documents, including its HA and written HACCP plan, and records documenting the

monitoring of CCPs, critical limits, verification activities, and the handling of processing deviations.
- Principle 7: Establish procedures to verify the HACCP system works as intended.

Validation ensures that the plans do what they were designed to do; that is, they are successful in ensuring the production of safe product. Plants will be required to validate their own HACCP plans. FSIS will not approve HACCP plans in advance, but will review them for conformance with the final rule. Verification ensures the HACCP plan is adequate, that is, working as intended. Verification procedures may include such activities as review of HACCP plans, CCP records, critical limits, and microbial sampling and analysis. Verification tasks are performed by plant personnel and also by FSIS inspectors.

HAZARD AND OPERABILITY (HAZOP) ANALYSIS

HAZOP analysis is a technique for identifying and analyzing hazards and operational concerns of a system. It is a very organized, structured, and methodical process for carrying out a hazard identification analysis of a system, from the concept phase through decommissioning. Although HAZOP is a relatively simple process in theory, the steps involved must be carefully observed in order to maintain the rigor of the methodology.

HAZOP analysis utilizes use key guidewords and system diagrams (design representations) to identify system hazards. Adjectives (guide words) such as *more, no,* and *less*, are combined with process/system conditions such as *speed, flow,* and *pressure*, in the hazard identification process. HAZOP analysis looks for hazards resulting from identified potential deviations in design operational intent. A HAZOP analysis is performed by a team of multidisciplinary experts in a brainstorming session under the leadership of a HAZOP team leader.

Some of the key components to a HAZOP analysis include:

- A structured, systematic, and logical process
- A multidisciplinary team with experts in many areas
- An experienced team leader
- The controlled use of system design representations
- The use of carefully selected system entities, attributes, and guide words to identify hazards

The purpose of HAZOP analysis is to identify the potential for system deviations from intended operational intent through the unique use of key guidewords. The potential system deviations then lead to possible system hazards. HAZOP analysis is applicable to all types of systems and equipment, with analysis coverage given to subsystems, assemblies, components, software, procedures, environment, and human error. HAZOP analysis can be conducted at different abstraction levels, such as conceptual design, top-level design, and detailed component design. HAZOP analysis has been successfully applied to a wide range of systems, such as chemical plants, nuclear power plants, oil platforms, and rail systems. The technique can be applied to a system very early in design development and thereby identify safety issues early in the design process. Early application helps system developers to design-in safety of a system during early development rather than having to take corrective action after a test failure or a mishap. The HAZOP analysis technique, when applied to a given system by experienced personnel, provide a thorough and comprehensive identification of hazards that exist in a given system or process.

HAZOP analysis was initially developed for the chemical process industry and its methodology was oriented around process design and operations. The methodology can be extended to systems and functions with some practice and experience. The HAZOP analysis technique provides for an effective HA. In essence, the HAZOP analysis is not much different than PHA or SSHA, except for the guidewords used. HAZOP analysis could be utilized for the PHA and/or SSHA techniques.

HAZOP analysis entails the investigation of deviations from design intent for a process or system by a team of individuals with expertise in different areas, such as engineering, chemistry, safety, operations, and maintenance. The approach is to review the process/system in a series of meetings during which the multidisciplinary team "brainstorm" the system design methodically by following a sequence based on prescribed guidewords and the team leader's experience. The guidewords are used to ensure that the design is explored in every conceivable manner. The HAZOP analysis is based on the principle that several experts with different backgrounds can interact and better identify hazards when working together than when working separately and combining their results.

The HAZOP analysis is conducted through a series of team meetings led by the team leader. The key to a successful HAZOP is selection of the right team leader and the selection of the appropriate team members. This HAZOP analysis is applied in a structured way by the team, and it relies upon their imagination in an effort to discover credible causes of deviations from design intent. In practice, many of the deviations will be fairly obvious, such as pump failure causing a loss of circulation in a cooling water facility; however, the great advantage of the technique is that it encourages the team to consider other less obvious ways in which a deviation may occur. In this way, the analysis becomes much more than a mechanistic checklist type of review. The result is that there is a good chance that potential failures and problems will be identified that had not previously been experienced in the type of plant/system being studied.

The HAZOP analysis is performed by comparing a list of system parameters against a list of guidewords. This process stimulates the mental identification of possible system deviations from design intent and resulting hazards. Establishing and defining the system parameters and the guidewords is a key step in the HAZOP analysis. The deviations from the intended design are generated by coupling the guideword with a variable parameter or characteristic of the plant, process, or system, such as reactants, reaction sequence, temperature, pressure, flow, and phase. In other words

$$\text{Guideword} + \text{Parameter} = \text{Deviation}$$

For example, when considering a reaction vessel in which an exothermic reaction is to occur and one of the reactants is to be added stepwise, the guideword "more" would be coupled with the parameter "reactant," and the deviation generated would be "thermal runaway." Systematic examinations are made of each part of a facility or system.

The HAZOP analysis technique is a detailed HA utilizing structure and rigor. It is typically performed using a specialized worksheet. Although the format of the analysis worksheet is not critical, generally, matrix or columnar-type worksheets are used to help maintain focus and structure in the analysis. The HAZOP analysis sessions are primarily reported in the HAZOP worksheets, in which the different items and proceedings are recorded. As a minimum, the following basic information is required from the HAZOP analysis worksheet:

1. Item under analysis
2. Guide words
3. System functions and parameters
4. System effect if guide word occurs
5. Resulting hazard or deviation (if any)
6. Risk assessment
7. Safety requirements for eliminating or mitigating the hazards

HAZOP is an analysis technique that is important and essential to system safety. For more detailed information on the HAZOP technique, see Clifton A. Ericson II, *Hazard Analysis Techniques for System Safety* (2005), chapter 21.

HAZARD CAUSAL FACTOR (HCF)

HCFs are the specific causes of a hazard. HCFs are the genesis of a mishap; they explain how a hazard will transform into a mishap, and they also explain what outcome to expect. Hazard–mishap theory states that hazards create the potential for mishaps, mishaps occur based on the level of risk involved, and risk is calculated from the hazard–mishap causal factors. HCFs are the unique factors that *create* the hazard, while mishap causal factors are the factors that *cause* the dormant hazard to become an active mishap event.

Just as hazard risk is the same as mishap risk, HCFs are the same as mishap causal factors.

For a hazard to exist, three hazard components must be present: (1) the HS which provides the basic source of danger, (2) the potential IMs that will transition the hazard from an inactive state to a mishap event, and (3) the TTO that will result from the expected mishap event. The HS and IM are the HCFs that are used to determine risk likelihood, and the TTO is the causal factor that establishes risk outcome and severity. The HS and IM hazard components can be broken into the major causal factor categories of hardware, software, humans, interfaces, functions, procedures, management safety culture, and the environment. Since hazard risk is the same as mishap risk, HCFs are also mishap causal factors because they are the same factors used to compute the risks.

If a hazard has multiple causal factors that can be ORed together, then there should really be multiple hazards, one hazard for each unique independent causal factor. If a hazard has multiple causal factors that must be ANDed together, then it is a single hazard with multiple causes that must occur together. It should be noted that AND and OR refer to Boolean logic used in FTA.

To illustrate the HCF concept, consider the hazard "Fire destroys a house due to ignition of a gas leak." The HCFs are flammable gas and faults that cause leaking and ignition of the gas. The house is destroyed because a fire resulted from ignited gas. The mishap is the destroyed house, caused by an explosion and fire, caused by gas ignition. For this particular hazard, all of the causal factors must be ANDed together; gas present in system, gas leak occurs, and gas ignition source occurs.

See *Hazard* and *Mishap* for additional related information.

HAZARD CHECKLIST

A hazard checklist is a generic list of items known to be hazardous or that might create potentially hazardous designs or situations. The hazard checklist should not be considered complete or all-inclusive. Hazard checklists are used in HA to assist in the identification of hazards. Hazard checklists help trigger the analyst's recognition of potential HSs, from past lessons learned.

Typical hazard checklist types include:

- Energy sources
- Hazardous functions
- Hazardous operations
- Hazardous components
- HAZMATs
- Lessons learned from similar type systems

- Known undesired mishaps
- Failure mode and failure state considerations

Hazard checklists provide a brainstorming process for recognizing hazard possibilities brought about by the unique system design and the hazard sources that are used in the system. A hazard checklist should never be considered a complete and final list but merely a mechanism or catalyst for stimulating hazard recognition.

HAZARD COMPONENTS

In order for a hazard to exist, three hazard components must be present in order to form the hazard: (1) the HS which provides the basic source of danger, (2) the potential IMs that will transition the hazard from an inactive state to a mishap event, and (3) the TTO that will be negatively impacted by the mishap event outcome. The existence of a hazard requires the existence of these three components as a prerequisite. These components are necessary to assess the risk and to know where and how to mitigate the hazard.

A hazard is an existing potential condition (inactive) that will result in a mishap when actualized. The hazard condition is a potential state formed by the HS (e.g., energy) and the potential IMs (e.g., failures) that will transform the hazard into a mishap and the hazard outcome. The hazard outcome is predicted in the expected mishap target (e.g., personnel) and the expected mishap threat (e.g., death or injury).

See *Hazard* and *Hazard Triangle* for additional related information.

HAZARD CONTROL

Hazard control is the process and methodology for reducing the mishap risk of a hazard when the hazard cannot be eliminated. Hazard control reduces the hazard mishap probability and/or the hazard mishap outcome severity. Hazard control involves taking specific design action to reduce the hazard mishap risk to an acceptable level. Design action is achieved through design measures, the use of safety devices, warning devices, training, or procedures. Hazard control is also known as mishap risk mitigation, since the process is reducing the potential mishap risk. Hazard control is identical to and the same as hazard mitigation; however, hazard mitigation is the preferred term.

HAZARD COUNTERMEASURE

A hazard countermeasure is a special and intentional feature in the design of a system or product employed specifically for the purpose of eliminating or mitigating the risk presented by an identified hazard. A hazard countermeasure may not be necessary for system function, but it is necessary for safety (i.e., risk reduction). A hazard countermeasure can be any device, technique, method, or procedure incorporated into the design to specifically eliminate or reduce the risk factors comprising a hazard.

See *Design Safety Feature (DSF)* and *Safety Order of Preference (SOOP)* for additional related information.

HAZARD DESCRIPTION

A hazard description is the brief narrative description of a potential hazard. A hazard description should contain the three elements that define a hazard: (1) the HS which provides the basic source of danger, (2) the potential IMs that will transition the hazard from an inactive state to a mishap event, and (3) the TTO that will be negatively impacted by the mishap event outcome.

HAZARD ELIMINATION

Hazard elimination means to completely remove the hazard from the system design. This generally involves completely removing the HS (e.g., fuel, explosives) from the system design and replacing it with another nonhazardous item. If the HS cannot be eliminated from the system, then hazards associated with it cannot be removed; they can only be mitigated.

See *Hazard Triangle* for additional related information.

HAZARD IDENTIFICATION

Hazard identification is the process of recognizing hazards that exist within a system design. This includes defining the hazards and describing their characteristics and components. Hazard identification is achieved through HA.

See *Hazard Analysis* for additional related information.

HAZARD LIKELIHOOD (HAZARD PROBABILITY)

Likelihood is one parameter in the risk equation. Risk is the safety measure of a potential future event, stated in terms of event likelihood and event severity. Hazard likelihood is the expected likelihood that the identified hazard will be activated and becomes an actual mishap. Hazard likelihood is the estimated likelihood of a hazard transitioning from a conditional state to an actual mishap event state, resulting in an actual mishap with undesired outcome.

In a risk assessment, hazard likelihood can be characterized in terms of probability, frequency, or qualitative criteria. Quite often, the term "hazard probability" is incorrectly used when the actual assessment is done in terms of frequency or qualitative criteria. This is why hazard likelihood is a more accurate term. Likelihood is a measure of how possible or likely it is that an event will occur, such as a hazard–mishap. Likelihood can typically be characterized in one of the following ways:

1. Probability—Probability is a number between zero and one. Although probability is a dimensionless number, it has to be calculated to some specific set of criteria, such as the probability of event X occurring in 700 h of operation.
2. Frequency—Frequency is the rate of occurrence of an event described in terms of occurrences per unit of time or operations, such as failures per million hours of operation or per million cycles. Frequency is quite often established from historical data.
3. Qualitative description—In this methodology, likelihood is classified in terms of different qualitative likelihood ranges. For example: "Likely to occur frequently in the life of an item," "Will occur several times in the life of an item," and "Likely to occur sometime in the life of an item."

Hazard probability is typically obtained from calculations made using the HCF failure rates and the exposure time involved. It should be noted that mishap likelihood and hazard likelihood are really the same entity, just viewed from two different perspectives.

See *Hazard Risk* and *Mishap Risk* for additional related information.

HAZARD LOG

A hazard log is a database of identified hazards. It is a formal record of all data and tasks associated with identifying and resolving hazards.

See *Hazard Tracking System (HTS)* for additional related information.

HAZARD MITIGATION

Hazard mitigation is the process and methodology for reducing the mishap risk of a hazard when the hazard cannot be eliminated. Mitigation reduces the hazard mishap probability and/or the hazard mishap outcome severity. Hazard mitigation involves taking specific design action to reduce the hazard mishap risk to an acceptable level. Design action is achieved through design measures, the use of safety devices, warning devices, training, or procedures. Hazard mitigation is also known as mishap risk mitigation, since the process is reducing the potential mishap risk.

See *Hazard Triangle* and *Risk Mitigation* for additional related information.

HAZARDOUS MATERIAL (HAZMAT)

A HAZMAT is any substance or compound that has the capability of producing adverse effects on the health of humans. As part of a safety program, a Hazardous Materials Management Program (HMMP) should be established for the identification and control of HAZMATs used and generated by the system. The HMMP should evaluate and manage the selection, use, and disposal of HAZMATs consistent with regulatory requirements and program cost, schedule, and performance goals. As alternative technology becomes available, the HMMP should replace HAZMATs in the system through changes in the system design, manufacturing, and maintenance processes, where technically and economically practicable. HAZMATs should be tracked within a HAZMAT database and handled within the framework of the HMMP. The HAZMAT database should list the material, quantities, location, any required special procedures (a.k.a. fire fighting), exposure levels, health hazard data, exposure times, disposal methods, and so on.

See *Material Safety Data Sheet (MSDS)* for additional related information.

HAZARDOUS

Something is hazardous if it involves risk or danger. For example, most energy sources are hazardous items because if the energy involved is uncontrolled, it provides the capability to cause damage and/or injury. For example, hydraulic pressure is a hazardous item because if the high pressure hydraulic hose ruptures, it could result in damage and/or injury.

Hazardous items provide the source component of a hazard. For example, gasoline is hazardous because it involves danger and risk and, it is the source component for many different types of gasoline related hazards, such as fire, explosion, and toxic vapors. Hazardous items are not "hazards" in themselves, but the source for hazards; broken glass is hazardous, but it is not a unique hazard, merely the source for many possible related hazards, such as the hazard of causing a flat tire resulting in a crash, or the hazard of cutting someone's feet that accidentally walks on the exposed glass pieces. The term *hazardous* is a problematic term for use because it involves a colloquial definition and a technical engineering definition for system safety purposes, similar to the definition for *hazard*.

See *Hazard* for additional related information.

HAZARDOUS CONDITION

A hazardous condition is a unique set of parameters, circumstances, and/or design factors that form a hazard. Essentially a hazardous condition is a hazard.
See *Hazard* for additional related information.

HAZARDOUS FUNCTION

A hazardous function is a system function that involves risk or danger should there be failure, error, or malfunction. A hazardous function is a hazard source for many different hazards associated with the function. For example, aircraft braking is a hazardous function for a jet aircraft. If the function fails or functions incorrectly, it opens the door for many different hazards associated with braking, such as failure to brake could result in collision, brakes failed "on" during landing could result in the aircraft skidding off the runway. System safety typically performs an FHA to identify system hazards and SCFs. FHA is particularly important for identifying SCFs in system software.
See *Functional Hazard Analysis (FHA)* and *Hazard* for additional related information.

HAZARDOUS OPERATION

A hazardous operation is a system operation that involves risk or danger. A hazardous operation is a hazard source for many different hazards associated with the operation. For example, a tethered walk in space from a spacecraft is a hazardous operation. If certain failures occur during the operation, the astronaut's life could be lost; each of the potential failure modes would constitute a different hazard. System safety typically performs an O&SHA to identify operational hazards.
See *Hazard* and *Operating and Support Hazard Analysis (O&SHA)* for additional related information.

HAZARD RISK

Risk is the safety measure of a potential future event, stated in terms of event likelihood and event severity. Likelihood can be characterized in terms of probability, frequency, or qualitative criteria, while severity can be characterized in terms of death, injury, dollar loss, and so on. Exposure is part of the likelihood component. *Hazard risk* is a safety metric characterizing the amount of danger presented by a hazard, where the likelihood of a hazard occurring and transforming into a mishap is combined with the expected severity of the mishap generated by the hazard's actuation. *Mishap risk* is the safety metric characterizing the amount of danger presented by a mishap, where the likelihood of the mishap's occurrence is multiplied by the resulting severity of the mishap.

It should be noted that hazard risk and mishap risk are really the same entity, just viewed from two different perspectives. HA looks into the future and predicts a potential mishap from a hazard perspective. Mishap risk can only be reached by first determining the hazard and its risk. Since a hazard merely predefines a potential mishap, the risk has to be the same for both. Hazard risk exposes the potential impact or threat presented by a hazard.

See *Mishap Risk* and *Risk* for additional related information.

HAZARD RISK INDEX (HRI)

The HRI is an index number indicating qualitatively the relative risk of a hazard. It is derived from the HRI matrix by identifying the matrix cell resulting from the intersection of the hazard likelihood and hazard severity values. In a typical HRI matrix, such as the matrix in MIL-STD-882, there are 20 cells created by a 4×5 matrix. The matrix cells are labeled with an index number of 1 through 20, where 1 represents the highest risk and 20 the lowest risk. The smaller the HRI number, the higher the safety risk presented by the hazard. The HRI number establishes the safety significance of a hazard and who can accept the risk for the hazard. It should be noted that the HRI is also often referred to as the mishap risk index (MRI). However, since hazard risk and mishap risk are really the same entity, then the HRI and the MRI are really the same entity, just viewed from two different perspectives.

See *Hazard Risk Index (HRI) Matrix* and *Mishap Risk Index (MRI)* for additional related information.

HAZARD RISK INDEX (HRI) MATRIX

The HRI matrix is a risk matrix that is utilized to establish the relative (vice absolute) risk of a hazard. The matrix maps hazard severity on one axis and hazard likelihood on the other axis. Once a hazard's severity and likelihood are determined, they are mapped to a particular HRI matrix cell (the likelihood–severity intersection), which yields the HRI risk level for that hazard. The likelihood and severity axes are broken into cells defined by qualitative and semiquantitative criteria.

The Hazard Risk Index (HRI) matrix is a risk management tool used by system safety for hazard/mishap risk assessment. The HRI matrix establishes the relative level of potential mishap risk presented by an individual hazard. By comparing the calculated qualitative severity and likelihood values for a hazard against the predefined criteria in the HRI matrix, a level of risk is

HAZARD RISK INDEX (HRI) MATRIX

determined by a derived index number. If the hazard can be controlled through mitigation, then the likelihood and/or severity can be changed, thereby affecting a positive change in the risk level. It should be noted that hazard risk and mishap risk are the same entity, just viewed from two different perspectives.

The HRI matrix was initially developed in MIL-STD-882 as a means for simple, quick, and efficient risk assessment. Variations of the HRI matrix are used in different industries and agencies. The basic HRI matrix in MIL-STD-882 is a 4×5 matrix with 20 index cells. It should be noted that tailoring of this matrix is allowed in order to meet the specific needs of a project.

Figure 2.37 presents the HRI matrix concept for establishing the level of potential mishap risk presented by a hazard. It can be seen from this figure that the HRI matrix concept essentially involves one matrix and three tables. The HRI matrix is the main component, which is based upon the combination of the hazard/mishap likelihood on one axis and hazard/mishap severity on the other axis. The hazard/mishap likelihood category is determined from the criteria stated in the Likelihood Table and the hazard/mishap severity category is determined from the criteria stated in the Severity Table. The Risk Level Table ranks each hazard into one of four risk levels (high, serious, medium, or low) based on the particular HRI matrix indices designated for the particular level.

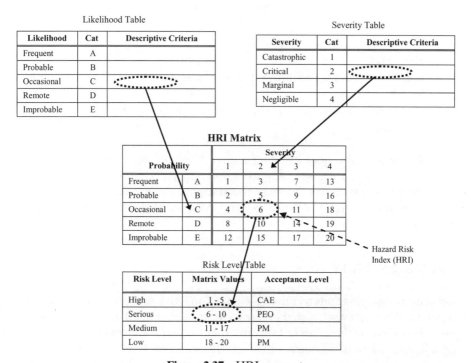

Figure 2.37 HRI concept.

The relative risk index (or HRI) is derived from the matrix cell resulting from the intersection of the likelihood and severity axes that represent a particular hazard. The HRI matrix maps hazard severity on one axis and hazard likelihood on the other axis. Once a hazard's severity and likelihood are determined, they are mapped to a particular HRI matrix cell, which yields the hazard index and the relative risk for that hazard. The hazard risk level establishes who can accept the risk by authority level.

Each matrix cell contains an HRI risk index number, indicating the relative (vice absolute) safety mishap risk presented by a particular hazard. Note that the cells for a 5 × 4 matrix are labeled with a risk index of 1 through 20, where 1 represents the highest risk and 20 the lowest risk. The smaller the HRI number, the higher the safety risk presented by the hazard. The HRI indices are divided into four groups which comprise the high, serious, medium, and low risk levels in the Risk Level Table.

The HRI matrix has several intertwined purposes that are useful to system safety. The HRI matrix allows the system safety analyst to:

1. Perform a risk assessment that is simple, efficient, and cost-effective
2. Determine the potential mishap risk presented by a hazard
3. Communicate hazard risk from a common framework
4. Rank hazards by risk index level
5. Prioritize highest risk hazards requiring immediate mitigation attention
6. Know how much mitigation is necessary to lower the risk to an acceptable level
7. Identify the level of authority that can accept the residual risk presented by a hazard
8. Compare hazards by risk level across a program and across various platforms

The HRI matrix also goes by other names; however, regardless of the name used, the matrices are essentially the same entity because hazard risk is the same entity as mishap risk. Some of the alternate names include: Risk Hazard Index (RHI) matrix, MRI matrix, and Mishap Risk Assessment Matrix (MRAM).

HAZARD SEVERITY

Severity is one parameter in the risk equation. Risk is the safety measure of a potential future event, stated in terms of event likelihood and event severity. Hazard severity is the expected severity that would result from the occurrence of a hazard when it becomes a mishap. Hazard severity is the predicted loss outcome resulting from a hazard given that it occurs and becomes a mishap.

It is the assessed consequences of the worst credible mishap that could be caused by a specific hazard. Severity can be stated in various terms, such as death, injury, system loss, system damage, environmental damage, and dollar loss. It should be noted that mishap severity and hazard severity are really the same entity, just viewed from two different perspectives.

See *Hazard Risk*, *Mishap Risk*, and *Mishap Severity* for additional related information.

HAZARD SEVERITY LEVELS

One method for measuring and ranking the severity of a hazard is to establish different hazard severity levels, with each level having an increasing amount of severity. MIL-STD-882 categorizes hazards into four severity levels, which are defined as follows:

- Category I (Catastrophic)—Could result in death, permanent total disability, loss exceeding $1 M, or irreversible severe environmental damage that violates law or regulation.
- Category II (Critical Catastrophic)—Could result in permanent partial disability, injuries, or occupational illness that may result in hospitalization of at least three personnel, loss exceeding $200 K but less than $1 M, or reversible environmental damage causing a violation of law or regulation.
- Category III (Marginal Catastrophic)—Could result in injury or occupational illness resulting in one or more lost work days(s), loss exceeding $10 K but less than $200 K, or mitigatible environmental damage without violation of law or regulation where restoration activities can be accomplished.
- Category IV (Negligible Catastrophic)—Could result in injury or illness not resulting in a lost work day, loss exceeding $2 K but less than $10 K, or minimal environmental damage not violating law or regulation.

These severity level definitions have been fairly stable in all versions of MIL-STD-882 but could conceivably change in future versions, particularly the dollar amounts.

HAZARDS OF ELECTROMAGNETIC RADIATION TO ORDNANCE (HERO)

HERO refers to the possible ignition of ordnance from EMR energy. HERO is concerned with EMR that can induce or otherwise couple currents and or voltages large enough to initiate electrically initiated devices or other sensitive explosive components. Analysis, test, and design measures are necessary to

ensure that ordnance ignition limits are not reached or exceeded during system operation.
See *Electromagnetic Radiation (EMR)* for additional related information.

HAZARDS OF ELECTROMAGNETIC RADIATION TO PERSONNEL (HERP)

HERP refers to the thermal effects of EMR on humans. Analysis, test, and design measures are necessary to ensure that personnel exposure limits (PEL) are not reached or exceeded during system operation. DoDINST 6055.11 provides the PELs for exposure of personnel to EMR energy, which is frequency dependent for personnel hazards.
See *Electromagnetic Radiation (EMR)* for additional related information.

HAZARD SYMBOL

Hazard symbols are recognizable symbols designed to warn about HAZMATs or locations. The use of hazard symbols is often regulated by law and directed by standards organizations. Hazard symbols may appear with different colors, backgrounds, borders, and supplemental information in order to signify the type of hazard. For example, the skull and crossbones is a common symbol for poisons and other sources of lethal danger.

HAZARD TRACKING

Hazard tracking is the process of systematically recording all identified hazards and the data associated with these hazards as they are mitigated to an acceptable level of risk. It is typically achieved through the use of a formal hazard tracking system (HTS). Hazard tracking is a basic required element of an effective SSP. Hazard tracking encompasses the HA process, the mishap risk management process, and the hazard mitigation process.
See *Hazard Tracking System (HTS)* for additional related information.

HAZARD TRACKING SYSTEM (HTS)

An HTS is a tool for formally tracking all identified hazards within a system. This involves ensuring identified hazards are properly mitigated and closed, and that all related actions are recorded. An effective HTS actually establishes how a process facilitates hazard control through the steps of mitigation design, mitigation verification, risk acceptance, and closure. Closed-loop hazard tracking is a basic required element of an effective SSP. An HTS enforces a formal

HAZARD TRACKING SYSTEM (HTS)   **199**

and systematic process that ensures identified hazards are resolved and which also provides a historical record. It encompasses the HA process, the mishap risk management process, and the hazard mitigation process. The HTS is sometimes referred to as a hazard log, hazard database, or hazard tracking database (HTDB).

An HTS does not imply that a hazard is just passively stored in a database and then forgotten. Hazard tracking is a dynamic process in which the SSP takes positive steps to eliminate or mitigate the hazard and record all actions. Hazards are tracked from inception (identification) to closure, with focus on reporting and acceptance of the final residual hazard–mishap risk. Hazard tracking should be a "closed-loop" process, meaning that the review and mitigation process is repeated iteratively, until final closure of the hazard is achieved.

The primary objectives of an HTS include:

1. Retain a record of all data and tasks associated with identifying and resolving hazards.
2. Provide a mechanism and discipline for tracking hazards from inception through closure.
3. Ensure that all identified hazards are adequately mitigated (i.e., none are lost).
4. Meet the SSP requirement for hazard tracking and closure.

Figure 2.38 shows the overall HTS concept. As hazards are identified, they entered into the HTS. As additional hazard related information is generated in the closed-loop process, it too is entered into the tracking system. Any desired information regarding the hazard can be obtainable from the HTS at any time during the closure process.

The HTS provides a process for risk mitigation, as well as maintaining a complete record and history of every identified hazard. The hazard status is maintained as "open" until it has been verified that the appropriate safety

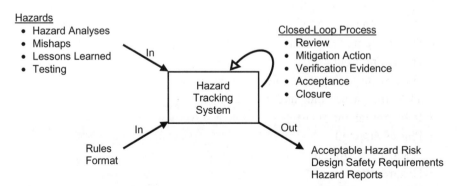

Figure 2.38   The hazard tracking system (HTS).

requirements for eliminating or controlling the hazard have been implemented and proven successful through testing. Following successful verification and validation of the SSRs for a hazard, and acceptance of the hazard risk level, the hazard's status can be changed to "closed."

In an HTS, the database can be a manual or a computerized system; however, it is highly recommended that an automated electronic database be utilized, particularly for medium and large system development programs. In addition, there are commercial electronic database software HTS packages available that are already set up specifically for hazard tracking. An automated electronic database provides many advantages, such as:

- Hazard data entry is easy and efficient.
- Data updates and changes are simple and efficient.
- Capability to search for specific items based on different queries.
- Capability to provide custom reports.
- Capability to place on Network or Internet for access by many users.
- Does not require programming if purchased.
- Can be utilized as a company standard for several different projects.

Some basic considerations to address when designing, procuring, and operating an HTS include:

- Rules for opening, monitoring, and closing a hazard
- A hazard numbering scheme
- A standard format that remains consistent
- Rules for who can enter data into the HTS
- Rules for who can modify or remove data from the HTS
- The capability to generate various reports

Data items typically contained in an HTS include:

- Program name
- System name
- Hazard number
- Hazard status (open, monitor, closed, etc.)
- TLM the hazard falls under
- Risk (initial and final)
- Phases affected
- Hazard description
- Hazard effect
- HCFs

- Recommended action
- Name of individual that identified hazard
- Name of individual assigned for action to resolve hazard
- Hazard mitigation verification requirements
- Hazard mitigation verification evidence
- Dates (open, close, action, meetings, etc.)
- Comments
- Lessons learned

HAZARD TRIANGLE

In order for a hazard to exist, three hazard components must be present to form the hazard: (1) the HS which provides the basic source of danger, (2) the potential IMs that will transition the hazard from an inactive state to a mishap event, and (3) the TTO that will be negatively impacted by the mishap event outcome. The existence of a hazard requires the existence of these three components as a prerequisite. These components are necessary to assess the risk and to know where and how to mitigate the hazard.

These three required components of a hazard form what is known as the *hazard triangle*. The hazard triangle conveys by the idea that a hazard consists of *three necessary* and *coupled* components, each of which forms the side of a triangle. All three sides of the triangle are essential and are required in order for a hazard to exist (i.e., HS, IM, and TTO). Remove any one of the triangle sides and the hazard is eliminated because it is no longer able to produce a mishap (i.e., the triangle is incomplete). For example, remove the human operator from an aircraft, and the hazard "aircraft crashes resulting in pilot death" is eliminated because the TTO is removed. Reduce the probability of the IM side and the mishap probability is reduced. Reduce an element in the HS or the TTO side of the triangle and the mishap severity is reduced. Refer back to Figure 2.35 for an illustration of the hazard triangle.

See *Hazard* for additional related information.

HAZARD TYPECAST

The hazard typecasts provide a taxonomy of different generic hazards that could be expected from a hazard source. For example, when electricity is used in a system, the various hazard typecasts include electrocution, shock, shock surprise, inadvertent power to critical circuits, power not available when required for safety, and ignition source. These are not fully identified and described hazards, only generic categories or typecasts. Each typecast should be analyzed in detail as applicable to the particular system design in order to identify specific hazards unique to the particular system.

HEALTH HAZARD ASSESSMENT (HHA)

The HHA is an analysis technique for evaluating the human health aspects of a system's design. These aspects include considerations for ergonomics, noise, vibration, temperature, chemicals, HAZMATs, and so on. The intent is to identify human health hazards during design and eliminate them through DSFs. If health hazards cannot be eliminated, then protective design measures must be used to reduce the associated risk to an acceptable level. Health hazards must be considered during manufacture, operation, test, maintenance, and disposal phases of the systems life cycle.

On the surface, the HHA appears to be very similar in nature to the O&SHA, and the question often arises as to whether they both accomplish the same objectives. The O&SHA evaluates operator tasks and activities for the identification of hazards, whereas the HHA focuses strictly on human health issues. There may occasionally be some overlap, but they each serve different interests. The objectives of the HHA are to provide a design safety focus from the human health viewpoint and to identify hazards directly affecting the human operator from a health standpoint. The intent of the HHA is to identify human health hazards and propose design changes and/or protective measures to reduce the associated risk to an acceptable level. Human health hazards can be the result of exposure to ergonomic stress, chemicals, physical stress, biological agents, HAZMATs, and the like.

The HHA is applicable to analysis of all types of systems, equipment, and facilities that include human operators. The HHA concentrates on human health hazards during the production, test, and operational phases of the system in order to eliminate or mitigate human health hazards through the system design. The HHA should be completed and system risk known prior to the conduct of any of the production or operational phases. Although some of the hazards identified through the HHA may have already been identified by the PHL, PHA, or SSHA analysis techniques, the HHA should not be omitted since it may catch hazards overlooked by these other analyses.

Typical health hazard checklist categories include, but are not limited to:

1. Ergonomic
2. Noise
3. Vibration
4. Temperature
5. Chemicals
6. Biological
7. HAZMATs
8. Physical stress

When performing the HHA the following factors should be given consideration:

1. Toxicity, quantity, and physical state of materials.
2. Routine or planned uses and releases of HAZMATs or physical agents.
3. Accidental exposure potentials.
4. Hazardous waste generated.
5. HAZMAT handling, transfer, and transportation requirements.
6. Protective clothing/equipment needs.
7. Detection and measurement devices required to quantify exposure levels.
8. Number of personnel potentially at risk.
9. Design controls that could be used, such as isolation, enclosure, ventilation, noise, or radiation barriers.
10. Potential alternative materials to reduce the associated risk to users/operators.
11. The degree of personnel exposure to the health hazard.
12. System, facility, and personnel protective equipment design requirements (e.g., ventilation, noise attenuation, radiation barriers) to allow safe operation and maintenance.
13. HAZMAT and long-term effects (such as potential for personnel and environmental exposure, handling and disposal issues/requirements, protection/ control measures, and life-cycle costs).
14. Means for identifying and tracking information for each HAZMAT.
15. Environmental factors that effect exposure (wind, temperature, humidity, etc.)

The HHA is a detailed HA utilizing structure and rigor when performed using an HHA worksheet. An example HHA worksheet is shown in Figure 2.39.

Although the format of the analysis worksheet is not critical, the following basic information is required from the HHA:

System: Subsystem: Operation: Mode:			Health Hazard Assessment				Analyst: Date:		
Hazard Type	No.	Hazard	Causes	Effects	IMRI	Recommended Action	FMRI	Comments	Status

Figure 2.39 Example HHA worksheet.

- Personnel health hazards
- Hazard effects (mishaps)
- HCFs (materials, processes, excessive exposures, etc.)
- Risk assessment (before and after DSFs are implemented)
- Derived safety requirements for eliminating or mitigating the hazards

HHA is an analysis technique that is important and essential to system safety. For more detailed information on the HHA technique, see Clifton A. Ericson II, *Hazard Analysis Techniques for System Safety* (2005), chapter 9.

HEAT STRESS

Heat stress is any combination of air temperature, thermal radiation, humidity, air flow, and workload which may stress the body as it attempts to regulate body temperature. Heat stress becomes excessive when the body's capability to adjust is exceeded, resulting in an increase of body temperature.

HEAT STROKE

Heat stroke is heat illness where the thermoregulatory system fails to function, so the main avenue of heat loss is blocked resulting in unconsciousness, convulsions, delirium, and possible death.

HERMETIC SEALING

Hermetic sealing is the process by which an item is totally enclosed by a suitable metal structure or case by fusion of metallic or ceramic materials. This includes the fusion of metals by welding, brazing, or soldering; the fusion of ceramic materials under heat or pressure; and the fusion of ceramic materials into a metallic support. Hermetic sealing can be used as a design safety mechanism for preventing explosive vapors from entering electronic equipment. This prevents the electronic equipment from providing an ignition source for the vapors.

HERTZ (HZ)

Hertz is the metric unit which expresses the frequency of a periodic oscillation in cycles per second. For example, 60 Hertz equals 60 cycles per second.

HETERARCHY

A heterarchy is an ordering of things in which there is no single peak or leading element, and which element is dominant at a given time depends on the total situation. This term is often used in contrast to hierarchy, which is also a vertical arrangement of entities (systems and their subsystems), but in an ordered format from the top downward.

HIERARCHY

Hierarchy refers to the organizational ordering of items or elements within a system. Hierarchy is an important aspect of a system that helps break down system complexity into manageable pieces that can be easily understood. System hierarchy is the layered structure defining dominant and subordinate relationships in the system, from the subsystems down to the lowest component or piece parts.

See *System Hierarchy* for additional related information.

HIGH-INTENSITY RADIO FREQUENCY (HIRF)

HIRF is EMR in the RF wavelength with sufficient field strength to interfere with other electronics. An RF EM wave has both an electric and a magnetic component (electric field and magnetic field), and it is often convenient to express the intensity of the RF environment at a given location in terms of units specific to each component. For example, the unit "volts per meter" (V/m) is used to express the strength of the electric field, and the unit "amperes per meter" (A/m) is used to express the strength of the magnetic. Another commonly used unit for characterizing the total EMF is "power density." Power density is most appropriately used when the point of measurement is far enough away from an antenna to be located in the far-field zone of the antenna.

HIRF is of system safety concern for safety-critical electronics because HIRF can cause these electronic systems to fail or operate incorrectly. Commercial aircraft avionics is a prime example of system requiring protection from HIRF interference. HIRF levels of radiation can be encountered near military activity with multiple emitters or if locked onto with a single fire control radar. The FAA cites 200 V/m and above as HIRF.

See *Electromagnetic Radiation (EMR)* for additional related information.

HIGH-LEVEL LANGUAGE

A high-level computer language is a computer language that is more readily readable and understandable to humans than a low-level language. Code written in a high-level language must be converted by an interpreter or compiler into computer-readable machine code for use by a computer. Example high-level computer languages include Basic, Fortran, Ada, C, C++, Modula, Jovial, and Java. A high-level computer language is also often called an HOL.

See *Compiler*, *High-Order Language (HOL)*, and *Interpreter* for additional related information.

HIGH-ORDER LANGUAGE (HOL)

An HOL is a high-level computer language that is more readily readable and understandable to humans than a low-level language. Code written in an HOL language must be converted by an interpreter or compiler into computer-readable machine code for use by a computer. Example HOL computer languages include Basic, Fortran, Ada, C, C++, Jovial, and Java. A high-level computer language is also often called an HOL. An HOL is a step above a low-level language such as assembler. A low-level language consists of somewhat cryptic machine instructions that are closer to the CPU instructions, but much more difficult for human understanding. Low-level languages typically run faster and take less memory space, but programmers tend to make more errors in these languages.

In computer science, a low-level programming language is a programming language that provides little or no abstraction from a computer's instruction set. The word *low* refers to the small or nonexistent level of abstraction between the language and machine language. Because of this, low-level languages are sometimes described as being close to the hardware. A low-level language does not need a compiler or interpreter to run; the processor for which the language was written is able to run the code without using either of these. In contrast, an HOL language isolates the execution semantics of a computer's instruction set from the specification of the program, making the process of developing a program simpler and more understandable.

See *Compiler* and *Interpreter* for additional related information.

HOLISM

Holism is a nonreductionist strategy for generating explanatory principles of whole systems. Attention is focused on the emergent properties of the whole rather than on the reductionist behavior of the isolated parts. The approach typically involves and generates empathetic, experiential, and intuitive understanding, not merely analytic understanding. Holism theory is the foundation for systems theory.

HUMAN ENGINEERING

Human engineering is the application of human factors knowledge to the design and development of systems. Human engineering provides a trade-off between human capabilities, limitations, reliability, skill, and proficiency with system performance, safety, and cost. Human engineering assures that the system or equipment design, required human tasks, and work environment are compatible with the sensory, perceptual, mental, and physical attributes of the personnel who will operate, maintain, control, and support it.

HUMAN ERROR

Human error is the incorrect or wrong execution of a required human action; that is, it is a human failure. Human error and mistakes typically result from the frailties of the human nature. Human errors can easily cause hazards and place people, equipment, and systems at risk. Human error is an act that through ignorance, deficiency, or accident departs from or fails to achieve what should be done. Errors can be predictable and random. Errors can also be categorized as primary or contributory. Primary errors are those committed by personnel immediately and are directly involved with an accident. Contributory errors result from actions on the part of personnel whose duties preceded and affected the situation during which the results of the error became apparent.

There are several ways to categorize human error, such as:

- By fault type
    - Omission
    - Commission
    - Sequence error
    - Timing error
- By situation assessment versus response planning
    - Errors in problem detection (see also signal detection theory)
    - Errors in problem diagnosis (see also problem solving)
    - Errors in action planning and execution (e.g., slips or errors of execution vs. mistakes or errors of intention)
- By level of analysis
    - Perceptual (e.g., optical illusions)
    - Cognitive
    - Communication
    - Organizational
- By exogenous versus endogenous source (i.e., originating outside vs. inside the individual)

Human error can be a significant cause or contributory factor in many hazards. For this reason, it is imperative that system safety evaluate all man–machine interfaces (MMIs) and human tasks for possible hazards, particularly in safety-critical applications. The most common human error causal categories are listed in Table 2.11.

TABLE 2.11 Typical Human Error Causal Categories

Human Error Cause	Description
Complexity	The size of the system, and/or number of items that must be checked, observed, noted, and/or acted upon as designed. Multitasking of tasks or thoughts.
Stress	Pressure on an operator, maintainer, or other personnel interfacing with the system to perform correctly, quickly, and safely.
Fatigue	Physical fatigue caused by lack of sleep, extended work hours, or little time off between shifts.
Psychological factors	Personality (risk taking, risk averse). Domestic situation. Psychological disorders. Tendency to assign blame; defensive reaction to a situation. Fear; lack of confidence that action is the correct one. Depression; state of anxiety and despair. Self-esteem; lack of personal worth.
Environment	Physical workplace conditions such as temperature, humidity, lighting and air quality, and the social/cultural environment (such as personnel morale and cooperation)
Training	The quality and quantity of training in general and on a particular system.
Experience	Total amount of experience that personnel have on a system greatly.
Skill-based lapses	Errors that occur while performing a learned task (one that is done with little or no conscious effort) usually as a result of a momentary memory lapse or distraction to the routine.
Rule-based lapses	Errors that occur when rules and/or procedures are applied or not applied, whether they are appropriate or not.
Knowledge-based mistakes	Errors that occur when performing new tasks or tasks in unfamiliar circumstances.
Unreliable behavior	Errors occur when the person performing the action cannot consistently perform in the manner prescribed/required.
Problem solving	Errors occur when either the problem is formulated (stated) incorrectly or the process to find a solution is in error.
Dysfunctional planning	Errors that occur due to improper and/or untimely planning.
Bias	Errors occur when the perception of a problem is warped or slanted due to personal feelings. Bias can be due to past history (endogenous) or due to external forces, expectations, and influence of others (exogenous).
Misunderstanding	Errors occur when the person performing a task, or people working together on separate but integrated tasks, do not fully comprehend what the total plan is or what each of the others is doing toward the ultimate goal.
Mismanagement/ organization	Errors occur due to inadequate funding of a project (with personnel and/or resources), poor planning, failure to implement plans, or from a poor safety culture. Errors resulting from workplace practices or pressures.

TABLE 2.11 *Continued*

Human Error Cause	Description
Environmental	Errors due to weather, internal heating, ventilation and air conditioning (HVAC) conditions or other uncontrollable location factors.
Reaction time	Response times not compatible with the system action/reaction.
Decisions by committee	Consensus designing.
Cognitive effect	Information overload, boredom, low morale, and information confusion. No conception of what an erroneous action has on the final outcome (big picture).
Maintenance	Incomplete, erroneous, or improper maintenance actions are human errors that can lead to subsequent human errors once malfunction occurs.
Perception of risk	Perceiving the risk of an action as lower or higher than it really is can lead to inaction, no action, or inappropriate action than can cause unsafe conditions and can lead to further human errors.
Situational awareness	Design and human factors leading to a degradation of operator's situational awareness.

In order to quantify human error risk, each task that a human performs in the operation, maintenance, or handling of the system must be reviewed. Each task must be analyzed to determine what effects a human error could have on the system and then quantify the probability that the error will occur. The probabilities are then combined to yield the risk attributable to human error on a system.

When performing a system safety analysis, some basic guidelines regarding human error include the following:

- The potential for human error will always exist, it cannot be avoided.
- The system's UI should always be analyzed for potential hazards that could be caused by human error.
- Human errors are quantifiable, for example Technique for Human Error Rate Prediction (THERP) is a tool commonly used.
- Complicating factors and multidiscipline effects must be included in any human factors/human error analysis. Feedback from the working environment contributes to the likelihood of error.
- Human error causes can be determined and eliminated, or the effects minimized.
- Since some potential human errors can only be identified or evaluated through testing, it is recommended that mockups and simulations be performed before the design phase is complete.
- Human error can be operator induced, system induced, or design induced.

HUMAN FACTORS

The term "human factors" refers to the body of scientific knowledge about human characteristics. The human factors field is a multidisciplinary field encompassing the methods, data, and principles of the behavioral sciences, engineering, physiology, anthropometry, and biomechanics. The field of human factors includes the MMI, as well as the larger socio-technical system, such as management, organization, and regulation.

Human factors covers all biomedical and psychosocial considerations; it includes, but is not limited to, principles and applications in the areas of human engineering, personnel selection, training, life support, job performance aids, and human performance evaluation. It is important to fully understand human factors in order to design systems that operate optimally and safely, considering the innate and error prone nature of the human operator.

HUMAN–MACHINE INTERFACE (HMI)

HMI is the human to machine interface in a system. It is the operator interface or UI, which in modern systems is typically a computer screen with computer input-output controls for the operator. HMI is also the application of human factors and human engineering to system design to ensure the safe and reliable operation of the system throughout its life cycle. Since personnel are a major component of any system, special design consideration must be given to human performance. HMI is also sometimes referred to as MMI. HMI is identical to the more recent term HSI.

See *Human Systems Integration (HSI)* for additional related information.

HUMAN RELIABILITY ANALYSIS (HRA)

HRA is the study of human performance and the reliability of the human in the system. The purpose is to render a complete description of the human contribution to safety and to identify ways to reduce the contribution of human error in system hazards. HRA can provide quantitative results which are useful in system safety studies and FTA. Human errors are quantifiable; for example, THERP is a tool commonly used.

HUMAN SYSTEMS INTEGRATION (HSI)

HSI is the application of human factors and human engineering to system design to ensure the safe and reliable operation of the system throughout its life cycle. Since personnel are a major component of any system, special design consideration must be given to human performance. The HMI, as well as the human influence on the system, must be part of all system design considerations. HSI includes consideration of software applications, graphics, and hardware that allow the operator to effectively give instructions to or receive data from the system.

HSI is a technical process, which ensures that the system is compatible with the operator's physical and cognitive attributes. It integrates the disciplines of Human Factors Engineering (HFE), system safety, manpower, personnel, training, habitability, survivability, and Environmental, Safety, and Occupational Health (ESOH) into the systems engineering of a material system to ensure safe, effective operability and supportability. HSI is a disciplined, unified, and synergistic approach to integrate human considerations into system design. Since personnel are a major component of any system, special design consideration must be given to human performance.

It is important to remember in any system humans can be a catalyst, a detriment, and/or a lifesaver. To implement complete system safety requires more than just evaluating the system hardware and software; the human factor element is also critical. The dependability to correctly perform tasks and make correct decisions at the correct time cannot be overemphasized. Because people are not components, they cannot be treated in terms of complete failure, as is done with hardware components. HSI is concerned with the degradation of human performance in subtle ways. In order for humans to operate or otherwise handle a system, system safety must include HSI in a single integrated program and not two separate programs. Safety must be foremost in the review whenever a human task is analyzed. A system must not only operate within its specified performance requirements, but it must also be safe, user friendly, and user safe. The HMI interface, as well as the human influence on the system must also be part of all analyses to improve human performance and human reliability from a safety standpoint.

An SSP cannot ignore the human element of the system without disastrous results. A well-rounded SSP must include the following HSI elements:

1. Human Factor Hazards—Ensure the SSP mitigates human physical and environmental hazards, such as exposure to electricity, HAZMATs, and air conditioning.
2. Man–Machine Hazards—Ensure the SSP mitigates human hazards caused by automation, such as workload stress, controls design, communications, and emergency egress.
3. Human Reliability Hazards—Ensure the SSP evaluates potential human error and mitigates hazards resulting from human error.
4. Occupational Hazards—Ensure the SSP meets all Safety and Occupational Health (SOH) requirements for humans as specified by local, state, and Federal agencies (i.e., OSHA).

HUMAN–ROBOT INTERACTION (HRI)

HRI is the study of interactions between humans and robots. It is often referred as HRI by researchers. HRI is a multidisciplinary field with contributions from human–computer interaction, AI, robotics, natural language understanding, and social sciences. HRI is the means by which humans interact with robots. It typically refers to interaction with anthropomorphic robots, as opposed to UMSs which are considered by some as robots. HRI has been a topic of both science fiction and academic speculation even before any robots existed. Because HRI depends on knowledge of human communication, many aspects of HRI are continuations of human communications topics that are much older than robotics per se.

The origin of HRI as a discrete safety concern was established by author Isaac Asimov in 1941, in his novel "I, Robot," where he states the Three Laws of Robotics as follows:

1. A robot may not injure a human being or, through inaction, allow a human being to come to harm.
2. A robot must obey any orders given to it by human beings, except where such orders would conflict with the First Law.
3. A robot must protect its own existence as long as such protection does not conflict with the First or Second Law.

The three laws of robotics determine the idea of safe interaction between humans and robots. The closer the human and the robot get, the higher the risk of injury. In industry, this is solved by not allowing the human and robot share the workspace at any time through the extensive use of work zones. The presence of a human is completely forbidden in some zones while the robot is working in that zone, and vice versa.

With the advances of AI, autonomous robots could eventually have more proactive behaviors, planning their motion in complex unknown environments. These new capabilities would have to be developed, keeping safety as a prime concern. The basic goal of HRI is to define a general human model that could lead to principles and algorithms allowing more natural and effective interaction between humans and robots.

HYPERGOLIC

Hypergolic is a property of various chemicals where the chemicals self-ignite upon contact with each other, without a spark or other external initiation source. Hypergolic liquids are sometimes used as rocket fuels.

HYPERTHERMIA

Hyperthermia is abnormally high body temperature. Hyperthermia has significant safety impact if not immediately corrected. It causes an individual to make poor decisions and leads to death.

HYPOTHERMIA

Hypothermia is abnormally low body temperature. Hypothermia has significant safety impact if not immediately corrected. It causes an individual to make poor decisions and leads to death.

ILLNESS

An illness or disease is nontraumatic physiological harm or loss of capacity produced by systemic, continued, or repeated stress or strain; exposure to toxins, poisons, fumes, and so on; or other continued and repeated exposures to conditions of the environment over a long period of time. For practical purposes, an occupational illness or disease is any reported condition that does not meet the definition of injury (from a mishap). Illness includes both acute and chronic illnesses, such as, but nor limited to, a skin disease, respiratory disorder, or poisoning.
See *Disease* for additional related information.

IMPORTANCE MEASURE

Importance measure is a term used in FTA that indicates the relative importance of an event or CS in the overall FT. There are several different types of importance measures, each measuring the event or CS contribution to the total FT probability using a different statistical method.

IMPROVISED EXPLOSIVE DEVICE (IED)

An IED is an explosive device that is improvised. It is typically used by terrorists as it is homemade, as opposed to being built by a manufacturer.

INADVERTENT FUNCTIONING

Inadvertent functioning is the accidental or unintentional initiation and execution of a system function. For example, inadvertent launch (IL) is an inadvertent functioning. An FHA identifies functions that are safety-critical when inadvertent function occurs.

INADVERTENT LAUNCH (IL)

IL is the unintentional launch of a weapon or missile. IL is typically the result of failures, errors, or combinations thereof. IL is a major system safety concern in the design of weapon systems; therefore, all functions associated with launch are typically considered to be safety-critical.

INADVERTENT ARMING

Inadvertent arming is the unintentional arming of a weapon, bomb, or missile. Inadvertent arming is typically the result of failures, errors, or combinations thereof. Inadvertent arming is a major system safety concern in the design of weapon systems; therefore, all functions associated with arming are typically considered to be safety-critical.

INADVERTENT RELEASE

Inadvertent release is the unintentional release of a weapon, bomb, or missile from the aircraft carrying it; it is typically the result of failures and/or errors.

INCIDENT

An incident is the occurrence of an unexpected event, which is generally undesired, where the outcome does not result in serious damage, injury, death, or loss. An incident is considered to be a *near miss* to a serious accident or mishap, which would have more serious consequential outcome. An incident is similar to a mishap in that it is the outcome resulting from an actualized hazard; however, in this case, the hazard is only partially actualized, thus precluding any loss or damage. If complete hazard actualization had taken place, a mishap would have resulted. Documenting and analyzing incident reports and near misses are important in system safety because an assessment may reveal a previously unidentified hazard.

INCREMENTAL DEVELOPMENT MODEL

In the incremental development process, a desired capability is identified, an end-state requirement is known, and that requirement is met over time by developing several increments, each dependent on available mature technology. This method breaks the development process into incremental stages, in order to reduce development risk. Basic designs, technologies, and methods are developed and proven before more detailed designs are developed. It is basically a form of progressive prototyping. Incremental development is a form of evolutionary acquisition.

See *Engineering Development Model* and *System Life-Cycle Model* for additional related information.

INDENTURE LEVEL

Indenture level, or indentured equipment list (IEL), is an engineering tool for identifying system components in a hierarchy that defines dominant and subordinate relationships between subsystems, down to the lowest piece part level. The indenture or hierarchy levels progress downward from the more complex high-level system elements to the simpler part/component elements. This methodology of system description is described in MIL-STD-1629A, Procedures for Performing a Failure Modes, Effects and Criticality Analysis. The purpose of indenture levels is to reduce system complexity into manageable pieces that can be easily understood. The use of indenture levels aids the safety analyst in ensuring that all of the system hardware and functions have been adequately covered by the appropriate HAs.

See *System Hierarchy* for additional related information.

INDENTURED EQUIPMENT LIST (IEL)

IEL is an engineering tool for identifying system components in a hierarchy that defines dominant and subordinate relationships between subsystems, down to the lowest piece part level. The indenture or hierarchy levels progress downward from the more complex high-level system elements to the simpler part/component elements. This methodology of system description is described in MIL-STD-1629A, Procedures for Performing a Failure Modes, Effects and Criticality Analysis. The purpose of indenture levels is to reduce system complexity into manageable pieces that can be easily understood and accounted for. The IEL aids the safety analyst in ensuring that all of the system hardware and functions have been adequately covered by the appropriate HAs.

See *System Hierarchy* for additional related information.

INDEPENDENCE (IN DESIGN)

Independence is a design concept that ensures the failure of one item does not cause the failure of another item (i.e., the items are independent). This concept is very important in system safety and reliability because it enhances system success since it removes many potential CCF sources. Event independence is also significant during quantitative analysis because the modeling and mathematics are less complicated when events are independent. Many models, such as FTA, assume failure event independence.

Independence is a design concept (and implementation) in which system components are independent of one another; that is, they have no effect on each other. Complete design independence is not practical, however, as many system components will naturally interface with one another, thus establishing some form of provisional effect on each other. Design independence is necessary for certain safety applications where it is undesirable to allow the failure of one component to directly cause the failure of another component.

Each design dependency is unique and may not be a safety concern; therefore, special safety analysis and testing may be required to ensure acceptable design independence where needed. Design dependence is where one component failure directly causes another component failure, which typically creates a safety concern. Thus, design independence is a DSF, while design dependence is a potential safety problem. CCFA is a safety tool for identifying design safety dependencies. For example, take the case where a resistor and transistor are linked together in a circuit. If the resistor and transistor fail due to its own inherent wearout failure rate, they are treated as independent failures. However, if the resistor fails shorted resulting in excessive current to the transistor causing the transistor to fail, this situation is treated as a dependent failure.

See *Common Cause Failure Analysis (CCFA)* and *Dependence (In Design)* for additional related information.

INDEPENDENT EVENT

Events are independent when the outcome of one event does not directly affect or influence the outcome of a second event. To say that two events are independent means that the occurrence of one event makes it neither more nor less probable that the other occurs. To find the probability of two dependent events both occurring, multiply the probability of A and the probability of B; $P(A \text{ and } B) = P(A) \cdot P(B)$. For example, find the probability of tossing two number dice and getting a 3 on each one. These events are independent, therefore $P(A \text{ and } B) = P(3) \cdot P(3) = (1/6) \cdot (1/6) = 1/36$.

INDEPENDENT FAILURE

A component failure is an independent failure is when the failure is solely due to an inherent failure mode of the component, and is not caused by the failure of a different component or an external event. In probability theory, events are independent when the outcome of one event does not influence the outcome of a second event. Components are independent when the outcome of one component failure does not influence the outcome of a second component failure. To say that two failures are independent means that the occurrence of one component failure makes it neither more nor less probable that the other component failure occurs. To find the probability of two independent failures both occurring, multiply the probability of the first failure by the probability of the second failure; $P(A \text{ and } B) = P(A) \cdot P(B)$. Most failure modes in an FTA are treated as independent failures.

See *Independent Event* for additional related information.

INDEPENDENT PROTECTION LAYER (IPL)

An IPL is a device, system, or action that is capable of preventing a scenario from proceeding to its undesired consequence independent of the IE or the action of any other layer of protection associated with the scenario. The effectiveness and independence of an IPL must be auditable. IPLs are safeguards (but not all safeguards are IPLs). IPLs are often depicted as an onion skin. Each layer is independent in terms of operation. The failure of one layer does not affect the next. IPLs provide defense in-depth; they can be thought of as layers of an onion skin, as shown in Figure 2.40.

An IPL is a layer of protection that will prevent an unsafe scenario from progressing regardless of the IE or the performance of another layer of protection. IPLs must be sufficiently independent so that the failure of one

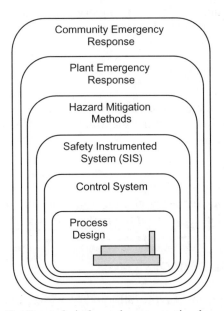

Figure 2.40 Example independent protection layers (IPLs).

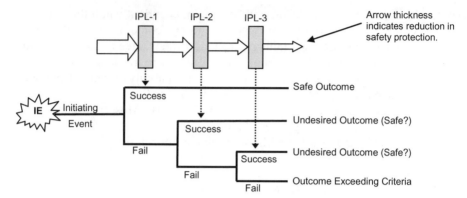

Figure 2.41 IPL evaluation using ETA.

IPL does not adversely affect the probability of failure of another IPL. IPLs are designed to prevent a hazardous event or mitigate the consequences of the event. An IPL is designed to perform its safety function during normal, abnormal, and design basis conditions. Figure 2.41 demonstrates the IPL concept, showing that an ET can be used for quantification.

See *Layers of Protection Analysis (LOPA)* for additional related information.

INDEPENDENT SAFETY FEATURE

A safety feature is independent if its integrity is not affected by the function or malfunction of another safety feature.

INDEPENDENT VARIABLE

An independent variable is the factor or element in an experiment that is manipulated in order to determine the affect dependent variables.

INDIVIDUAL RISK

Individual risk is the frequency at which an individual may be expected to sustain a given level of harm from the realization of a specified hazard.

INDOOR AIR QUALITY (IAQ)

IAQ is a term referring to the air quality within and around buildings and structures, especially as it relates to the health and comfort of building occupants. IAQ can be affected by microbial contaminants (e.g., mold, bacteria), gases (e.g., carbon monoxide, radon, volatile organic compounds [VOCs]), particulates, or any mass or energy stressor that can induce adverse health conditions. Indoor air is becoming an increasingly more concerning health hazard than outdoor air. Using ventilation to dilute contaminants, filtration, and source control are the primary methods for improving IAQ in most buildings. Determination of IAQ involves the collection of air samples, monitoring human exposure to pollutants, collection of samples on building surfaces, and computer modeling of air flow inside buildings. Houseplants, together with the medium in which they are grown, can reduce components of indoor air pollution, particularly VOCs such as benzene, toluene, and xylene. The compounds are removed primarily by soil microorganisms. Plants can also remove CO_2, which is correlated with lower work performance, from indoor areas. Plants also appear to reduce airborne microbes and increase humidity.

INDUCTIVE REASONING

Inductive reasoning is a logical process in which a conclusion is proposed that contains more information than the observation, data, or experience on which it is based. For example, every crow ever seen was black; therefore, all crows are black. The truth of the conclusion is verifiable only in terms of future experience, and certainty is attainable only if all possible instances have been examined. In the example, there is no certainty that a white crow will not be found tomorrow, although past experience would make such an occurrence seem unlikely. In inductive reasoning, the conclusion is broader and may imply more than the known premises can guarantee with the data available.

See *Deductive Reasoning* and *Inductive Safety Analysis* for additional information.

INDUCTIVE SAFETY ANALYSIS

Inductive safety analysis is an analysis that reasons from the specific to the general to determine what overall effect could result. For example, an FMEA inductively analyzes from a component failure to a system-level effect. An inductive HA might conclude more than the given data intend to yield. This is useful for general hazard identification. It means a safety analyst might try to identify (or conclude) a hazard from limited design knowledge or information. Inductive analysis tends to be a bottom-up approach (going from the specific to the general), such as is done in the FMEA.

As an example, when analyzing the preliminary design of a high-speed subway system, a safety analyst might conclude that the structural failure of a train car axle is a potential hazard that could result in car derailment and passenger injury. The analyst does not know this for sure, there is no conclusive evidence available, but the conclusion appears reasonable from past knowledge and experience. In this case, the conclusion seems realistic but is beyond any factual knowledge or proof available at the time of the analysis; however, a hazard has been identified that now needs to be proven credible.

In system safety, inductive analysis tends to be for hazard identification (when the specific root causes are not known or proven), and deductive analysis for root cause identification (when the hazard is known). Obviously, there is a fine line between these definitions because sometimes the root causes are known from the start of an inductive HA. This is why some analysis techniques can actually move in both directions. The PHA is a good example of this. Using the standard PHA worksheet, hazards are identified inductively by asking what if this component fails, and hazards are also identified by deductively asking how can this UE happen.

See *Deductive Reasoning*, *Deductive Safety Analysis*, and *Inductive Reasoning* for additional information.

INFORMAL SPECIFICATION

An informal specification is a specification where a natural language is used to express the requirements of a system or components. This is opposed to a formal specification that utilizes mathematical formalism and notation.

INFRARED

Infrared is EMR with wavelengths which lie within the range 0.7 mm–1 mm.

INHERENT HAZARD

An inherent hazard is a hazard that is naturally inherent within a system due to the unique system design. For example, electrocution by contact with high voltage is an inherent hazard for any system using high voltage. This is not necessarily a recommended term as it often causes some confusion.

INITIAL RISK

Initial risk is the level of risk presented by a hazard when the hazard has been first identified and assessed, and prior to any risk mitigation effort. Note that if the decision is made to not perform mitigation, then the residual risk is the same as the initial risk.

INITIATING EVENT (IE)

An IE is an event that commences a sequence of events that ultimately lead to the occurrence of a mishap or UE. Initiating event analysis (IEA) utilizes the concept of scenarios for identifying hazards, risk, and risk mitigations. In the scenario approach to mishap analysis, the scenario consists of a series of events that begins with the IE, which is then followed by one or more critical PEs until a final consequence has resulted. Typically, barriers are designed into the system to protect against IEs and the PEs are failure or success of these barriers.

IEs are one of the fundamental building blocks in the scenario approach to HA and PRA; they are the "mishap starters" that typically initiate a mishap scenario. An IE is a postulated event that could occur in a system, thereby putting the system into a hazardous state. It is an event occurrence that creates a disturbance in the system and has the potential to lead to a mishap.

IEs are usually categorized as internal or external initiators, reflecting the origin of the events. Internal IEs are the hardware failures, software failures, or operator errors that arise during system operation. External IEs are those events encountered outside the domain of the normal system boundary; events associated with the occurrence of natural phenomena (e.g., earthquakes, lightning, tornadoes, fires and floods); or external man-made physical conditions (e.g., RF radiation, water leak).

In hazard theory, an IE would be the event that begins the IM aspect of the hazard (recall that a hazard has three components: an HS, an IM, and a TTO). In IE analysis, a hazard or undesired state would occur when an IE occurs, followed by one or more PEs.

See *Event Tree Analysis (ETA)* and *Initiating Event Analysis (IEA)* for additional related information.

INITIATING EVENT ANALYSIS (IEA)

IEA is a methodology for identifying hazards, risk, and risk mitigations that utilizes the concept of mishap scenarios. The mishap scenario concept is shown in Figure 2.42.

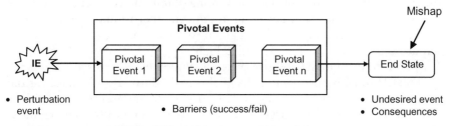

Figure 2.42 Mishap scenario breakdown.

In the scenario approach to mishap analysis, the scenario begins with the IE, and it ends after the PEs have occurred along with the resulting consequences. The key to this methodology is identifying the correct IEs and all of the relevant PEs. Typically, barriers are designed into the system to protect against IEs resulting in end states with undesirable consequences. As each barrier fails, the outcome consequence worsens. Refer back to Figure 2.41 for a diagram of the IPLs concept using barriers as PEs.

See *Event Tree Analysis (ETA)* and *Initiating Event (IE)* for additional related information.

INJURY

An injury is a traumatic wound or other condition of the body caused by external force including stress or strain. The injury is identifiable as to time and place of occurrence and the part or function of the body affected, and is caused by a specific event or series of events within a single day or work shift. Injuries include cases such as, but not limited to, a cut, fracture, sprain, or amputation.

INSENSITIVE MUNITIONS (IMS)

Insensitive munitions are munitions which are stable enough to withstand mechanical shocks, fire, and impact by shrapnel, but that are still able to explode as intended in order to destroy their targets. Insensitive munitions will only burn (rather than explode) when subjected to fast or slow heating, bullets, shrapnel, shaped charges, or the detonation of another nearby munition. The term typically refers to warheads, bombs, and rocket motors. Insensitive munitions are munitions that are less sensitive to unplanned stimuli that might cause unplanned ignition of the munition.

The insensitive munitions program considers the explosive component as part of a total system, which includes the munitions casing, liner material, the insensitive explosive composition, the booster formulation, the detonator mechanism, the energy source which is matched to the detonator, and any other components. The munitions, as a total system, is to be made insensitive. This may include features such as the venting of munitions to relieve pressure during thermal cook-off or shielding in the container to mitigate the effects of bullet or fragment impact or sympathetic detonation. At the same time, the output performance of the insensitive high explosive must equal or exceed that of the conventional explosive which it is intended to replace.

Threat Hazard Assessment (THA) helps in establishing guidelines for the sensitivity assessment of munitions. The purpose of an insensitive munitions program is to increase the survivability of ships and aircraft by making munitions less sensitive to unplanned stimuli. MIL-STD-2105 (Series) provides a series of tests to assess the reaction of energetic materials to external stimuli representative of credible exposures in the life cycle of a weapon and requires the use of a THA in developing test plans. Although the THA may identify additional required tests, the minimum set of insensitive munitions tests required by MIL-STD-2105 includes the following:

1. Fast cook-off tests
2. Slow cook-off tests
3. Bullet impact tests
4. Fragment impact tests
5. Sympathetic detonation tests
6. Shaped charge jet impact test
7. Spall impact tests
8. 40-foot drop test
9. 28-day temperature and humidity test
10. Vibration tests
11. 4-day temperature and humidity test

INSPECTION

Inspection is the process of measuring, examining, gauging, or otherwise comparing an article or service with specified requirements. Inspection is one method for determining the quality of a product. CSIs may require a more rigorous inspection process during manufacture in order to ensure the part performs as required for SR applications.
See *Critical Safety Item (CSI)* for additional information.

ISSUE

An "issue" is a product of an audit; it is a concern. It may not be specific to compliance or process improvement but may be a safety, system, program management, organizational, or other concern that is detected during the audit review. Typical categories of items resulting from an audit include compliance, finding, observation, issue, and action.
See *Audit* for additional related information.

INSTRUMENT

An instrument is a piece of equipment consisting of sensors, hardware, and/or software for making measurements or observations. For most purposes, an instrument is considered to be a subsystem. In many situations, system safety must evaluate the safety of instruments, such as when being used aboard an aircraft during flight testing or to ensure it is properly calibrated for taking safety-critical measurements.

INTERFACE

An interface is a point of contact (POC) and interaction between two interfacing system elements, such as systems, subsystems, and software modules. The interface between a human and a computer is called a UI. Interfaces between hardware components are physical interfaces. A software interfaces between separate software components provides a mechanism by which the components can communicate.

Functional and physical interfaces would include mechanical, electrical, thermal, data, control, procedural, and other interactions. Interfaces may also be considered from an internal/external perspective. Internal interfaces are those that address elements inside the boundaries established for the system addressed. External interfaces, on the other hand, are those which involve entity relationships outside the established boundaries. In system development, interfaces are generally identified and controlled through the use of interface requirements.

See *Interface Requirement* and *System* for additional related information.

INTERFACE REQUIREMENT

In engineering, a requirement is a singular documented need of what a particular product or service should be or perform. It is most commonly used in a formal sense in systems engineering or software engineering. It is a statement that identifies a necessary attribute, capability, characteristic, or quality of a system in order for it to have value and utility to a user. An interface requirement is a requirement that specifies the necessary attributes between two interfacing system elements, such as systems, subsystems, and software modules.

See *Requirement* for additional related information.

INTERLOCK

An interlock is a design safety arrangement whereby the operation of one control or mechanism allows, or prevents, the operation of another function. The safety interlock is a special safety device used in a system design to increase the level of safety of a specific function; it is a DSF. The primary purpose of an interlock is to provide a mechanism to make or break an SR function, based upon a set of predetermined safety criteria. It should be noted that interlocks are absolutely not necessary for the operational functionality of a system. An interlock is a device added to the design in order to achieve the needed safety required of the system, not for the operational effectiveness of the system. Interlocks control state transitions in an attempt to prevent the system from entering an unsafe state or to assist in exiting from an unsafe state.

An interlock can be used in either of two ways: (a) to break a function when necessary, and (b) to make a function complete when necessary. The break-function interlock interrupts a critical system function when a known hazardous state is about to be entered, thereby preventing a mishap. For example, if a hazardous laser is being operated in a locked room with no personnel in the room, a sensor-switch interlock on the door automatically removes, or breaks, power from the laser system when the door is opened by someone inadvertently entering the room. This interlock breaks the laser operation function and prevents a potential eye injury mishap. The system cannot be initiated and operated unless the "door is closed" safety criteria is satisfied. In this case, an unsafe system state is prevented by forcing the system into a known safe state (i.e., laser off state). Once the door is opened, the system must be re-initialized, with assurance that no one is in the room. In another example, an interlock could be used to automatically remove electrical power from contacts inside an electrical panel when the panel door is opened, thereby eliminating the possibility of accidental personnel electrocution. The make-function interlock prevents an SR function from being executed until the interlock safety criteria become valid. For example, in a certain missile system design, the power to launch a missile would not reach the missile until three separate and independent switches are intentionally closed. The switches are interlocks and switch closures are based on timing and passing the safety criteria at each point in time. Each switch requires certain safety criteria to be true before the switch is closed. This type of safety interlock protects against inadvertent or premature function that could result from possible system failure modes.

Figure 2.43 provides a pictorial representation of a missile launch interlock example using three switches in the functional launch path. These switches are considered as three independent interlocks that significantly reduce the probability of missile launch power inadvertently reaching the missile due to random failures in the system. The hazardous launch state cannot be entered until all three interlocks have passed their safety criteria and closed. The decision logic box represents the methodology selected for monitoring and judging the safety criteria. It is important to ensure that there is valid justification for each interlock, and that there is a suitable safety case showing that the interlocks are completely independent and not subject to common mode or CCFs.

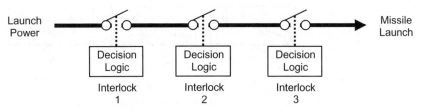

Figure 2.43 Interlock example.

226 SYSTEM SAFETY TERMS AND CONCEPTS

It is also important to ensure that the combined probability of failure for the three interlocks is adequately small; if parts with high failure rates are used, the safety advantage of the design may be negated by an overall high probability of activation.

Figure 2.44 contains an FT for the interlock design shown in Figure 2.43. This FT illustrates how the interlock design protects against SPFs. It also shows how the design introduces AND gates into the FT logic, which help to reduce the overall top probability of the UE.

For normal system operation missile launch would be expected when all of the following conditions are successful: (a) launch power is made available, (b) interlock 1 closure occurs, (c) interlock 2 closure occurs, *and* (d) interlock 3 closure occurs. The FT models these normal intended events occurring prematurely or inadvertently due to system failures.

In Figure 2.44, the input events are shown as diamonds to indicate that they can be further expanded via FT logic when the design is analyzed in more detail. It should be noted from this FT that if the interlocks were not in the system design, IL would occur as soon as event A occurred, and the resulting top probability would be about 1.0E-3. It is clear from the FT that the top probability is directly affected by the numbers of interlocks utilized, as shown in Table 2.12. Changes in the individual interlock probabilities will also influence the top probability.

A safety interlock is a single device that is part of a larger system function. Its purpose is to prevent the overall system function from being performed until a specified set of safety parameters are satisfied. An interlock can be implemented in either hardware or software. If implemented in software, it is done as a function in the software code. It should be noted that software

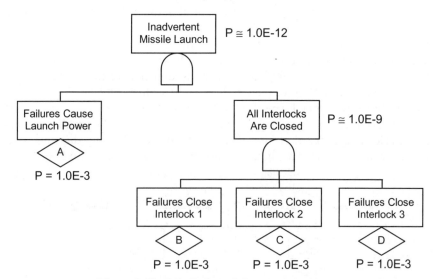

Figure 2.44 Missile launch interlock fault tree.

TABLE 2.12 Interlock Influence on Fault Tree Top Probability

Number of Interlocks	Top Probability
0	$P \cong 1E\text{-}3$
1	$P \cong 1E\text{-}6$
2	$P \cong 1E\text{-}9$
3	$P \cong 1E\text{-}12$

interlocks are more vulnerable to common mode and CCF. Interlocks should not be confused with enables. An enable is primarily required for system functionality. An example of an enable is an ignition switch in a car; the ignition switch is used to start and stop the car engine. It is not for added safety of the system, it is needed as a mechanism to turn the system on and off. However, the enable appears similar to the make-function interlock in a system design where many interlocks are utilized. For example, a light switch is a system-enabling device that prevents the light function until the light-on is desired. In this case, the enable does perform a make-function interlock like operation, but it is somewhat necessary for system operation.

INTEROPERABILITY

Interoperability is the ability of systems to provide data, information, commands, materiel, and services to other systems and to accept the same from other systems. The purpose is to allow diverse systems to effectively operate together for a common goal. Interoperability allows U.S. services to operate together and also with coalition forces, when they work together with their own individually provided equipment. Interoperability embraces the system-of-systems concept. Interoperability increases the complexity, difficulty and cost for system safety. In addition to ensuring that each individual system is safe, safety must ensure that a combined and much larger system-of-systems is safe. The safety complexity is increased due to factors such as multiple networks, hardware interfaces, different languages, different protocols, and timing constraints.

INTERPRETER

An interpreter reads the program instructions written in a computer language and executes the computer language statements directly. This is as opposed to a compiler, which translates the statements into a binary format that is then independently available for execution. Interpreted code tends to run slower than compiled code. The interpreter must always be present in order to run the code, whereas the compiler does not have to be present after the code is compiled. The Basic language is an example of an interpreter and interpreted language.

See *Compiler* for additional related information.

INTRINSIC SAFETY (IS)

Intrinsic safety (IS) is a protection technique for safe operation of electronic equipment in explosive atmospheres. IS involves ensuring that the available electrical and thermal energy in the system is always low enough that ignition of the hazardous atmosphere cannot occur. This is achieved by ensuring that only low voltages and currents enter the hazardous area. For example, intrinsically safe electrical lanterns are used in coal mines to prevent ignition of explosive vapors that may emanate from the mine. In intrinsic designs the maximum energy level is below hazardous levels. No single field device or wiring is intrinsically safe by itself (except for battery-operated, self-contained devices; see http://en.wikipedia.org/wiki/Battery_%28electricity%29), but is intrinsically safe only when employed in a properly designed IS system.

In normal uses, electrical equipment often creates internal tiny sparks in switches, motor brushes, connectors, and in other places. Such sparks can ignite flammable substances present in air. A device termed "intrinsically safe" is designed to not contain any components that produce sparks or that can hold enough energy to produce a spark of sufficient energy to cause an ignition. For example, during marine transfer operations when flammable products are transferred between the marine terminal and tanker ships or barges, two-way radio communication needs to be constantly maintained in case the transfer needs to stop for unforeseen reasons such as a spill. The U.S. Coast Guard requires that the two-way radio must be certified as intrinsically safe.

IONIZING RADIATION

Ionizing radiation is EMR in the upper frequency end of the EM spectrum. EM waves carry energy, and the higher the frequency, the greater the energy. EM waves from the upper end of the spectrum are referred to as ionizing because they are powerful enough to break the internal bonds of atoms and molecules, creating charged particles called ions. Ionizing radiation is EMR having sufficiently large photon energy to directly ionize atomic or molecular systems with a single quantum event.

Ionizing radiation comes from radioactive materials, X-ray tubes, particle accelerators, and is present in the environment. It is invisible and is not directly detectable by human senses, so instruments such as Geiger counters are usually required to detect its presence. It has many practical uses in medicine, research, construction, and other areas, but presents a health hazard if used improperly. Exposure to radiation causes damage to living tissue, resulting in skin burns, radiation sickness, and death at high doses, and cancer, tumors, and genetic damage at low doses.

See *Electromagnetic Radiation (EMR)* for additional related information.

ISHIKAWA DIAGRAM

Ishikawa diagrams are diagrams that show the causes of a certain event. Common uses of the Ishikawa diagram are product design and quality defect prevention, to identify potential factors causing an overall effect. Each cause or reason for imperfection is a source of variation. Ishikawa diagrams are also called fishbone diagrams or cause-and-effect diagrams.

Causes are typically grouped into major categories to identify these sources of variation. The categories typically include:

- People—Anyone involved with the process
- Methods—How the process is performed and the specific requirements for doing it, such as policies, procedures, rules, regulations, and laws
- Machines—Any equipment, computers, tools, and so on, required to accomplish the job
- Materials—Raw materials, parts, pens, paper, and so on, used to produce the final product
- Measurements—Data generated from the process that are used to evaluate its quality
- Environment—The conditions, such as location, time, temperature, and culture in which the process operates

Figure 2.45 shows the Ishikawa diagram concept. Notice that it tends to look like a fish bone, explaining the derivation of its alternate name. The primary and secondary causes are listed in the fishbone lines.

Ishikawa diagrams are used to help analyze problems (or mishap) and determine the causes for the problem. It provides a methodology for including

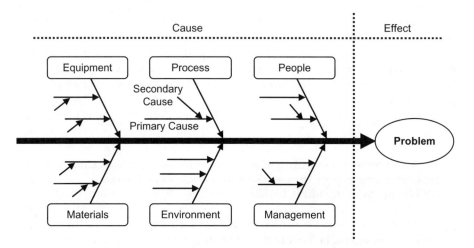

Figure 2.45 Ishikawa diagram (or fish bone diagram).

all possible considerations. Although it looks slightly different, it is very similar to mind mapping where brain-storming ideas are grouped together.

See *Fishbone Diagram* for additional related information.

JAVA

Java is a computer programming language. It is known as a high-level-type language or HOL. Java has some unique features which are not desirable in safety-critical applications.

JEOPARDY

Jeopardy represents danger and the possibility of incurring loss, harm, or misfortune. Someone or something is in jeopardy when there is a risk of it being involved in a mishap resulting in death, injury, or loss. Jeopardy is the condition of being susceptible to harm or injury, that is, susceptible to a mishap. Since a hazard presents danger and is a precursor to a mishap, jeopardy must be exposure to a hazard and the risk of a potential mishap. Jeopardy is not a recommended term for use in system safety because it lacks firm definition for precise technical decision-making usage; it tends to be an emotional term that evokes misunderstanding and can generate disagreements.

See *Hazard* and *Risk* for additional related information.

JOB HAZARD ANALYSIS (JHA)

JHA is an HA technique for identifying and mitigating job related hazards. JHA analyzes the tasks associated with performing a certain job in order to identify hazards associated with the job tasks and to establish corrective action (mitigation) methods for reducing the risk presented by the hazards. The intent of JHA is to ensure a safe workplace for employees by eliminating or mitigating hazards in any of the job tasks the employees will perform.

The Federal Occupational Safety and Health Administration (OSHA) Act states:

> Employers must furnish a place of employment free of recognized hazards that are causing or are likely to cause death or serious physical harm to employees. (OSHA Act of 1970, 29 CFR 1903.2)

This OSHA requirement provides motivation for employers to perform JHA. All employers should be conducting JHA and JHA training in order to keep the workplace and employees safe. JHA covers all industries and all type of jobs and tasks. It should go without saying that JHA be extensively conducted on hazardous jobs and job sites, such as welding, machining, and construction. It should be noted that a JHA is not the same as an O&SHA. Although they are similar in nature, they each have slightly different purposes and focuses. The JHA applies to OSHA-related activities, whereas the O&SHA evaluates tasks for system in development in order to affect the system design. The primary purpose of a JHA is to keep employees safe and free from the possibility of injuries on the job. Job safety also improves employee morale, absenteeism, and medical days lost. In general, a JHA provides the following major benefits:

- Protects employees from accidents/mishaps
- Reduces injuries
- Increases morale
- Reduces absenteeism
- Increases productivity
- Assists in OSHA safety compliance

Hazards, whether in a system design or job process, are always both existing and potential at the same time. If a hazard is there, then it exists. And, by definition, a hazard is a potential condition that can result in a mishap; therefore, it is always a potential. The objectives of JHA are to identify hazards in the job process, assess the hazard–mishap risk, prioritize corrective actions, and reduce and/or eliminate the hazards.

JHA involves a multistep technique that includes the following primary steps:

- Identify the job, tasks, and steps to be performed
- Identify tools and equipment to be used
- Identify proper procedures and permits
- Identify how injury or illness can occur—what can go wrong
- Determine the consequences
- Determine how it could happen
- Identify all contributing factors
- Establish the risk—likelihood the hazard and injury will occur
- Implement proper controls—determine a safer way to do the job
- Have the JHA reviewed by employees and management

The JHA analysis technique is a detailed HA utilizing structure and rigor. It is typically performed using a specialized worksheet which helps to add focus and thoroughness to the analysis, record the process and data, and help support justification for the identified hazards and controls. Although the format of the analysis worksheet is not critical, generally, matrix or columnar-type worksheets are used for clarity in recording the analysis information.

As a minimum, the following basic information should be obtained from the JHA analysis worksheet:

- Job/tasks/steps under analysis
- Hazards
- Hazard effects (mishaps)
- HCFs (to subsystem identification)
- Safety-critical factors or parameters
- Risk assessment (before and after DSFs are implemented)
- Derived safety requirements for eliminating or controlling the hazards

An example JHA worksheet is shown in Figure 2.46. When performing a JHA some example hazard sources to consider include:

- Heat
- Impact
- Sharp edges
- Penetration
- Compression
- Chemical exposures
- Repetitive motions
- Optical radiation
- Hazardous movements
- "Struck by" hazards
- Worker posture/balance
- Physical pinch points
- Harmful airborne contaminants
- Employee jewelry
- Suspended loads
- Environmental hazards

Job Steps	Hazard	Effect	Causal Factors	Risk	Recommended Controls

Figure 2.46 Example JHA worksheet.

Parameters to consider when performing a JHA include:

- Is personal protective equipment available?
- Has worker been properly trained?
- Is the worker position/posture proper?
- Is lockout/tagout used?
- What is the flow of work?
- What are the sources and levels of chemicals, noise, vibration, and so on?
- Are slips, trips, and falls a possibility?

When controlling or mitigating JHA hazards, the preferred control hierarchy is as follows: (1) engineering controls, (2) administrative controls, (3) procedures, and (4) PPE. It is important to remember that it is not the JHA form that will keep workers safe on the job, but rather the process it represents. It is of little value to identify hazards and devise controls if the controls are not put in place. Everyone in the workforce should be involved in creating the JHA. The more minds and the more years of experience applied to analyzing job hazards, the more successful the work group will be in controlling them. Remember that a JHA can become a quasi-legal document and can be used in incident investigations, contractual disputes, and court cases.

See *Hazard* for additional related information.

JOB SAFETY ANALYSIS (JSA)

JSA is a safety HA technique for identifying job-related hazards. JSA is synonymous with JHA; they are the same analysis with different names.

See *Job Hazard Analysis (JHA)* for additional related information.

JOULE

A Joule is a unit of energy; 1 joule = 1 watt × 1 second. One joule is defined as the amount of work done by a force of 1 Newton moving an object through a distance of 1 m. Other relationships are:

- The work required to move an electric charge of one coulomb through an electrical potential difference of one volt; or one coulomb volt (C·V). This relationship can be used to define the volt;
- The work required to continuously produce one watt of power for one second; or one watt second (W·s) (compare kilowatt hour). This relationship can be used to define the watt.

LABEL

A label is an identifying or descriptive marker that is attached to an object. Labels are used to identify equipment with information, such as manufacturer name, model number, and serial number. For transportation purposes, the U.S. Dept. of Transportation requires labels of certain sizes, colors, and shapes, particularly for shipping HAZMATs. See 49 CFR 172 for label information.

LASER

Laser is an acronym for light amplification by stimulated emission of radiation. A laser is a device that produces radiant energy predominantly by stimulated emission. It is any device that can be made to produce or amplify EMR in the X-ray, UV, visible, and infrared, or other portions of the spectrum by the process of controlled stimulated emission of photons. Laser radiation may be highly coherent temporally, or spatially, or both.

A laser typically operates in one of three output modes: single pulse, repetitive pulse, or CW. In the single pulse mode, the laser outputs a single short pulse, where a pulse is defined as <0.25 s. A repetitively pulsed laser outputs a train of multiple pulses occurring in a sequence. In the CW mode, the laser operates with a continuous output for a period >0.25 s.

The radiation produced by a laser has three unique properties that ordinary light sources do not have. These unique properties are as follows:

1. The light emitted from a laser is *monochromatic*; that is, it is of one color or wavelength. The laser radiation actually consists of a very narrow band of wavelengths or possibly several narrow bands based on the atomic structure of the lasing medium. The optical cavity is usually tuned to output a specific wavelength. In contrast, ordinary white light is a combination of many wavelengths of light.
2. Lasers emit radiation that is *highly directional*; that is, the laser radiation is emitted as a relatively narrow beam in a specific direction. Ordinary light, such as from a light bulb, is emitted in many directions away from the light source.
3. The radiation from a laser is *coherent*, meaning that the waves of the laser radiation are in phase, thus producing a resultant wave much stronger than any single wave. The light rays from an ordinary light source have a random phase.

See *Laser Safety* for additional related information.

LASER DIODE

A laser diode is a laser employing a forward-biased semiconductor junction as the active medium; sometimes referred to as a semiconductor laser.

LASER FOOTPRINT

A laser footprint is the projection of the laser beam, and buffer zone, onto the ground or target area. The laser footprint may be part of the laser surface danger zone if the laser footprint lies within the nominal ocular hazard distance (NOHD) of the laser.

LASER SAFETY

In the United States, laser safety is a matter of U.S. law; the Food and Drug Administration (FDA) is the regulatory authority for laser products. Within the FDA, the Center for Devices and Radiological Health (CDRH) has the charter to protect the public health by eliminating unnecessary human exposure to laser radiation emitted from electronic products. The FDA enforces 21 Code of Federal Regulations (CFR) Parts 1040.10 and 1040.11, Federal Performance Standard for Light-Emitting Products. This federal standard became effective in August 1976 and has been amended several times. All laser products distributed in the United States after August 1, 1976 must comply with the requirements of this federal standard. Additionally, this federal standard describes the methods by which lasers must be classified and defines the safety features required for each laser classification.

In 1976, the FDA issued Exemption Number 76EL-OIDOD to the DoD, which permits the DoD to grant manufacturers of military laser products an exemption from the FDA laser safety standards. The exemption encompasses laser products that are intended for use by the U.S. government or whose function or design cannot be divulged by the manufacturer for reasons of national security, as evidenced by government security classification. This includes lasers that are either designed expressly for combat and combat training, or lasers that are classified in the best interest of national security. However, the FDA stipulated that in lieu of full compliance to the federal standards, DoD must establish alternative control measures that would ensure the safety of public health.

Laser safety is typically part of the SSP. Laser safety involves ensuring the safe design, implementation, and operation of lasers in a system. In addition, laser safety involves ensuring that all regulations are followed. For Navy programs, if a system utilizes a laser, the system must have a laser safety program and the system design must be reviewed and approved by the Laser Safety Review Board (LSRB).

There are two laser classification systems, the "old system" used before 2002, and the "revised system" being phased in since 2002. The latter reflects the greater knowledge of lasers that has been accumulated since the original classification system was devised, and permits certain types of lasers to be recognized as having a lower hazard than was implied by their placement in the original classification system. The revised system is part of the revised IEC

60825 standard. From 2007, the revised system is also incorporated into the US-oriented American National Standards Institute (ANSI) Laser Safety Standard (ANSI Z136.1).

The maximum permissible exposure (MPE) for lasers is the highest power or energy density (in W/cm^2 or J/cm^2) of a light source that is considered safe, that is, that has a negligible probability for creating damage. It is usually about 10% of the dose that has a 50% chance of creating damage under worst-case conditions. The MPE is measured at the cornea of the human eye or at the skin, for a given wavelength and exposure time.

A calculation of the MPE for ocular exposure takes into account the various ways light can act upon the eye. For example, deep-UV light causes accumulating damage, even at very low powers. Infrared light with a wavelength longer than about 1400 nm is absorbed by the transparent parts of the eye before it reaches the retina, which means that the MPE for these wavelengths is higher than for visible light. In addition to the wavelength and exposure time, the MPE takes into account the spatial distribution of the light (from a laser or otherwise). Collimated laser beams of visible and near-infrared light are especially dangerous at relatively low powers because the lens focuses the light onto a tiny spot on the retina. Light sources with a smaller degree of spatial coherence than a well-collimated laser beam lead to a distribution of the light over a larger area on the retina. For such sources, the MPE is higher than for collimated laser beams. In the MPE calculation, the worst-case scenario is assumed, in which the eye lens focuses the light into the smallest possible spot size on the retina for the particular wavelength and the pupil is fully open. Although the MPE is specified as power or energy per unit surface, it is based on the power or energy that can pass through a fully open pupil ($0.39\,cm^2$) for visible and near-infrared wavelengths. This is relevant for laser beams that have a cross-section smaller than $0.39\,cm^2$. IEC-60825-1 and ANSI Z136.1 standards include methods of calculating MPEs.

Laser Classification (old system). There are four classes of lasers for safety purposes, with Class 1 lasers being the least hazardous and Class 4 lasers being the most hazardous. Proper laser classification is very important because it provides the basis for laser safety design. The required safety features for a given laser are based on its classification. The laser safety community has developed this laser classification system based on injury potential. The classification levels are dependent upon the lasers wavelength, power or energy, exposure duration, pulse repetition frequency (PRF), and the size of the laser beam.

The four laser hazard classifications determine the required extent of radiation safety controls. These range from Class 1 lasers that are safe for direct beam viewing under most conditions to Class 4 lasers that require the strictest of controls. Laser product classification pertains to intended use only. When a laser product is disassembled for maintenance, and pro-

TABLE 2.13 Laser Classifications (Old System)

Class	Safety Description
1	Lasers, which by inherent design, cannot emit radiation levels in excess of the permissible exposure limits. Not hazardous under any operational or viewing condition. Requires no controls.
2	Low-powered lasers and laser system's which emit less than 1 mW visible continuous wave (CW) radiation. Not considered hazardous for momentary unintentional exposure. These lasers carry a CAUTION label.
2a	Low-powered lasers and laser systems which emit less than 1 mW visible continuous wave (CW) radiation. Not considered hazardous for momentary unintentional exposure. These lasers carry a CAUTION label.
3a	Low-powered laser systems which emit 1–5 mW visible CW radiation. Lasers or laser systems of less than 2.5 mW/cm^2 are not considered to be hazardous for momentary unintentional exposures unless the beam is viewed with magnifying optics. These lasers carry a CAUTION label. Lasers which exceed 2.5 mW/cm^2 carry a DANGER label and should not be viewed even momentarily.
3b	Medium-powered lasers or laser systems considered to be potentially hazardous when the direct or specularly reflected beam is viewed without protection. Special care is required to prevent intrabeam viewing and to control specular reflections from mirror-like surfaces. These lasers carry a DANGER label and require the use of protective eyewear.
4	High-powered lasers or laser systems which can be hazardous to the eye from intrabeam viewing, specular reflections, or diffuse reflections. They may also be hazardous to the skin or flammable materials, causing a fire. These lasers carry a DANGER label. Strict controls are required, including use of protective eyewear and door interlocks.

tective features are removed, the laser classification may change to a more hazardous class.

These laser classification levels in the old system are shown in Table 2.13.

Revised Laser Classification (new system). Table 2.14 describes the main characteristics and requirements for the new classification system as specified by IEC 60825-1.

Table 2.15 lists some typical laser related hazards. Lasers always have two primary hazards associated with them: eye damage and high-energy heat source. Collateral laser hazards involve hazards from the laser system components, primarily from the lasing medium materials and the energy source used for excitation.

TABLE 2.14 Laser Classifications (New System)

Class	Safety Description
1	A Class 1 laser is safe under all conditions of normal use. This means the maximum permissible exposure (MPE) cannot be exceeded. This class includes high-power lasers within an enclosure that prevents exposure to the radiation and that cannot be opened without shutting down the laser. For example, a continuous laser at 600 nm can emit up to 0.39 mW, but for shorter wavelengths, the maximum emission is lower because of the potential of those wavelengths to generate photochemical damage. The maximum emission is also related to the pulse duration in the case of pulsed lasers and the degree of spatial coherence.
1M	A Class 1M laser is safe for all conditions of use except when passed through magnifying optics such as microscopes and telescopes. Class 1M lasers produce large-diameter beams or beams that are divergent. The MPE for a Class 1M laser cannot normally be exceeded unless focusing or imaging optics are used to narrow the beam. If the beam is refocused, the hazard of Class 1M lasers may be increased and the product class may be changed. A laser can be classified as Class 1M if the total output power is below Class 3B but the power that can pass through the pupil of the eye is within Class 1.
2	A Class 2 laser is safe because the blink reflex will limit the exposure to no more than 0.25 s. It only applies to visible-light lasers (400–700 nm). Class 2 lasers are limited to 1 mW continuous wave, or more if the emission time is less than 0.25 s or if the light is not spatially coherent. Intentional suppression of the blink reflex could lead to eye injury. Many laser pointers are Class 2.
2M	A Class 2M laser is safe because of the blink reflex if not viewed through optical instruments. As with Class 1M, this applies to laser beams with a large diameter or large divergence, for which the amount of light passing through the pupil cannot exceed the limits for Class 2.
3R	A Class 3R laser is considered safe if handled carefully, with restricted beam viewing. With a Class 3R laser, the MPE can be exceeded, but with a low risk of injury. Visible continuous lasers in Class 3R are limited to 5 mW. For other wavelengths and for pulsed lasers, other limits apply.
3B	A Class 3B laser is hazardous if the eye is exposed directly, but diffuse reflections such as from paper or other matte surfaces are not harmful. Continuous lasers in the wavelength range from 315 nm to far infrared are limited to 0.5 W. For pulsed lasers between 400 and 700 nm, the limit is 30 mJ. Other limits apply to other wavelengths and to ultrashort pulsed lasers. Protective eyewear is typically required where direct viewing of a Class 3B laser beam may occur. Class 3B lasers must be equipped with a key switch and a safety interlock.
4	Class 4 lasers include all lasers with beam power greater than Class 3B. By definition, a Class 4 laser can burn the skin, in addition to potentially devastating and permanent eye damage as a result of direct or diffuse beam viewing. These lasers may ignite combustible materials, and thus may represent a fire risk. Class 4 lasers must be equipped with a key switch and a safety interlock. Most entertainment, industrial, scientific, military, and medical lasers are in this category.

TABLE 2.15 Typical Laser System Hazards

Hazard	Causal Factors
Eye damage	Eye exposure to energy from laser beam
Burn	Skin exposure to high-energy laser beam
Ignition source	High-energy beam ignites explosive vapors or materials
Cutting source	High-energy laser beam cuts solid materials
Fire	Laser system components cause fire
Explosion	Laser system components cause explosion
Toxic chemicals	Exposure to chemicals in lasing medium
Electrocution	Exposure to high-voltage electronics
Noise	Excessive noise
Cryogenics	Spills; failure to cool system components

LASER SAFETY OFFICER (LSO)

In many jurisdictions, organizations that operate lasers are required to appoint an LSO. The LSO is responsible for ensuring that safety regulations are followed by all workers in the organization.

See *Laser System Safety Officer (LSSO)* for additional related information.

LASER SAFETY REVIEW BOARD (LSRB)

The LSRB provides a systems safety review of all Department of the Navy (DON) lasers used in combat, combat training, or classified in the interest of national security, and all lasers capable of exceeding Class 3a levels, including those used in optical fiber communication systems. If a system utilizes a laser, the system design must be reviewed and approved by the LSRB. It is the responsibility of the SSP to initiate contact with the LSRB and ensure all applicable requirements are satisfied.

LASER SYSTEM

A laser system is an assembly of electrical, mechanical, and optical components which includes a laser.

LASER SYSTEM SAFETY OFFICER (LSSO)

LSSO refers to personnel functioning as a laser safety authority on a Navy program. The LSSO has authority to monitor, evaluate, and enforce the control of laser hazards. The LSSO must have successfully completed approved laser safety courses.

LATENT FAILURE (OR LATENCY)

A latent failure refers to a component that is not checked for operability before the start of a mission; thus, it could unknowingly be failed when required for use. When failure of the component occurs, it is not detected or annunciated. Certain latent faults can be of system safety concern because they are involved in system designs where operation is critical. Latency can significantly increase the potential safety risk because this situation effectively increases the component exposure time. The latent time period is the time between maintenance checks, which can often be significantly greater than the mission time. This large exposure time can make a large impact on the probability. Latency is also sometimes referred to as dormancy or dormant failure.

Figure 2.47 provides an example FTA showing how significant the numerical error can be if latency is not properly handled in the safety analysis and the system design.

In this example, the FTA models an uncontained fire in the lower bay of a commercial aircraft. There is a fire detection/suppression system in the bay to detect and extinguish a fire. An uncontained fire will result if the fire detection/suppression fails, causing loss of aircraft. FTA-A evaluates the design ignoring the effect of latency. It assumes both components are checked for failed state prior to flight. However, in reality, the fire detection/suppression may be checked only every 6000h (contrived number) because it can only be done in the maintenance shop. FTA-B shows the correct calculation which accounts for the fact that the fire detection/suppression system cannot be checked for failed state prior to flight. Note the two orders of magnitude difference in the probability calculations, and that FTA-A understates the actual risk.

LAYERS OF PROTECTION (LOP)

LOP are the layers of mitigation that are implemented to mitigate a hazard. LOP are also referred to as IPLs.

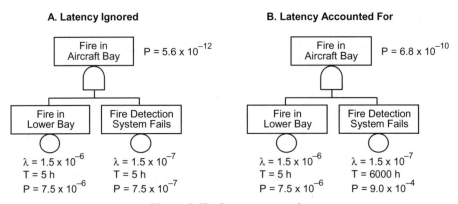

Figure 2.47 Latency example.

LAYERS OF PROTECTION ANALYSIS (LOPA)

LOP are the layers of mitigation that are implemented to mitigate a hazard. LOPA is a semiquantitative methodology that can be used to identify safeguards that meet the IPL criteria. LOPA provides specific criteria and restrictions for the evaluation of IPLs. LOPA is limited to a single cause–consequence pair as a scenario. LOPA is a recognized technique that can establish a proper safety integrity level (SIL) of the process. LOPA is a methodology for hazard evaluation and risk assessment, and lies between simple qualitative and more elaborate quantitative analysis techniques.

See *Independent Protection Layer (IPL)* for additional related information.

LEAD EXPLOSIVE

Lead and booster explosives are compounds and formulations which are used to transmit and augment the detonation reaction. They are typically used in the beginning stage of an explosive train, for example, in a fuze.

LEVEL OF ASSEMBLY

Level of assembly refers to the natural hierarchy of elements within a system. A system is composed of subsystems, assemblies, subassemblies, components, and parts, all of which can be categorized by level of assembly or system hierarchy. Level of assembly is a decomposition of a system into a hierarchy of levels that incrementally reduce the complexity description of the system. Levels of assembly are typically used for describing development, analysis, and test configurations of a system. This type of system breakdown is also referred to as system indenture levels and can also be part of a system work breakdown structure (WBS).

The following breakdown describes typical levels of hardware assembly in a system:

1. Part—A hardware element that is not normally subject to further subdivision or disassembly without destruction of design use. Examples include resistor, IC, relay, connector, bolt, and gaskets.
2. Subassembly—A subdivision of an assembly. Examples are wire harnesses and printed circuit boards.
3. Assembly—A functional subdivision of a component consisting of parts or subassemblies that perform functions necessary for the operation of the component as a whole. Examples are a power amplifier, gyroscope, radar antenna, and so on.
4. Component or unit—A functional subdivision of a subsystem and generally a self-contained combination of items performing a function

necessary for the subsystem's operation. Examples are electronic box, transmitter, gyro package, actuator, motor, and battery. For most purposes, "component" and "unit" are used interchangeably.
5. Section—A structurally integrated set of components and integrating hardware that forms a subdivision of a subsystem, module, and so on. A section forms a testable level of assembly, such as components/units mounted into a structural mounting tray or panel-like assembly, or components that are stacked.
6. Subsystem—A functional subdivision of a system consisting of two or more components. Examples are structural, attitude control, electrical power, and communication subsystems.
7. Instrument—A spacecraft subsystem consisting of sensors and associated hardware for making measurements or observations in space. For most purposes, an instrument is considered a subsystem.
8. Module—A major subdivision of the system that is viewed as a physical and functional entity for the purposes of analysis, manufacturing, testing, and record keeping. Examples include spacecraft bus, science payload, and upper-stage vehicle.
9. Payload—An integrated assemblage of modules, subsystems, and so on, designed to perform a specified mission, such as a satellite carried in a spacecraft or a weapon aboard an aircraft.

See *System Hierarchy* for additional related information.

LIFE CYCLE

Life cycle refers to the life cycle of a system which includes the following phases:

- Conceptual design
- Development (includes preliminary design, final design, and test)
- Production
- Operation
- Disposal

See *System Life-Cycle Model* for additional related information.

LIFE SUPPORT ITEM

All man-mounted or aircraft installed equipment and components designed to protect, sustain, or save human lives are categorized as life support. This includes, but is not limited to, ejection systems, crew seats, passenger seats, emergency escape slides, parachutes, life rafts and preservers, survival kits, emergency radios and beacons, aircrew helmets, oxygen masks, goggles, visors, chemical defense equipment, and selected clothing and uniform items.

LIGHT

Light is EMR in the visible range, which can be detected by the human eye. This term is commonly used to describe wavelengths which lie in the range 0.4–0.7 nm. The primary properties of light include intensity, frequency or wavelength, polarization, and phase.

LIKELIHOOD

Likelihood is the probability of a specified outcome or event; a measure of how likely it is that some event will occur. It can be stated in qualitative or quantitative terms.

LIMITED LIFE ITEMS

Limited life items are hardware items that exhibit the following:

- An expected failure-free life that is less than the projected mission life, when considering cumulative test, storage, and operation.
- Limited shelf-life material used to fabricate the hardware.

LINE-OF-SIGHT (LOS)

LOS refers to a visual condition that exists when there is no obstruction between the viewer and the object being viewed. In RF communications, a condition that exists when there is no intervening object, such as dense vegetation, terrain, man-made structures, or the curvature of the Earth, between the transmitting and receiving antennas, and transmission and reception would be impeded.

LINE REPLACEABLE UNIT (LRU)

An LRU indicates the operations or maintenance level at which a system element can be replaced. An LRU is typically a subsystem that can be replaced on the operational line from a spare that is in inventory. It is

usually a functional subsystem that is at a higher level in the system hierarchy, such as an HF radio in an aircraft or a helicopter transmission unit.
See *System Hierarchy* for additional related information.

LOCKIN

A lockin is a protective device that restricts personnel inside specific limits to prevent contact with a hazard outside those limits or that maintains the hazard inside those limits so that it cannot affect anything outside.

LOCKOUT

A lockout is a protective device that restricts personnel outside specific limits to prevent contact with a hazard inside those limits or that maintains the hazard outside those limits so it cannot affect anything inside. For example, a padlock can be used as a lockout device on a gate to prevent unauthorized entry into an area.

LOWER EXPLOSIVE LIMIT (LEL)

LEL is the concentration of vapor or dust in air below which an explosion cannot occur.

LOWER FLAMMABLE LIMIT (LFL)

LFL is the concentration of a vapor or dust in air below which a burning reaction cannot be sustained.

LOW RATE INITIAL PRODUCTION (LRIP)

LRIP is a term commonly used in DoD projects/programs to designate the phase of initial, small-quantity production of a weapons systems. LRIP gives the system buyer time to thoroughly test the system before going into full-scale production. This allows time to gain a reasonable degree of confidence as to whether the system actually performs to the agreed-upon requirements before contracts for mass production are signed.

MAIN CHARGE

The explosive or pyrotechnic charge which is provided to accomplish the end result in the munition, such as producing blast and fragments, dispensing submunitions, or producing other effects for which it was designed.

MAINTAINABILITY

Maintainability is the relative ease and economy of time and resources with which an item can be retained in, or restored to, a specified condition when maintenance is performed by personnel having specified skill levels, using prescribed procedures and resources, at each prescribed level of maintenance and repair. Maintainability is the probability that an item can be retained in, or restored to, a specified condition when maintenance is performed by personnel having specified skill levels, using prescribed procedures and resources, at each prescribed level of maintenance and repair.

MAINTENANCE

Maintenance involves the actions necessary for retaining an item in, or restoring it to, a specified condition. Maintenance includes inspection, testing, servicing, repair, rebuilding, and reclamation.

MALFUNCTION

A malfunction is the occurrence of unsatisfactory or incorrect performance of a component or function. There can be many different forms of malfunction, depending upon the item and the system involved. For example, the following are some typical forms of malfunction:

- Fails to operate or function
- Operates out of specified tolerances
- Functions early
- Functions prematurely
- Functions late
- Partially functions

See *Failure* for additional related information.

MANAGEMENT OVERSIGHT AND RISK TREE (MORT) ANALYSIS

MORT is an analysis technique for identifying SR oversights, errors, and/or omissions that led to the occurrence of a mishap. MORT is primarily a reactive analysis tool for accident/mishap investigation, but it can also be used for the proactive evaluation and control of hazards. MORT analysis is used to trace out and identify all of the causal factors leading to a mishap or UE. MORT utilizes the logic tree structure and rules of FTA, with the incorporation of some new symbols. Although MORT can be used to generate risk probability calculations like FTA, quantification is typically not employed. MORT analysis provides decision points in a safety program evaluation where design or program change is needed. MORT attempts to combine design safety with management safety.

MORT is a root-cause analysis tool that provides a systematic methodology for planning, organizing, and conducting a detailed and comprehensive mishap investigation. It is used to identify those specific design control measures and management system factors that are less than adequate (LTA) and need to be corrected to prevent the reoccurrence of the mishap, or prevent the UE. The primary focus of MORT is on oversights, errors, and/or omissions and to determine what failed in the management system. MORT analysis is applicable to all types of systems and equipment, with analysis coverage given to systems, subsystems, procedures, environment, and human error. The primary application of MORT is in mishap investigation to identify all of the root causal factors and to ensure that corrective action is adequate. MORT analysis is capable of producing detailed analyses of root causes leading to a UE or mishap. By meticulously and logically tracking energy flows within and out of a system, MORT analysis compels a thorough analysis for each specific energy type. The degree of thoroughness depends on the self-discipline and ability of the analyst to track logically the flows and barriers in the system. Although simple in concept, the MORT process is labor intensive and requires significant training.

The theory behind MORT analysis is fairly simple and straightforward. The analyst starts with a predefined MORT graphical tree that has been developed by the original MORT developers. The analyst works through this predefined tree comparing the management and operations structure of his/her program to the ideal MORT structure, and develops a MORT diagram modeling the program or project. MORT and FTA logic and symbols are used to build the program MORT diagram. The predefined tree consists of 1500 basic events, 100 generic problem areas, and a large number of judging criteria. This diagram can be obtained from the MORT User's Manual, Department of Energy (DOE) SSDC-4, 1983. The concept emphasizes energy-related hazards in the system design and management errors. MORT analysis is based on energy transfer and barriers to prevent or mitigate mishaps. Consideration is given to management structure, system design, potential human error, and environmental factors.

MORT analysis is essentially an FTA that asks *what* oversights and omissions could have occurred to cause the UE or mishap, and *why* in terms of the management system. In some ways, MORT analysis is like using the basic MORT diagram as a checklist to ensure everything pertinent is considered. Figure 2.48 shows the top level of the ideal MORT analysis from the MORT User's Manual.

MORT is an analysis technique that is useful in some system safety applications. For more detailed information on the MORT technique, see Clifton A. Ericson II, *Hazard Analysis Techniques for System Safety* (2005), chapter 24.

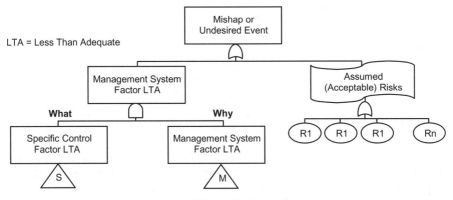

Figure 2.48 MORT top tiers.

MANAGING ACTIVITY (MA)

The MA is the organization responsible for the procurement of a product or system. Typically, the MA is a government agency; however, it could also be a private sector enterprise or corporation purchasing products or systems from another contractor.

See *Contracting Process* and *Contractor* for additional related information.

MAN-PORTABLE

Man-portable refers to the capability of an item being carried by one man. Typically, the upper weight limit is approximately 31 lb. In land warfare, equipment which can be carried by one man over long distance without serious degradation of the performance of normal duties is man-portable.

MAN-TRANSPORTABLE

Man-transportable refers to the capability of items that are usually transported on wheeled, tracked, or air vehicles, but have integral provisions to allow periodic handling by one or more individuals for limited distances (100–500 m). The upper weight limit is approximately 65 lb per individual.

MARGINAL FAILURE

A marginal failure is a failure that can degrade performance or result in degraded operation. Special operating techniques or alternative modes of operation involved by the loss can be tolerated throughout a mission but should be corrected upon its completion.

MARGINAL HAZARD

A marginal hazard is a hazard that has a Category III (Marginal) severity level, as defined by the hazard severity criteria in MIL-STD-882.
　See *Hazard Severity Levels* for additional related information.

MARGIN OF SAFETY

Margin of safety is a term describing the cushion between expected load and actual design strength in a component or product design. A margin of safety is typically the ratio of the safety design to the required design. The required design is what the item is required to be able to withstand for expected conditions, whereas the margin of safety includes a safety margin that provides a measure of how much more the actual design can tolerate. The terms SF, factor of safety, margin of safety, and safety margin are synonymous.
　See *Safety Factor (SF)* for additional related information.

MARKOV ANALYSIS (MA)

Markov analysis (MA) is an analysis technique for modeling system state transitions and calculating the probability of reaching various system states from the model. MA is a tool for modeling complex system designs involving timing, sequencing, repair, redundancy, and fault tolerance. MA is accomplished by drawing system state transition diagrams and examining these diagrams for understanding how certain undesired states are reached, and their relative probability. MA can be used to model system performance, dependability, availability, reliability,, and safety. MA describes failed states and degraded states of operation where the system is either partially failed or in a degraded mode where some functions are performed while others are not.
　Markov chains are random processes in which changes occur only at fixed times. However, many of the physical phenomena observed in everyday life are based on changes that occur continuously over time. Examples of these continuous processes are equipment breakdowns, arrival of telephone calls, and radioactive decay. Markov processes are random processes in which changes occur continuously over time, where the future depends only on the present state and is independent of history. This property provides the basic framework for investigations of system reliability, dependability, and safety. There are several different types of Markov processes. In a semi-Markov process, time between transitions is a random variable that depends on the transition. The discrete and continuous-time Markov processes are special cases of the semi-Markov process (as will be further explained).
　MA can be applied to a system early in development and thereby identify design issues early in the design process. Early application will help system

MARKOV ANALYSIS (MA) 249

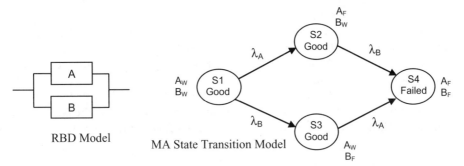

Figure 2.49 MA model for one component system with repair.

Figure 2.50 MA model for two component parallel system with no repair.

developers to design-in safety and reliability of a system during early development rather than having to take corrective action after a test failure, or worse yet, a mishap. MA is a somewhat difficult technique to learn, understand, and master. A high-level understanding of mathematics is needed to apply the methodology.

Figure 2.49 shows an example MA state transition model for a one component system with repair. The reliability block diagram (RBD) shows the system design complexity. In this MA model, only two states are possible, the operational state S1 and the failed state S2. The starting state is S1, in which the system is operational (good). In state S2, the system is failed. The transition from state S1 to state S2 is based on the component failure rate λ_A. Note that A_W indicates component A working and AF indicates component A failed. The connecting edge with the notation λ_A indicates the transitional failure of component A. Note how component A can return from the failed stated to the operational state at the repair transition rate μ_A.

Figure 2.50 shows an example MA model for a two component parallel system with no repair. The RBD indicates that successful system operation only requires successful operation of either component A or B. Both components must fail to result in system failure.

In this MA model, four states are possible. The starting state is S1, whereby the system is good (operational) when both A and B are working. Based on A failure rate λ_A, it transitions to the failed state S2. In state S2, A is failed, while B is still good. In state S3, B is failed, while A is still good. In state S4, both A and B are failed. In states S1, S2, and S3 the system is good, while in state S4 the system is failed.

When larger, more complex systems are modeled, the MA model and corresponding equations become larger and more complex. For complex models, the equations between MA and FTA will differ. However, many demonstrations have shown that the FTA probability approximations are typically close enough to the MA results for most safety risk assessments, and the FTA models are easier to comprehend and mathematically compute. MA is a tool for modeling complex system designs involving timing, sequencing, repair, redundancy, and fault tolerance. MA provides both graphical and mathematical (probabilistic) system models. MA models can easily become too large in size for comprehension and mathematical calculations, unless the system model is simplified. Computer tools are available to aid in analyzing more complex systems. MA is typically utilized when very precise mathematical calculations are necessary.

MA is an HA technique that is important and essential to system safety and reliability engineering. For more detailed information on the MA technique, see Clifton A. Ericson II, *Hazard Analysis Techniques for System Safety* (2005), chapter 18.

MASTER EQUIPMENT LIST (MEL)

The MEL is the systems engineering tool for exhibiting both system components and system hierarchy relationships. The MEL is a system hierarchy that establishes nomenclature and terminology that support clear, unambiguous communication and definition of the system, its functions, components, operations, and associated processes. A system hierarchy refers to the organizational structure defining dominant and subordinate relationships between subsystems, down to the lowest component or piece part level. The MEL should include all system hardware and equipment. The purpose of indenture or hierarchy levels is to reduce system complexity into manageable pieces that can be easily understood and accounted for. The MEL aids the safety analyst in ensuring that all of the system hardware and functions have been adequately covered by the appropriate HAs.

See *System Hierarchy* for additional related information.

MASTER LOGIC DIAGRAM (MLD)

An MLD is a hierarchical, top-down display of IEs, showing general types of UEs at the top, proceeding to increasingly detailed event descriptions at lower tiers, and displaying IEs at the bottom. The goal is not only to support identification of a comprehensive set of IEs, but also to group them according to the challenges that they pose (the responses that are required as a result of their occurrences). IEs that are completely equivalent in the challenges that they pose, including their effects on subsequent PEs, are equivalent in the risk model.

MLDs are useful in development IE because they facilitate organizing thoughts and ideas into a comprehensive list of candidate IEs. An MLD resembles an FT, but it lacks explicit logic gates. An MLD also differs from an FT in that the initiators defined in the MLD are not necessarily failures or basic events. Specifically, MLDs are a hierarchical depiction of ways in which system perturbations occur. For example, in a nuclear power plant system, these perturbations involve failure to contain, failure to control, and failure to cool or otherwise maintain temperatures within acceptable ranges. An MLD shows the relationship of lower levels of assembly to higher levels of assembly and system function. The top event in each MLD is an end state. Events that are necessary but not sufficient to cause the top event are enumerated in even more detail as the lower levels of the MLD are developed. For complex missions, it may be necessary to develop phase-specific MLDs since threats and IEs may change as the mission progresses.

A key concept in MLD development is the pinch point; without some termination criterion, an MLD could be developed endlessly. The pinch point is the termination criterion applied to each MLD branch. A pinch point occurs when every lower level of the branch has the same consequence (relative to system response) as the higher levels. Under such conditions, more detailed MLD development will not contribute further insights into IEs capable of causing the end state being investigated. Figure 2.51 illustrates the MLD concept.

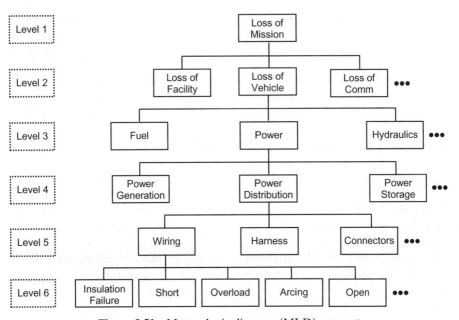

Figure 2.51 Master logic diagram (MLD) concept.

MATERIAL SAFETY DATA SHEET (MSDS)

An MSDS is a form containing data regarding the properties of a particular substance. An important component of workplace safety, it is intended to provide workers and emergency personnel with procedures for handling or working with that substance in a safe manner, and includes information such as physical data (melting point, boiling point, flash point, etc.), toxicity, health effects, first aid, reactivity, storage, disposal, protective equipment, and spill-handling procedures. MSDS are a widely used system for cataloging information on chemicals, chemical compounds, and chemical mixtures. MSDS information may include instructions for the safe use and potential hazards associated with a particular material or product. MSDS formats can vary from source to source within a country depending on national requirements.

An MSDS for a substance is not primarily intended for use by the general consumer, focusing instead on the hazards of working with the material in an occupational setting. In some jurisdictions, the MSDS is required to state the chemical's risks, safety, and effect on the environment. It is important to use an MSDS specific to both country and supplier, as the same product (e.g., paints sold under identical brand names by the same company) can have different formulations in different countries. The formulation and hazard of a product using a generic name (e.g., sugar soap) may vary between manufacturers in the same country. In the United States, the OSHA requires that MSDS be available to employees for potentially harmful substances handled in the workplace under the Hazard Communication regulation. The MSDS is also required to be made available to local fire departments and local and state emergency planning officials under Section 311 of the Emergency Planning and Community Right-to-Know Act. The American Chemical Society defines Chemical Abstracts Service (CAS) Registry Numbers which provide a unique number for each chemical and are also used internationally in MSDSs.

MAXIMUM PERMISSIBLE EXPOSURE (MPE)

MPE is the highest power or energy density (in W/cm^2 or J/cm^2) of a light source that is considered safe for human exposure. The MPE for laser safety is the level of laser radiation to which a person may be exposed without hazardous effect or adverse biological changes in the eye or skin. It is usually about 10% of the dose that has a 50% chance of creating damage under worst-case conditions. The MPE is measured at the cornea of the human eye or at the skin, for a given wavelength and exposure time. A calculation of the MPE for ocular exposure takes into account the various ways light can act upon the eye.

See *Laser Safety* for additional related information.

MAY

With regard to design requirements, the term "may" indicates optional action in a requirement. It is typically used in a requirement when the application of the requirement is desired but optional.

MEAN TIME BETWEEN FAILURES (MTBFs)

MTBF is a basic measure of reliability for repairable items. The mean number of life units during which all parts of the item perform within their specified limits, during a particular measurement interval under stated conditions. MTBF is the predicted elapsed time between inherent failures of a system, component, or product during operation. MTBF can be calculated as the arithmetic mean (average) time between failures of an item. The MTBF is typically part of a model that assumes the failed item is immediately repaired (zero elapsed time), as a part of a renewal process. This is in contrast to the mean time to failure (MTTF), which measures average time between failures with the modeling assumption that the failed item is not repaired. Reliability increases as the MTBF increases. The MTBF is usually specified in hours but can also be used with other units of measurement such as miles or cycles.

Formulas for calculating MTBF are:

$MTBF = $ (total hours of operation)/(total number of failures)

$MTBF = 1/\lambda$, where λ is the failure rate

Reliability is quantified as MTBFs for repairable products and MTTF for nonrepairable products, where:

$R(t) = e^{-T/MTBF} = e^{-\lambda T}$ (where T is the exposure time)

MEAN TIME TO FAILURE (MTTF)

MTTF is a basic measure of reliability for nonrepairable items. It is the total number of life units of an item population divided by the number of failures within that population, during a particular measurement interval under stated conditions. MTTF measures the average time between failures with the modeling assumption that the failed item is not repaired. Reliability increases as the MTTF increases. The MTTF is usually specified in hours, but can also be used with other units of measurement such as miles or cycles. MTBFs measure the average time between failures with the modeling assumption that the failed item is repaired.

MEAN TIME TO REPAIR (MTTR)

MTTR is a basic measure of maintainability. It is the sum of corrective maintenance times at any specific level of repair, divided by the total number of failures within an item repaired at that level, during a particular interval under stated conditions.

MICROBURST

A microburst is also known as wind shear. Wind shear is a sudden and violent change in wind speed and/or direction. Low-altitude wind shear is caused by the strong downdraft of a shower or thunderstorm, which as it hits the ground, spreads out in all directions, producing dangerous air currents. Low-altitude wind shear can have devastating effects on an aircraft during taking or landing because there is usually insufficient altitude for the pilot to prevent hitting the ground.

MILESTONE DECISION AUTHORITY (MDA)

The MDA is the individual designated in accordance with criteria established by the Undersecretary of Defense for Acquisition, Technology and Logistics (USD[AT&L]) to approve entry of an acquisition program into the next phase.

MINIMAL CUT SET (MCS)

A CS is a set of events (typically failures) that together cause the top UE of an FT to occur. A CS can have any number of events in it; for example, a CS could be composed of one event or 10 events. Multiple events in a CS indicate that the events are ANDed together. An MCS is a CS where none of the set elements can be removed from the set and still cause the top event to occur. It is also referred to as a min CS.

See *Cut Set* for additional related information.

MISFIRE

A misfire is the failure of an explosive device to fire or explode properly; failure of a primer or the propelling charge of a round or projectile to function wholly or in part.

MISHAP

An industry standard definition of mishap, provided in MIL-STD-882D, is "An unplanned event or series of events resulting in death, injury, occupational illness, damage to or loss of equipment or property, or damage to the environment." It should be noted that in system safety, the term mishap is synonymous with accident. A mishap is effectively an actualized hazard, whereby the hazard transitions from the dormant conditional state to the active mishap event state. The three required components of a hazard predefine the mishap. A mishap would not be possible without the preexistence of a hazard. A mishap is an actual event that has occurred and has resulted in an undesired outcome. Mishaps and a hazards are directly related, they are linked together by risk and state space. A mishap is an actuated hazard; it is the direct result of a potential hazard, when the hazard's IMs (or causal factors) occurs, transitioning the hazard from a potential condition state to a mishap event state with loss outcome.

Mishaps are assumed by many to be stochastic events, that is, random, haphazard, unpredictable. However, a mishap is more than a random unplanned event with an unpredictable free will. Mishaps are not events without apparent reason; they are the result of actuated hazards. Hazards are predictable and controllable and they occur randomly based on their statistical predilection, which is typically controlled by a failure or error rate. Thinking of a mishap as a chance event without justification gives one the sense that mishaps involve an element of destiny and futility. System safety, on the other hand, is built upon the premise that mishaps are not just chance events; instead they are seen as deterministic, predictable, and controllable events (in the disguise of hazards).

Mishaps involve a set of causal factors that lead up to the final mishap event, and these factors are the actuated hazard conditions. Mishap causal factors can be identified prior to an actual mishap through the application of HA. Mishaps are an inevitable consequence of antecedent causes and, given the same causal factors, the same mishap is repeatable, with the frequency based on the component probabilities. Mishaps can be predicted via hazard identification, and they can be prevented or controlled via hazard elimination or hazard control methods. This safety concept demonstrates that we do have control over the potential mishaps in the systems we develop and operate. We are not destined to face an unknown suite of undesired mishaps, unless we allow it to be so (by not performing adequate system safety). In the safety sense, mishaps are preplanned events in that they are actually created through poor design and/or inadequate design foresight.

Figure 2.52 depicts the hazard–mishap relationship, whereby a hazard and a mishap are two separate states of the same phenomenon, linked by a state transition that must occur. You can think of these states as the *before* and *after* states. A hazard is a "potential event" at one end of the spectrum, which may be transformed into an "actual event" (the mishap) at the other end of the spectrum, based upon the state transition.

256 SYSTEM SAFETY TERMS AND CONCEPTS

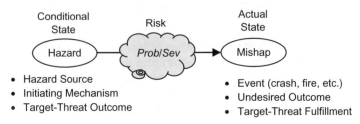

Figure 2.52 Hazard–mishap relationship.

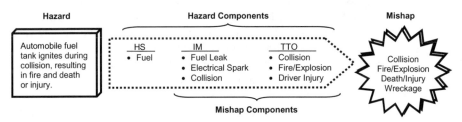

Figure 2.53 Hazard–mishap example components.

Mishaps are the immediate result of actualized hazards. The state transition from a hazard to a mishap is based on two factors: (1) the unique set of *hazard components* involved and (2) the *risk* presented by the hazard components. The hazard components are the items comprising a hazard, and the risk is the probability of the hazard–mishap occurring combined with the severity of the resulting outcome loss.

Figure 2.53 contains an example hazard, which is broken down into its constituent parts. The corresponding mishap is also broken down into its constituent parts. A hazard is composed of three required components: (1) an HS, (2) an IM, and (3) a TTO. A mishap is composed of (1) mishap trigger, (2) event, and (3) undesired outcome. Note in this example that "collision" falls into both the HS and TTO categories, which is not uncommon for certain hazards.

A hazard is a wrapper containing (or describing) all of the latent conditions (or components) necessary to result in a mishap, when the latent factors actualize. And, the hazard wrapper also describes the mishap outcome to be expected. The major difference between a hazard and mishap description is how they are stated; a hazard "could happen" whereas a mishap "did happen." In addition, a hazard is stated with all the necessary elements as a potential condition, while a mishap is only stated as the final outcome (which is defined in the hazard statement). It is interesting to note that when a mishap actually does occur, it is the job of the accident investigation team to discover all of the mishap causal factors, which in effect describes the hazard components that should have originally been identified.

See *Hazard* for additional related information.

MISHAP CAUSAL FACTOR

Mishap causes are conditions or events that explain why a mishap occurred. Events within a mishap may have multiple causes. Causes are the genesis of the mishap, not the reason that damage or injury occurred. For example, a fire may have damaged a house, but the mishap was not caused by the fire, it was caused by an ignition source combined with a flammable material, resulting in a fire.

Hazard–mishap theory states that hazards create the potential for mishaps, mishaps occur based on the level of risk involved, and risk is calculated from the hazard–mishap causal factors. HCFs are the factors that create the hazard, while mishap causal factors are the factors that cause the dormant hazard to become an active mishap event. Just as hazards and mishaps are directly linked, so are their causal factors. For a hazard to exist, three hazard components must be present: (1) the HS which provides the basic source of danger, (2) the potential IMs that will transition the hazard from an inactive state to a mishap event, and (3) the TTO that will result from the expected mishap event. The HS and IM are the HCFs that are used to determine risk probability, and the TTO is the causal factor that establishes risk severity. The HS and IM hazard components can be broken into the major causal factor categories of hardware, software, humans, interfaces, functions, procedures, management safety culture, and the environment. Since hazard risk is the same as mishap risk, HCFs are also mishap causal factors.

See *Hazard* and *Mishap* for additional related information.

MISHAP LIKELIHOOD (MISHAP PROBABILITY)

Likelihood (or probability) is one parameter in the risk equation. Risk is the safety measure of a potential future event, stated in terms of event likelihood and event severity. Hazard likelihood is the expected likelihood that the identified hazard will be activated and becomes an actual mishap. Hazard likelihood is the estimated likelihood of a hazard transitioning from a conditional state to an actual mishap event state, resulting in an actual mishap with undesired outcome.

In a risk assessment, hazard likelihood can be characterized in terms of probability, frequency, or qualitative criteria. Quite often, the term "hazard probability" is incorrectly used when the actual assessment is done in terms of frequency or qualitative criteria. This is why hazard likelihood is a more accurate term. Likelihood is a measure of how possible or likely it is that an event will occur, such as a hazard–mishap. Likelihood can typically be characterized in one of the following ways:

1. Probability—Probability is a number between zero and one. Although probability is a dimensionless number, it has to be calculated to some

258 SYSTEM SAFETY TERMS AND CONCEPTS

> specific set of criteria, such as the probability of event X occurring in 700 hours of operation.
> 2. Frequency—Frequency is the rate of occurrence of an event described in terms of occurrences per unit of time or operations, such as failures per million hours of operation or per million cycles. Frequency is quite often established from historical data.
> 3. Qualitative description—In this methodology, likelihood is classified in terms of different qualitative likelihood ranges. For example, "Likely to occur frequently in the life of an item," "Will occur several times in the life of an item," and "Likely to occur sometime in the life of an item."

It should be noted that mishap likelihood and hazard likelihood are really the same entity, just viewed from two different perspectives.

See *Hazard Risk* and *Mishap Risk* for additional related information.

MISHAP RISK

Risk is the safety measure of a potential future event, stated in terms of event likelihood and event severity. Likelihood can be characterized in terms of probability, frequency, or qualitative criteria, while severity can be characterized in terms of death, injury, dollar loss, and so on. Exposure is part of the likelihood component. Mishap risk is the safety metric characterizing the amount of risk damage presented by a postulated mishap, where the likelihood of the mishap's occurrence is multiplied by the resulting severity of the mishap. Hazard risk is a safety metric characterizing the amount of danger presented by a hazard, where the likelihood of a hazard occurring and transforming into a mishap is combined with the expected severity of the mishap generated by the hazard's actuation. It should be noted that mishap risk and hazard risk are really the same entity, just viewed from two different perspectives. HA looks into the future and predicts a potential mishap from a hazard perspective. Mishap risk can only be reached by first determining the hazard and its risk. Since a hazard merely predefines a potential mishap, the risk has to be the same for both hazard risk and mishap risk.

See *Hazard Risk* for additional related information.

MISHAP RISK INDEX (MRI)

The MRI is an index number indicating qualitatively the relative risk of a hazard. It is derived from the MRI matrix by identifying the matrix cell resulting from the intersection of the hazard likelihood and hazard severity values. The MRI number establishes the safety significance of a hazard and who can accept the risk for the hazard. It should be noted that hazard risk and mishap risk are the same entity, just viewed from two different perspectives. Therefore, the MRI and the Hazard Risk Index (HRI) are essentially identical tools;

therefore, a more detailed definition is provided under the Hazard Risk Index (HRI) definition.

See *Hazard Risk Index (HRI)* for additional related information.

MISHAP RISK INDEX (MRI) MATRIX

The MRI matrix is a risk matrix that is utilized to establish the relative (vice absolute) risk of a hazard. The matrix maps hazard severity on one axis and hazard likelihood on the other axis. Once a hazard's severity and likelihood are determined, they are mapped to a particular MRI matrix cell (the likelihood–severity intersection), which yields the MRI risk level for that hazard. The likelihood and severity axes are broken into cells defined by qualitative and semiquantitative criteria.

The MRI matrix is a risk management tool used by system safety for hazard/mishap risk assessment. The MRI matrix establishes the relative level of potential mishap risk presented by an individual hazard. By comparing the calculated qualitative severity and likelihood values for a hazard against the predefined criteria in the MRI matrix, a level of risk is determined by a derived index number. If the hazard can be controlled through mitigation, then the likelihood and/or severity can be changed, thereby affecting a positive change in the risk level.

It should be noted that hazard risk and mishap risk are the same entity, just viewed from two different perspectives. Therefore, the MRI matrix and the Hazard Risk Index (HRI) matrix are essentially identical tools; thus, a more detailed definition is provided under the Hazard Risk Index (HRI) matrix definition.

See *Hazard Risk Index (HRI) Matrix* for additional related information.

MISHAP RISK ANALYSIS

Risk is the safety measure of a potential future event, stated in terms of event likelihood and event severity. Likelihood can be characterized in terms of probability, frequency, or qualitative criteria, while severity can be characterized in terms of death, injury, dollar loss, and so on. Mishap risk analysis is the process of identifying and evaluating the risk presented by a system hazard. HA is an integral part of risk analysis since safety risk can only be determined via the identification of hazards and risk assessment of those identified hazards.

See *Hazard Risk* and *Mishap Risk* for additional related information.

MISHAP SEVERITY

Severity is one parameter in the risk equation. Risk is the safety measure of a potential future event, stated in terms of event likelihood and event severity. Mishap severity is the expected severity that would result from the occurrence of a hazard when it becomes a mishap. Hazard severity is the predicted loss outcome resulting from a hazard given that it occurs and becomes a mishap. It is the assessed consequences of the worst credible mishap that could be caused by a specific hazard. Severity can be stated in various terms, such as death, injury, system loss, system damage, environmental damage, and dollar loss. It should be noted that mishap severity and hazard severity are really the same entity, just viewed from two different perspectives.

See *Hazard Risk* and *Mishap Risk* for additional related information.

MISSION CRITICAL

Mission critical is a designation given to the minimum subset of equipment, tasks, and assets necessary to achieve completion of a mission. This includes operation under both normal and abnormal conditions. The failure or malfunction of this equipment, tasks, or assets would preclude safe and successful mission completion. Depending on the type of system and the system definitions, flight-critical equipment could be the same mission-critical equipment. As a general rule, should a contingency necessitate shutting down certain equipment, the mission critical must not be shut down.

See *Mission Essential* for additional related information.

MISSION ESSENTIAL

Mission essential is a designation given to equipment, tasks, and assets necessary to achieve completion of a mission. Typically, this subset of equipment applies to normal operation (i.e., no failures) and is therefore larger than the subset of equipment designated as mission critical.

See *Mission Critical* for additional related information.

MODE

Modes and states are terms used to divide a system into segments which can explain different physical and functional configurations of the system that occur during the operation of the system. Although the two terms are often confused with one another and are sometimes used interchangeably, there are some clear definitions that make the two terms distinct and separate. The purpose of modes and states is to simplify and clarify system design and architecture for a more complete understanding of the system design and operation. A mode is a functional capability, whereas a state is a condition that characterizes the behavior of the functional capability.

A system mode is the functional configuration of the system, which established the system manner of operation. A mode is a set of functional capabilities that allow the operational system to accomplish tasks or activities. Modes tend to establish operational segments within the system mission. A system can have primary modes and sub-modes of operation. The system can only be in one mode at any one time. Correctly defining the mode establishes all of the modal constraints, which impact both design and operation.

A system state is the physical configuration of system hardware and software. Hardware and/or software are configured in a certain manner in order to perform a particular mode. A system state characterizes the particular physical configuration at any point in time. States identify conditions in which a system or subsystem can exist. A system or subsystem may be in only one state at a time. There is a connection between modes and states in that for every operational mode, there is one or more states the system can be in. For example, a stopwatch is a system that typically has modes such as on, off, timing, and reset. The "on" mode may go into the initialization state, which requires the initialization software package.

Modes and states are important to system safety because some modes and states are safety-critical and should not be performed erroneously or inadvertently. Evaluating system modes and states is an important factor during HA.

See *State* for additional related information.

MODE CONFUSION

Mode confusion is when the system or product operator is unaware that the current mode of operation is not the desired or intended mode. It involves incorrect or erroneous awareness of the operational mode. The confusion could be due to many different factors, such as human error, the system inadvertently jumping into the wrong mode, and display errors or failures. System safety is concerned about mode confusion because the operator may execute potentially hazardous commands if the system is in the incorrect mode. In critical applications, the operator can be forced into mode confusion due to poor system design or poor UI design.

See *Graphical User Interface (GUI)* and *User Interface (UI)* for additional information.

MODE OF CONTROL

Mode of control is the means by which a UMS receives instructions governing its actions and feeds back information to the operator. Modes of control include remote control, tele-operation, semiautonomous, and fully autonomous.

MODIFIED CONDITION/DECISION COVERAGE (MC/DC)

MC/DC is an exhaustive form of software testing, where independence of a condition is shown by proving that only one condition changes at a time. In MC/DC all of the following must be true at least once:

- Each decision tries every possible outcome
- Each condition in a decision takes on every possible outcome
- Each entry and exit point is invoked
- Each condition in a decision is shown to independently affect the outcome of the decision

In MC/DC, every point of entry and exit in the program has been invoked at least once, every condition in a decision in the program has taken on all possible outcomes at least once, and each condition has been shown to affect that decision outcome independently. A condition is shown to affect a decision's outcome independently by varying just that condition while holding fixed all other possible conditions. The condition/decision criterion does not guarantee the coverage of all conditions in the module because in many test cases, some conditions of a decision are masked by the other conditions. Using the modified condition/decision criterion, each condition must be shown to be able to act on the decision outcome by itself, everything else being held fixed. MC/DC criterion is thus much stronger than the condition/decision coverage. MC/DC testing (or equivalent) is required by RTCA/DO-178B for Level A software certification.

See *Code Coverage* for additional related information.

MODULE

A module is a major subdivision of the system that is viewed as a physical and functional entity for the purposes of analysis, manufacturing, testing, and record keeping. Examples include system power bus, payload, upper stage vehicle, and so on. A module represents a level in a system hierarchy.

See *System Hierarchy* for additional related information.

MONITOR

A monitor is a system component that watches the health and operation of another system component. It is generally a part of fault-tolerant design or architecture. A monitor is typically a piece of electronic equipment that keeps track of the operation of a system component and warns of trouble, or is part of a larger subsystem that automatically switches to an alternate component. A monitor is a safety design feature.

MORAL HAZARD

A moral hazard occurs when an individual insulated from risk may behave differently than if he were fully exposed to the risk. Moral hazard is a special case of information asymmetry, a situation in which one party in a transaction has more information than another. The party that is insulated from risk generally has more information about its actions and intentions than the party paying for the negative consequences of the risk. More broadly, moral hazard occurs when the party with more information about its actions or intentions has a tendency or incentive to behave inappropriately from the perspective of the party with less information. Moral hazard arises because an individual or institution does not take the full consequences and responsibilities of its doings, and therefore has a tendency to act less carefully than it alternately would, leaving another party to hold some responsibility for the consequences of those actions. For example, a person with insurance against automobile theft may be less cautious about locking his or her car, because the negative consequences of vehicle theft are (partially) the responsibility of the insurance company. Early usage of the term carried negative connotations, implying fraud or immoral behavior (usually on the part of an insured party).

MULTIPLE OCCURRING EVENT (MOE)

In FTA an MOE is a failure or fault that occurs in multiple places within the FT; it is a repeated event. This happens merely due to the system design and the FT logic. When this happens in an FT, the repeated events must be correctly reduced from the final Boolean equation in order to avoid adding the same event more than once in the probability calculation.

In one sense, an MOE can be thought of as a CCF because it is the same cause for more than one event in the FT. Some rules of thumb for MOEs are as follows:

- A CCF is always an MOE, but an MOE is not always a CCF.
- CCF typically applies to the loss of redundant items, but it can also apply to the loss of multiple items from a single source.
- If a CCF occurs in parallel redundant items, it eliminates the redundancy causing system loss.
- If a CCF occurs in several elements in series, then it is really an MOE as any one failure causes system loss.

Figure 2.54 shows the application of an MOE and CCF for an example system. In this example, the event "X1—Power Supply Fails" is the common fault that causes loss of components A1, A2, B, and C. Since CCF primarily applies to redundancy elimination, X1 is both a CCF and MOE for causing loss of A1

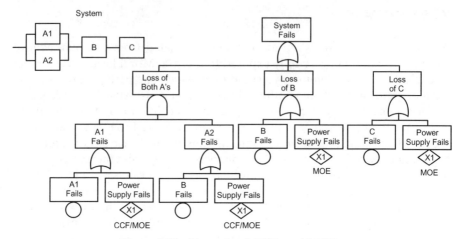

Figure 2.54 Example of CCFs and MOEs.

and A2. Because components B and C are in series and are not redundant, event X1 is considered an MOE in causing their failure.

See *Common Cause Failure (CCF)* and *Common Mode Failure (CMF)* for additional related information.

MULTIPLE-VERSION DISSIMILAR SOFTWARE

In multiple-version dissimilar software, a set of two or more computer programs are developed separately (and independently) to satisfy the same functional requirements. Errors specific to one of the versions should be detected by comparison of the outputs between the different versions. When multiple-version dissimilar software is used in redundant computer systems, the likelihood of the same errors in both systems is theoretically significantly reduced, thereby achieving a higher level of safety for the system. Multiple-version dissimilar software involves applying the dissimilar design concept to software.

See *Dissimilar Software* for additional related information.

MUNITION

A munition is an assembled ordnance item that contains explosive material(s) and is configured to accomplish its intended mission (MIL-STD-2105C, Hazard Assessment Tests for Non-Nuclear Munitions, July 23, 2003). The term munitions refers to all non-nuclear energetic devices, including bombs, missiles, torpedoes, mines pyrotechnics devices, demolition charges, rocket motors, and other devices that utilize energetic materials.

NEAR MISHAP

A near mishap is the occurrence of an unplanned event that did not result in injury, illness, or damage, but had the potential to do so.
  See *Near Miss* for additional related information.

NEAR MISS

A near miss is the occurrence of an unplanned event that did not result in injury, illness, or damage, but had the potential to do so. Only a fortunate break in the chain of events prevented an injury, fatality, or damage. Related terms are "incident," "close call," "near mishap" or "near collision." The term is often misunderstood and misused. An event is called a near miss to stress that not only did things go wrong, but that a catastrophe was barely missed. In the airline industry, if two airliners pass within a quarter mile of each other, this is by definition a "near miss" event. Some individuals feel it is a euphemistic term for a "near hit."
  A near miss could be viewed as the partial actualization of a hazard, resulting in an incident rather than a mishap. It could also be a situation where only part of a hazard occurs for various reasons, such as operator alertness and counteraction, thus preventing a mishap event. For example, a collision avoidance system intended to prevent a collision between two aircraft in the same airspace may have failed without any warning, but an alert pilot saw the situation and took countermeasures to prevent a collision.

NEED NOT

With regard to design requirements, the term "need not" indicates action that does not necessarily have to be performed. It is typically used in a requirement when the application of the requirement or procedure is optional.

NEGATIVE OBSTACLE

A negative obstacle refers to terrain below the horizontal plane; a terrain feature that presents a negative deflection relative to the horizontal plane. For a UGV, this can be safety-related as it prevents the UGVs continuation on its original path. Examples are depressions, canyons, creek beds, ditches, bomb craters, and so on.

NEGLIGIBLE HAZARD

A negligible hazard is a hazard that has a Category IV (Negligible) severity level, as defined by the hazard severity criteria in MIL-STD-882.
  See *Hazard Severity Levels* for additional information.

NET CENTRIC

Net centric is relating to or representing the attributes of net-centricity. Net-centricity is a robust, globally interconnected network environment (including infrastructure, systems, processes, and people) in which data are shared timely and seamlessly among users, applications, and platforms. Net-centricity enables substantially improved military situational awareness and significantly shortened decision-making cycles.

NET EXPLOSIVES WEIGHT (NEW)

NEW is an explosives metric based on explosives compounds that are equal to one pound of TNT. A compound may weigh 2 lb but have the blast effects of only 1 lb of TNT. It is then said to have a NEW of 1 lb. If the compound weighs 1 lb but has the blast effects of 2 lb of TNT, the NEW is considered as 2 lb NEW.

NIT

Nit is a unit of brightness, equal to one candela per square meter, commonly used to describe display brightness. Not having the correct amount of display brightness could be an HCF.

NOMINAL HAZARD ZONE (NHZ)

The NHZ describes the space within which the level of the direct, reflected, or scattered EM or laser radiation during normal operation exceeds the applicable MPE level. Exposure levels beyond the boundary of the NHZ are below the appropriate MPE level.

NOMINAL OCULAR HAZARD DISTANCE (NOHD)

NOHD is the distance along the axis of the laser beam beyond which the irradiance (W/cm^2) or radiant exposure (J/cm^2) is not expected to exceed the appropriate personnel exposure level, that is, the safe distance from the laser. The NOHD-O is the NOHD when viewing with optical aids.

NOISE

Noise consists of sounds or signals having a complex character with numerous separate frequency components extending over a wide range of frequencies and not generated to convey meaning or information. Excessive noise or high-volume noise are a personnel health hazard source. Noise can damage physiological and psychological health. Noise pollution can cause annoyance and aggression, hypertension, high stress levels, tinnitus, hearing loss, sleep disturbances, and other harmful effects. Chronic exposure to noise may cause noise-induced hearing loss. High noise levels can contribute to cardiovascular effects and exposure to moderately high levels during a single 8-h period causes a statistical rise in blood pressure of 5–10 points and an increase in stress and vasoconstriction leading to the increased blood pressure noted above, as well as to increased incidence of coronary artery disease.

NOISE POLLUTION

Noise pollution (or environmental noise) is displeasing human, animal, or machine-created sound that disrupts the activity or balance of human life. The source of most outdoor noise worldwide is transportation systems, including motor vehicle noise, aircraft noise, and rail noise. Poor urban planning may give rise to noise pollution, since side-by-side industrial and residential buildings can result in noise pollution in the residential area. Indoor and outdoor noise pollution sources include car alarms, emergency service sirens, mechanical equipment, fireworks, compressed air horns, grounds keeping equipment, barking dogs, appliances, lighting hum, audio entertainment systems, electric megaphones, and loud people.

NONDEVELOPMENTAL ITEM (NDI)

An NDI is an item that is used in a new system development program, but the item is not developed as part of the program. NDIs are provided to a program, are purchased by a program, for as-is use in a new system, as opposed to being designed and developed as part of the program. NDIs have already been developed for previous purposes and exist external to the program.

NDIs are typically developed for other programs under different design requirements than the requirements that may be in effect for the current program they are being used on. They were not developed under the program design requirements and are procured as a "black box" in the system design, which means that often very little is known about their internal workings and pedigree. For this reason, NDIs can present many problems for safety, particularly when used in SCFs or applications. NDIs are essentially the same as COTS items, except NDIs may not be commercially available, but are provided from another program.

Projects must understand that often the use of NDIs is cheaper only because standard development tasks have not been performed, such as system safety.

When system safety is performed, it may become obvious that the NDI is not the best alternative. Specifically, hazards must be identified, risks assessed, and the risk made acceptable regardless if the component/function is provided as NDI. The decision to use NDI or COTS items does not negate SSRs. A very important rule-of-thumb for NDI safety is that the use of NDIs does not eliminate the requirement for a safety analysis and risk assessment; COTS/NDI items must be evaluated for safety as an integral part of the entire system. COTS/NDI items are *not* exempt from the system safety process. The decision to use COTS/NDI items does not negate the need for SSRs and processes for COTS/NDI applications.

See *COTS* and *GFE* for additional related information.

NONCONFORMANCE

Nonconformance refers to the condition where a product or system is developed with one or more characteristics that do not conform to the development requirements and/or specifications. Nonconformance is a departure from, or a failure to conform to, design requirements or specifications. Nonconformance results in failure of an item to perform its required function as intended. An item may be a system, subsystem, product, software, human, or component. As applied in quality assurance, nonconformance falls into two categories: (a) discrepancies and (b) failures.

See *Deviation*, *Discrepancy*, and *Failure* for additional information.

NONFUNCTIONAL REQUIREMENT

A nonfunctional requirement is a requirement that specifies criteria that can be used to judge the operation of a system, rather than specific behaviors. This should be contrasted with functional requirements that define specific behavior or functions. In general, functional requirements define what a system is supposed to do whereas nonfunctional requirements define how a system is supposed to be. Nonfunctional requirements are often called qualities of a system. Other terms for nonfunctional requirements are "constraints," "quality attributes," "quality goals" and "quality of service requirements," and "nonbehavioral requirements." Nonfunctional requirements address the hidden areas of the system that are important but not always immediately obvious to the user. They do not deal with functionality; they tend to be fuzzy and global factors necessary for system success. They tend to be "ility"-type requirements, such as safety, reliability, maintainability, and scalability.

See *Requirement* and *Specification* for additional related information.

NONIONIZING RADIATION

Nonionizing radiation is EMR in the lower frequency end of the EM spectrum. Nonionizing EMR refers to the weaker EM waves that cannot alter the bonds of molecules. Nonionizing radiation is radiation that is not capable of stripping electrons from atoms in the media through which it passes. Examples include radio waves, microwaves, visible light, and UV radiation.

See *Electromagnetic Radiation (EMR)* for additional related information.

NON-LINE-OF-SIGHT (NLOS)

NLOS refers to a visual condition that exists when there is an obstruction between the viewer and the object being viewed. In RF communications, it is a condition that exists when there is an intervening object, such as dense vegetation, terrain, man-made structures, or the curvature of the Earth, between the transmitting and receiving antennas, and transmission and reception would be impeded. An intermediate ground, air, or space-based retransmission capability may be used to remedy this condition.

NORMAL DISTRIBUTION

The normal distribution is a symmetrical distribution of values or measures that approximate a bell-shaped curve, with as many cases above the mean as there are below the mean, according to a precise mathematical equation. A normal distribution is described by its mean (x) and standard deviation (σ).

Figure 2.55 is an example of the normal distribution curve, which always has the following characteristics:

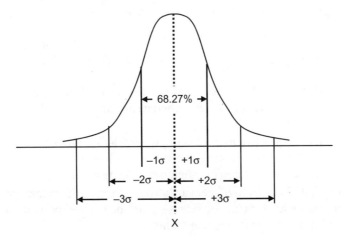

Figure 2.55 Normal distribution curve.

- 68.27% of the data points lie within +/− 1σ of the mean
- 95.45% of the data points lie within +/− 2σ of the mean
- 99.73% of the data points lie within +/− 3σ of the mean

NORMAL OPERATION

Normal operation (of a system) is system behavior which is in accordance with the documented requirements and expectations under all normally conceivable conditions. This is essentially operation where everything behaves properly and nothing goes wrong; no human errors, faults, failures, or other perturbations occur.

See *Abnormal Operation* for additional related information.

NOTE

A note is a procedure, practice, condition, and so on, that is essential to emphasize in a manual or procedures document. Typically, notes stress important information in operations, test, and maintenance manuals.

NUCLEAR WEAPON SAFETY

Nuclear weapon safety covers those safety aspects involved with a nuclear weapon system. A nuclear weapon system is a high-consequence system requiring an extensive SSP and Nuclear Safety Program. Nuclear safety must be planned in the conceptual phase, designed into components in the development phase, and continually examined throughout the test and operational phases of each device. The DoD has established four safety standards that are the basis for nuclear weapon system design and the safety rules governing nuclear weapon system operation. These standards require that, as a minimum, the system design shall incorporate the following positive safety measures:

1. There shall be positive measures to prevent nuclear weapons involved in accidents or incidents, or jettisoned weapons, from producing a nuclear yield.
2. There shall be positive measures to prevent deliberate prearming, arming, launching, firing, or releasing of nuclear weapons, except upon execution of emergency war orders or when directed by competent authority.
3. There shall be positive measures to prevent inadvertent prearming, arming, launching, firing, or releasing of nuclear weapons in all normal and credible abnormal environments.
4. There shall be positive measures to ensure adequate security of nuclear weapons.

Standard hazard analysis apply to nuclear weapon systems; however, because of the political and military consequences of an unauthorized or accidental nuclear or high-explosive detonation, additional analyses are necessary to demonstrate positive control of nuclear weapons in all probable environments. The following analyses, in whole or in part, are performed:

1. A quantitative analysis to assure that the probability of inadvertent nuclear detonation, inadvertent programmed launch, accidental motor ignition, inadvertent enabling, or inadvertent prearming meets the numerical requirements specified in applicable nuclear safety criteria documents.
2. An unauthorized launch analysis to define the time, tools, and equipment required to accomplish certain actions leading to unauthorized launch. The results of this analysis are used by the nuclear safety evaluation agency in determining which components require additional protection, either by design or procedural means.
3. A Nuclear Safety Cross-Check Analysis of software and certain firmware, which directly or indirectly controls or could be modified to control critical weapon functions. This analysis, by an independent contracting agency, must determine that the final version of software or firmware is free from programming, which could contribute to unauthorized, accidental, or inadvertent activation of critical system function.
4. A safety engineering analysis of all tasks in modification or test programs at operational sites. This analysis is specifically oriented toward identifying hazards to personnel and equipment in the work area and is in addition to the analysis of the safety impact of the change to the weapon system.

N-VERSION PROGRAMMING

N-version programming is a system design technique that involves producing two or more software components that provide the same function, but in different ways. This approach is intended to avoid sources of common errors between redundant components. N-version programming is also referred to as multi-version software, dissimilar software, or software diversity.

The degree of dissimilarity and hence the degree of safety protection is not usually measurable. In this approach, software modules for redundant subsystems are developed to the same design requirements using design diversity. The probability of loss of system function will decrease if the module results are compared during operation and adjustments are made when error differences are discovered. Software verification is the first line of defense against errors, while dissimilar software provides a second line of defense. For example, an aircraft flight control system that has two redundant computers might develop the software for one computer in the C language and the software for the second computer in the Ada language.

See *Design Diversity* and *Dissimilar Software* for additional related information.

OBSERVATION

An "observation" is a product of an audit; is the identification of a potential improvement. An observation is not a compliance issue and does not need to be addressed before approval. Typical categories of items resulting from an audit include: compliance, finding, observation, issue, and action.

See *Audit* for additional related information.

OFFGASSING

Offgassing is the same as outgassing. It is the slow release of a gas that was trapped, frozen, absorbed, or adsorbed in some material.

See *Outgassing* for additional related information.

OPEN ARCHITECTURE

Open architecture is a type of computer hardware or software architecture that allows users access to all or parts of the architecture without any proprietary constraints. Open architecture allows adding, upgrading, modifying, and swapping components. Typically, an open architecture publishes all or parts of its architecture that the developer or integrator wants to share. The business process involved with an open architecture may require some license agreements between entities sharing the architecture information.

Open architecture systems are systems that provide a varied combination of interoperability, portability, and open software standards. It can also mean systems configured to allow unrestricted access by people or other computers. Open architecture will most likely apply primarily to system backbones, such as the network structure, computer structure, and operating systems. Application hardware and software will plug into these backbone structures, but will themselves probably not be open. For example, the IBM PC hardware has an open architecture, whereas the Apple home computer has a closed architecture. The open architecture allows third-party vendors to develop hardware for the PC. Linux is software that is an open architecture operating system that allows users to modify it and make their own enhancements, whereas Microsoft Windows is not open architecture.

Open architecture can have both a positive and a negative impact on system safety and SwS. The open feature gives the safety analyst access to the design, which can then be evaluated for safety impact. However, the open feature also gives designers and users access to the design, which might make configuration control a safety concern, especially if software can be modified by users. Open source software (OSS) presents the same dangers as COTS software, with the exception that the source code is available for analysis.

OPEN SOURCE SOFTWARE (OSS)

OSS is software that is licensed to users with the following basic freedoms:

- To not have to pay a royalty or fee for the software
- To run the software for any purpose
- To study and modify the software
- To freely redistribute copies of either the original or modified software without royalty payments or other restrictions

OSS is free to procure; however, when considering the total cost of ownership, it is not entirely free. There is a cost associated with understanding and modifying the code, but it may provide the lowest cost for a particular situation. Extant OSS is considered to be COTS software for various reasons. Examples of OSS include the Linux operating system and the GNAT Ada compiler. OSS presents the same dangers as COTS software, with the exception that the source code is available for analysis.

See *Commercial Off-the-Shelf (COTS)* for additional related information.

OPEN SYSTEM

An open system is a system which continuously interacts with its environment. The system involves input, throughput, and output as it interacts with and reacts to its environment. The interaction can take the form of information, energy, or material transfers into or out of the system boundary, depending on the discipline which defines the concept. An open system should be contrasted with the concept of a closed system which does not exchange energy, matter, or information with its environment.

See *Closed System* for additional information.

OPERATING AND SUPPORT HAZARD ANALYSIS (O&SHA)

The O&SHA is an analysis technique for identifying hazards in system operational tasks. The O&SHA is an analysis technique for specifically assessing the safety of operations by integrally evaluating operational procedures, the system design, and the HSI interface. The scope of the O&SHA includes normal operation, test, installation, maintenance, repair, training, storage, handling, transportation, and emergency/rescue operations. Consideration is given to system design, operational design, hardware failure modes, human error, and task design. Human factors and HSI design considerations are a large factor in system operation, and therefore also in the O&SHA. The O&SHA is conducted during system development in order to affect both the design and operational procedures for future safe operations.

The purpose of the O&SHA is to ensure the safety of the system and personnel in the performance of system operation. Operational hazards can be introduced by the system design, procedure design, human error, and/or the environment. The overall O&SHA goal is to:

1. Provide safety focus from an operations and operational task viewpoint
2. Identify task or operational-oriented hazards caused by design, hardware failures, software errors, human error, timing, and so on.
3. Assess the operations mishap risk
4. Identify design SSRs to mitigate operational task hazards
5. Ensure all operational procedures are safe

The O&SHA is conducted during system development and is directed toward developing safe design and procedures to enhance safety during operation and maintenance. The O&SHA identifies the functions and procedures that could be hazardous to personnel or, through personnel errors, could create hazards to equipment, personnel, or both. Corrective action resulting from this analysis is usually in the form of design requirements and procedural inputs to operating, maintenance, and training manuals. Many of the procedural inputs from system safety are in the form of caution and warning notes.

The O&SHA is applicable to the analysis of all types of operations, procedures, tasks, and functions. It can be performed on draft procedural instructions or detailed instruction manuals. The O&SHA is specifically oriented toward the HA of tasks for system operation, maintenance, repair, test, and troubleshooting. The O&SHA evaluates the system design and operational procedures to identify hazards and to eliminate or mitigate operational task hazards. The O&SHA also provides insight into design changes that might adversely affect operational tasks and procedures. The O&SHA effort should start early enough during system development to provide inputs to the design, and prior to system test and operation. The O&SHA worksheet provides a format for entering the sequence of operations, procedures, tasks, and steps necessary for task accomplishment. The worksheet also provides a format for analyzing this sequence in a structured process that produces a consistent and logically reasoned evaluation of hazards and controls.

When performing an O&SHA, some considerations for hazard identification include the following:

1. Potentially hazardous system states under operator control
2. Operator hazards resulting from system design (hardware aging and wear, distractions, confusion factors, worker overload, operational tempo, exposed hot surfaces, environmental stimuli, etc.)
3. Operator hazards resulting from potential human error
4. The safety effect of concurrent tasks and/or procedures
5. Errors in procedures and instructions
6. Activities which occur under hazardous conditions, their time periods, and the actions required to minimize risk during these activities/time periods
7. Changes needed in functional or design requirements for system hardware/software, facilities, tooling, or support/test equipment to eliminate or control hazards or reduce associated risks
8. Requirements for safety devices and equipment, including personnel safety and life support equipment
9. Warnings, cautions, and special emergency procedures (e.g., egress, rescue, escape, render safe, EOD, backout, etc.), including those necessitated by failure of a computer software-controlled operation to produce the expected and required safe result or indication
10. Requirements for packaging, handling, storage, transportation, maintenance, and disposal of HAZMATs
11. Requirements for safety training and personnel certification
12. The safety effect of COTS items
13. Requirements for support equipment, tools, tool calibration, and so on
14. The safety effect of user interfaceUI design

The O&SHA is a detailed hazard analysis utilizing structure and rigor. It is desirable to perform the O&SHA using a specialized worksheet. Although the specific format of the analysis worksheet is not critical, as a minimum, the following basic information is required from the O&SHA:

- Tasks and task steps to be performed
- Tools, support equipment, and protective equipment involved
- UI complexity, criticality, modes, and so on
- Identified hazard, hazard effect, and hazard casual factors
- Recommended mitigating action (DSF, safety devices, warning devices, special procedures and training, caution and warning notes, etc.)
- Risk assessment (initial and final)

System: Operation:			Operating and Support Hazard Analysis				Analyst: Date:			
Task	Hazard No.	Hazard	Causes	Effects	IMRI	Recommended Action	FMRI	Comments	Status

Figure 2.56 Example O&SHA worksheet.

Figure 2.56 shows an example columnar format O&SHA worksheet which has proven to be useful and effective in many applications; it provides all of the information necessary from an O&SHA.

Note that in this analysis methodology, each and every procedural task is listed and analyzed. For this reason, not every entry in the O&SHA form will constitute a hazard, since not all tasks will be hazardous. This process documents that the O&SHA considered all tasks and identifies which tasks are hazardous and which are not.

O&SHA is an analysis technique that is important and essential to system safety. For more detailed information on the O&SHA technique, see Clifton A. Ericson II, *Hazard Analysis Techniques for System Safety* (2005), chapter 8.

OPERATIONAL ENVIRONMENT

An operational environment is the composite of the conditions, circumstances, and influences affecting a system during operation of the system.

OPERATIONAL READINESS REVIEW (ORR)

The ORR examines the system characteristics and the procedures used in the system or end product's operation to ensure that all system and support (flight and ground) hardware, software, personnel, procedures, and user documentation accurately reflect the deployed state of the system.

See *Critical Design Review (CDR)*, *Preliminary Design Review (PDR)*, and *Systems Engineering Technical Reviews (SETR)* for additional information.

OPERATIONAL RISK MANAGEMENT (ORM)

ORM is a decision-making tool to identify operational risks and determine the best course of action for any given situation. ORM is usually performed after system deployment during operational use. For example, an ORM might be performed by the military or a commercial air carrier prior to each flight or mission. This risk management process is designed to minimize risks in order to reduce mishaps, preserve assets, and safeguard the health and welfare of personnel.

Appropriate use of ORM increases both an organization's and individual's ability to accomplish their operational mission. Whether it is flying an airplane, commercial or military, in peacetime or combat, loading a truck with supplies, planning a joint service exercise, establishing a computer network, or driving home at the end of the day, the ORM process, when applied properly, can reduce risk effectively. Application of the ORM process ensures more consistent results, while ORM techniques and tools add rigor to the traditional approach to mission accomplishment, thereby directly strengthening our defense posture as well as nonmilitary applications.

ORM is simply a formalized way of thinking about operational hazards and risk. ORM involves a simple six-step process that identifies operational hazards and takes reasonable measures to reduce risk to personnel, equipment, and the mission. These steps include:

Step 1: Identify the Hazard

A hazard or threat is defined as any real or potential condition that can cause degradation, injury, illness, death, or damage to or loss of equipment, or property, or the environment. Experience, common sense, and specific analytical tools help identify risks.

Step 2: Assess the Risk

The assessment step is the application of quantitative and/or qualitative measures to determine the level of risk associated with specific hazards. This process defines the probability of occurrence and severity of an accident that could result from the hazards based upon the exposure of humans or assets to the hazards.

Step 3: Analyze the Risk Control Measures

Investigate specific strategies and tools that reduce, mitigate, or eliminate the risk. All risks have two components—(1) likelihood of occurrence, and (2) severity of the consequence, where the consequence involves exposure of people and equipment to the risk. Effective control measures reduce or eliminate at least one of these components of risk. The analysis must take into account the overall costs and benefits of remedial actions, providing alternative choices if possible.

Step 4: Make the Control Decisions

After identifying the appropriate decision maker, that decision maker must choose the best control or combination of controls, based on the analysis completed in step 3.

Step 5: Implement the Risk Controls

Management must formulate a plan for applying the controls that have been selected, and then must provide the time, materials, and personnel needed to put these measures in place.

Step 6: Supervise and Review the ORM Process

Once controls are in place, the process must be periodically reevaluated to ensure its effectiveness. Workers and managers at every level must carry out their respective roles to assure that the controls are maintained over time.

ORM must be a fully integrated part of planning and executing any operation, and routinely applied by management, not just a way of reacting when some unforeseen problem occurs. Careful determination of risks, along with the analysis and control of the hazards they create, results in a plan of action that anticipates difficulties that might arise under varying conditions and predetermines ways of dealing with these difficulties. Managers are responsible for the routine use of risk management at every level of activity, starting with the planning of that activity and continuing through its completion. The risk management process continues throughout the life cycle of the system, mission, or activity.

Four precepts govern all actions associated with ORM. These continuously employed principles are applicable before, during, and after all tasks and operations, by individuals at all levels of responsibility. The four ORM precepts include:

- Accept no unnecessary risk
- Make risk decisions at the appropriate level
- Accept risk when benefits outweigh the costs
- Integrate ORM into planning at all levels

ORM is very similar to the basic system safety process of identifying hazards, assessing risk, and mitigating risk to an acceptable level via design safety measures. It appears that ORM is being supplanted by a newer but similar tool called safety management system (SMS).

See *Safety Management System (SMS)* for additional related information.

OPERATIONAL SAFETY PRECEPT (OSP)

An OSP is a safety precept that that is directed at the operational phase. These precepts are operational safety rules that must be adhered to during system operation. One aspect of OSPs is that they may directly spawn the need for DSPs that will help realize the OSP via design methods. Although the development PM has no influence during system operation, the OSPs will provide guidance for many design features that will need to be implemented in order to meet the OSPs. An example OSP might be "The unmanned system shall be considered unsafe until a safe state can be verified."

See *Safety Precept*s for additional related information.

OPERATOR ERROR

Operator error is an inadvertent action by an operator, an operation that is not performed or is performed incorrectly. Operator errors generally have safety significance; they can eliminate, disable, or defeat safety features. They can also perform steps out of sequence or forget to perform steps, thereby creating hazards.

ORDNANCE

Ordnance is a term used for military material used in all kinds of combat weapons. Ordnance includes all the things that make up ground, ship, or aircraft armament. Ordnance items include items containing explosives, nuclear fission or fusion materials, and biological and chemical agents. This includes bombs and warheads; guided and ballistic missiles; artillery, mortar, rocket, and small arms ammunition; mines, torpedoes and depth charges, and demolition charges; pyrotechnics; clusters and dispensers; cartridge and propellant actuated devices; EEDs; clandestine and IEDs; and all similar or related items or components explosive in nature.

ORGANIZATION

An organization is a social arrangement which pursues collective goals, controls its own performance, and has a boundary separating it from its environment. An organization is a system in which people are the most numerous component and which is structured so as to create products and services of value to others. An organization generally consists of structure, functional purpose, operational policies, guidelines, and an inherent culture. By coordinated and planned cooperation of the elements, the organization is able to solve tasks (and build systems) that lie beyond the abilities of the single elements.

A system safety organization is a key component of an SSP; it is responsible for performing the necessary system safety tasks to design and build safety into a product or system design. The number of individuals in the organization depends upon many factors, such as project size, system complexity, system safety-criticality, and funding.

See *System Safety Program (SSP)* for additional related information.

OR GATE

An OR gate is a logic gate used in FTA. The OR gate logic states that a gate output occurs when any one, or more, of the gate inputs occur. See Figure 2.27 for an example of an OR gate.

See *Fault Tree Symbols* for additional related information.

ORIGINAL EQUIPMENT MANUFACTURER (OEM)

An OEM is the individual, activity, or organization that performs the physical fabrication processes that produce a deliverable part, product, or system. An OEM produces the part or product in-house. For example, the engines that Boeing might use on a commercial aircraft would be from an OEM such as General Electric or Rolls Royce.

OUTGASSING

Outgassing is the slow release of a gas that was trapped, frozen, absorbed, or adsorbed in some material. It can include sublimation and evaporation which are phase transitions of a substance into a gas, as well as desorption, seepage from cracks or internal volumes, and gaseous products of slow chemical reactions. Boiling is generally considered a separate phenomenon from outgassing because it occurs much more rapidly. Outgassing is sometimes called offgassing, particularly when in reference to air quality. Outgassing is also emanation of volatile materials under vacuum conditions resulting in a mass loss and/or material condensation on nearby surfaces. Outgassing is a concern in spacecraft as outgassing products can condense onto optical elements, thermal radiators, or solar cells, and obscure them. Outgassing can also affect air quality in a spacecraft.

OVERRIDE

An override is the forced bypassing of prerequisite checks on the operator-commanded execution of a function. Overrides always have potential safety consequence, and this should be considered in HAs.

OXIDIZER

An oxidizer is a substance that is capable of reacting with a fuel and gaining electrons in an oxidation reaction.

OZONE

Ozone (O_3) is a triatomic form of oxygen that is a bluish, irritating gas of pungent odor. It is an allotrope of oxygen that is much less stable than the diatomic allotrope (O_2). Ozone in the lower atmosphere is an air pollutant with harmful effects on the respiratory systems of animals and will burn sensitive plants; however, the ozone layer in the upper atmosphere is beneficial, preventing potentially damaging UV light from reaching the Earth's surface. Ozone is present in low concentrations throughout the Earth's atmosphere. Although ozone has many industrial and consumer applications, critical products like gaskets and O-rings may be attacked by ozone produced within compressed air systems. Fuel lines are often made from reinforced rubber tubing and may also be susceptible to attack, especially within engine compartments where low levels of ozone are produced from electrical equipment. Storing rubber products in close proximity to DC electric motors can accelerate the rate at which ozone cracking occurs. The commutator of a motor creates sparks which in turn produce ozone. Ozone is a potential hazard source for many different types of hazards.

PARADIGM

A paradigm is the set of fundamental beliefs, axioms, and assumptions that order and provide coherence to our perception of what is and how it works; a basic worldview; also, example cases and metaphors. A paradigm is the generally accepted perspective of a particular discipline at a given time. When a paradigm's philosophy changes, and becomes universally accepted, then a *paradigm shift* has taken place.

PARETO PRINCIPLE

The Pareto principle is a general observation, which states that for many events, roughly 80% of the effects come from 20% of the causes. This principle can be applied to many different fields, for example, 80% of a product's sales will typically come from 20% of the clients. Some software developers feel that by fixing the top 20% of the most reported bugs, 80% of the errors and crashes will be eliminated.

PART

A part is the lowest level of separately identifiable items in a system. A part is a hardware element that is not normally subject to further subdivision or disassembly without destruction of design use. Examples include resistor, IC, relay, connector, bolt, and gaskets. A part is one level in a system hierarchy, typically the lowest level. To some extent parts are synonymous with components, or parts may be combined together to form a component. Parts and components are combined together to create subassemblies or assemblies. A part is an entity that is a portion of a component or a subassembly.

In system safety, parts and components are of prime interest because it is often their unique failure modes, within unique system architectures, that provide the IM for certain hazards within a system design. FMEA and FTA typically deal with the system at the part or component level in order to determine the risk presented by a particular hazard. When an FTA is performed to determine the causal factors for a particular hazard or UE, the FTA is generally conducted to the part level. Failure rates can be obtained for parts, which can be used in the FTA to generate a quantitative result.

See *System Hierarchy* for additional related information.

PARTIAL DETONATION

With regard to explosives, this is the second most violent type of explosive event. In a partial detonation, some, but not all, of the energetic material reacts like a full detonation. An intense shock wave is formed; some of the case is broken into small fragments; a ground crater can be produced; adjacent metal plates can be damaged as in a detonation; and there will be blast overpressure damage to nearby structures. A partial detonation can also produce large case fragments as in a violent pressure rupture (brittle fracture). The amount of damage, relative to a full detonation, depends on the portion of material that detonates.

See *Explosive Event* for additional related information.

PARTICULAR RISK

Particular risk refers to risk associated with those events or influences that are outside the system(s) and item(s) concerned, but which may violate failure independence claims.

PARTICULAR RISK ASSESSMENT

A particular risk assessment examines those common events or influences that are outside the system(s) concerned but which may violate independence requirements. These particular risks may also influence several zones at the same time, whereas zonal safety analysis (ZSA) is restricted to each specific zone. Some of these risks may also be the subject of specific airworthiness requirements. Some of particular risks result from airworthiness regulations, while others arise from known external threats to the aircraft or systems.

Typical particular risks include, but are not limited to the following:

1. Fire
2. High energy devices (engines, motors, fans)
3. Leaking fluids (fuel, hydraulic, battery acid, water)
4. Hail, ice, snow
5. Bird strike
6. Tread separation from tire
7. Wheel rim release
8. Lightning
9. High-intensity radiated fields
10. Flailing Shafts
11. Bulkhead rupture

Having identified the appropriate risks with respect to the design under consideration, each risk should be the subject of a specific study to examine and document the simultaneous or cascading effect(s) of each risk. The objective is to ensure that any SR effects are either eliminated or the risk is shown to be acceptable. A particular risk assessment is required for aircraft airworthiness certification by the FAA; the process is documented in SAE/ARP-4761, Guidelines and Methods for Conducting the Safety Assessment Process on Civil Airborne Systems and Equipment, 1996.

PARTITIONING

Partitioning is the act of segregating the functions of a system into verifiably distinct, separate, and protected collections of functions.

PASCAL

Pascal is a computer programming language developed by Nicolas Wirth. It is known as a high-level type language or HOL. The Ada language borrowed a lot of concepts from Pascal.

PATCH

A patch is a modification to a computer module that is separately compiled and inserted into the machine code of a host or parent program. This avoids modifying the source code of the host or parent program. Consequently, the parent or host source code no longer corresponds to the combined object code.

Software patches present a significant concern to SwS because the patched "fix" is typically not evaluated for safety impact. There is no assurance that the modified object code is entirely correct or safe. In addition, there is no guarantee that the patch safety fits into the overall source code. In safety-critical applications, patches are typically prohibited.

PAYLOAD

A payload is an integrated assemblage of modules, subsystems, and so on, designed to perform a specified mission, such as a satellite carried in a spacecraft or a weapon aboard an aircraft.

PERCEPTION

Perception is the act of being aware of objects or data through any of the senses. In system design, it is often critical that the design take into account operator perception and all of the ramifications involved, particularly the effect of erroneous perception of information.

PERFORMANCE SHAPING FACTOR (PSF)

PSFs are the set of influences on the performance of an operating crew resulting from the human-related characteristics of the system, the crew, and the individual crew members. The characteristics include procedures, training, and other human factors aspects (e.g., MMIs in a control room, cockpit, alarm station, and operations center) of the working environment.

Risk analysts describe these influential factors as PSFs, and they are used to describe any factor that influences human performance. PSFs are essential to determining and understanding the root causes of errors (and accidents, as errors can be linked to accidents) and for improving human system effectiveness. PSFs are typically classified as external, internal, or team. External PSFs include factors such as lighting conditions and temperature. Internal PSFs include factors such as high stress, excessive fatigue, and deficiencies in knowledge and skill. Lastly, team PSFs include lack of communication, inappropriate task allocation, and excessive authority gradient. The importance of PSFs is not trivial and can significantly influence the performance of tasks. Research has demonstrated the relationship between PSFs and performance; PSFs can be used as a predictive measure of performance.

PERFORMANCE TESTING

Performance testing covers a broad range of engineering or functional evaluations where a material, product, system, or person is tested to determine and assess the final measurable performance characteristics. Performance testing is evaluating performance requirements, as opposed to detailed design requirements. Performance testing can refer to the assessment of human performance; for example, a behind-the-wheel driving test is a performance test of whether a person is able to perform the functions required of a competent driver of an automobile.

Software performance testing is used to determine the speed or effectiveness of a computer, network, software program, or device. This process can involve quantitative tests done in a lab, such as measuring the response time or the number of millions of instructions per second (MIPS) at which a system functions. Qualitative attributes such as reliability, scalability, and interoperability may also be evaluated. Performance testing is often done in conjunction with stress testing.

PERFORMANCE VALIDATION

In spaceflight development, performance validation is the determination by test, analysis, or inspection (or a combination of these) that the payload element can operate as intended in a particular mission; this includes being satisfied that the design of the payload or element has been qualified and that the particular item has been accepted as true to the design and ready for flight operations.

PETRI NET ANALYSIS (PNA)

PNA is a graphical and mathematical modeling tool. It consists of places, transitions, and arcs that connect them. Input arcs connect places with transitions, while output arcs start at a transition and end at a place. There are other types of arcs, for example, inhibitor arcs. Places can contain tokens; the current state of the modeled system (the marking) is given by the number (and type if the tokens are distinguishable) of tokens in each place. Transitions are active components. They model activities which can occur (the transition fires), thus changing the state of the system (the marking of the Petri net). Transitions are only allowed to fire if they are enabled, which means that all the preconditions for the activity must be fulfilled (there are enough tokens available in the input places). When the transition fires, it removes tokens from its input places and adds some at all of its output places. The number of tokens removed/added depends on the cardinality of each arc. The interactive firing of transitions in subsequent markings is called a token game.

PNA is an analysis technique for identifying hazards dealing with timing, state transitions, sequencing, and repair. PNA consists of drawing graphical Petri Net (PN) diagrams and analyzing these diagrams to locate and understand design problems. Models of system performance, dependability, and reliability can be developed using PN models. PNA is very useful for analyzing properties such as reachability, recoverability, deadlock, and fault tolerance. The biggest advantage of PNs, however, is that they can link hardware, software, and human elements in the system. PNA may be used to evaluate safety-critical behavior of control system software. In this situation, the system design and its control software is expressed as a timed PN. A subset of the PN states are designated as possible unsafe states. The PN is augmented with the conditions under which those states are unsafe. A PN reachability graph will then determine if those states can be reached during the software execution.

Figure 2.57 shows an example PN model with three transition states. In state 1, Place 1 has a token but Place 2 does not. Nothing can happen until Place 2 receives a token. In state 2, Place 2 receives a token. Now Transition D1 has both inputs fulfilled, so after delay D1 it fires. State 3 shows the final transition, whereby D1 has fired, it has removed the two input tokens (Places 1 and 2) and has given an output token to Place 3. Note that the "D1" text is removed after completion of the process.

PNs can be used to model an entire system, subsystem, or system components at a wide range of abstraction levels, from conceptual to detailed design. When a PN model has been developed for analysis of a particular abstraction, its mathematical representation can support automation of the major portions of the analysis. PNA is a good tool for modeling and understanding system operation. PNA is a limited HA tool because it only identifies system hazards dealing with timing and state change issues, and it does not identify root causes. PNA models quickly become large and complex; thus, it is more suitable to small systems or high-level system abstractions. PNA can easily become too large in size for understanding, unless the system model is simplified. PNA is

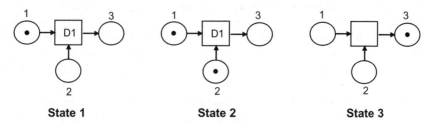

Figure 2.57 Example PN model with three transition states.

a tool for identifying a special class of hazards, such as those dealing with timing, state transitions, sequencing, and repair. PNA provides both a graphical and mathematical model. For system safety applications, PNA is not a general purpose HA tool and should only be used in situations to evaluate suspected timing, state transition, sequencing, and repair type hazards.

PNA is an analysis technique that is important and essential to system safety and reliability engineering. For more detailed information on the PNA technique, see Clifton A. Ericson II, *Hazard Analysis Techniques for System Safety* (2005), chapter 17.

PHYSICAL CONFIGURATION AUDIT (PCA)

A PCA is a formal audit that establishes the product baseline as reflected in an early production CI. The audit determines whether the system was built in accordance with the design package reviewed at the CDR. PCA is a physical examination to verify that the CI(s) "as built" conform to the technical documentation which defines the item. Approval of the CI product specification and satisfactory completion of this audit establishes the product baseline. PCA may be conducted on first full production or first LRIP item. The PCA is a CM activity and is conducted following procedures established in the CM plan. As part of a PCA, system safety typically makes a presentation summarizing the safety of the functional configuration. This may include a safety assessment of the DSFs in the system design, verification of the safety design requirements, and the current level of mishap risk which the design presents.

PITOT TUBE

A pitot tube is a short right-angled tube with an open end that is used with a manometer to measure the velocity of fluids or air by means of pressure differentials. Pitot tubes are used in aircraft to measure aircraft speed. Pitot tube errors feeding into automated flight control systems have caused aircraft crashes; therefore, they are a potential hazard source.

PIVOTAL EVENT (PE)

In the scenario approach to mishap analysis, the scenario consists of a series of events that begins with the IE, which is then followed by one or more critical PEs until a final consequence has resulted. Typically, barriers are designed into the system to protect against IEs, and the PEs are failure or success of these barriers. PEs are the intermediary events between an IE and the final resulting mishap. These are the failure/success events of the design safety methods established to prevent the IE from resulting in a mishap. If a PE works successfully, it stops the accident scenario and is referred to as a mitigating event. If a PE fails to work, then the accident scenario is allowed to progress and is referred to as an aggravating event. The PE is primarily used in ETA and CCA. Refer back to Figure 2.42 which shows the PE concept.

See *Event Tree Analysis (ETA)* and *Initiating Event (IE)* Analysis for additional related information.

PLASTICIZER

A plasticizer is a liquid that is mixed with a polymer to change its mechanical properties and make the polymer more rubbery and less stiff or brittle. Plasticizers are sometimes used in energetic materials.

POINT OF CONTACT (POC)

A POC is the single individual that is the most knowledgeable on a subject or is responsible for something, such as a task and a product. The POC is typically the technical authority on a subject, that is, SME.

POSITIVE CONTROL

Positive control refers to ensuring that command and control functions are performed in a safe manner. This safety assurance is primarily required in UMSs. Positive control requires the completion of the following functions:

1. A valid command is issued
2. The command authority is authenticated
3. The command is received
4. The command is acknowledged back to sender
5. The command is verified for correctness by sender and acknowledged
6. The command is executed

See *Valid Command* for additional related information.

POSITIVE MEASURE

Positive measure refers to the positive safety action taken to control or mitigate hazards. Measures include design features, safety equipment, procedures, safety rules, or other controls including physical security and coded control systems, used collectively or individually, to enhance safety and to reduce the likelihood, severity, or consequences of a mishap. Positive measures do not provide absolute assurance against a mishap or unauthorized act, but they do provide an improved system design that presents acceptable risk assurance for continuing safe operation of the system.

POTENTIAL HAZARD

This term is meant to indicate that a hazard is a potential; however, by definition, a hazard is a potential condition. Thus, the term adds unnecessary redundancy. It is recommended that this term not be used, as it is redundant (a hazard is a potential event) and causes some confusion.

See *Hazard* for additional related information.

POWER

Power is the rate at which energy is emitted, transferred, or received. Power can be produced and utilized in many different forms, such as electrical power, hydraulic power, and nuclear power. Power in these different forms is what makes a system work. System safety is concerned about power because most power sources are also hazard sources. One of the first steps in an HA is to identify all of the power sources and then consider all the possible ways these energy sources can contribute to different hazards. For example, electrical power can be a hazard source in hazards such as electrocution, failure to be present when needed, out of tolerance, applied prematurely or inadvertently.

PRELIMINARY DESIGN REVIEW (PDR)

The PDR demonstrates that the preliminary design meets all system requirements with acceptable risk and within the cost and schedule constraints and establishes the basis for proceeding with detailed design. It will show that the correct design options have been selected, interfaces have been identified, and verification methods have been described. The PDR precedes the CDR. As a rough rule of thumb, about 15% of production drawings are released by PDR. This rule is anecdotal and is only a guidance relating to an "average" defense hardware program.

System safety should be involved in the PDR, typically making a presentation summarizing the safety effort to date, the DSFs in the system design, and the current level of mishap risk which the design presents. System safety provides, as a minimum, an FHA, PHA and SAR for this review.

See *Preliminary Design Review (PDR)*, *Systems Engineering Technical Reviews (SETR)*, and *System Requirements Review (SRR)* for additional information.

PRELIMINARY HAZARD ANALYSIS (PHA)

PHA analysis is both a system safety analysis type and technique for identifying for the early identification of hazards and potential mishap risk. The PHA provides a methodology for identifying and collating hazards in the system and establishing the initial SSRs for design from preliminary and limited design information. The intent of the PHA is to affect the DFS as early as possible in the development process. The PHA normally does not continue beyond the SSHA time frame.

The purpose of the PHA is to analyze identified hazards, usually provided by the PHL, and to identify previously unrecognized hazards early in the system development. The PHA is performed at the preliminary design level, as its name implies, when detailed design information is usually not available. In addition, the PHA identifies HCFs, consequences, and relative risk associated with the initial design concept. The PHA provides a mechanism for identifying initial design SSRs that assist in designing-in safety early in the design process. The PHA also identifies SCFs and TLMs that that provide a safety focus during the design process.

The PHA is applicable to the analysis of all types of systems, facilities, operations, and functions; the PHA can be performed on a unit, subsystem, system, or an integrated set of systems. The PHA is generally based on preliminary or baseline design concepts and is usually generated early in the system development process, in order to influence design and mishap risk decisions as the design is developed into detail. The PHA is easily and quickly performed; it is comparatively inexpensive, yet provides meaningful results. A PHA worksheet provides rigor for focusing for the identification and evaluation of hazards; it provides a methodical analysis technique that identifies majority of system hazards and provides an indication of system risk. Commercial software is available to assist in the PHA process.

The PHA is probably the most commonly performed HA technique. In most cases, the PHA identifies the majority of the system hazards. The remaining hazards are usually uncovered when subsequent HAs are generated and more design details are available. Subsequent HAs refine the hazard cause–effect relationship, and uncover previously unidentified hazards and refine the design safety requirements. In the PHA, hazards are identified based on hypothesis or theory rather than experiment; they are derived by logic, without observed facts. A hazard checklist is a basic part of the analysis, which is compared against the system design, which should include hardware equipment, energy sources, system functions, software functions, and so on.

Output from the PHA includes identified and suspected hazards, HCFs, the resulting mishap effect, mishap risk, SCFs, and TLMs. PHA output also includes design methods and SSRs established to eliminate and/or mitigate identified hazards. It is important to identify SCFs because these are the areas that generally affect design safety and that are usually involved in major system hazards. Since the PHA is initiated very early in the design phase, the data available to the analyst may be incomplete and informal (i.e., preliminary). Therefore, the analysis process should be structured to permit continual revision and updating as the conceptual approach is modified and refined. When the subsystem design details are complete enough to allow the analyst to begin the SSHA in detail, the PHA is generally terminated.

When performing a PHA, the following factors should be considered, as a minimum:

1. Hazardous components (e.g., energy sources, fuels, propellants, explosives, and pressure systems).
2. SCFs.
3. Subsystem interfaces (e.g., signals, voltages, timing, human interaction, and hardware).
4. System compatibility constraints (e.g., material compatibility, EMI, transient current, and ionizing radiation).
5. Environmental constraints (e.g., drop, shock, extreme temperatures, noise and health hazards, fire, electrostatic discharge, lightning, X-ray, EMR, and laser radiation).
6. Undesired states (e.g., inadvertent activation, fire/explosive initiation and propagation, and failure to safe).
7. Malfunctions to the system, subsystems, or computing system.
8. Software errors (e.g., programming errors, programming omissions, and logic errors).
9. Operating, test, maintenance, and emergency procedures.
10. Human error (e.g., operator functions, tasks, and requirements).
11. Crash and survival safety (e.g., egress, rescue, and salvage).
12. Life-cycle support (e.g., demilitarization/disposal, EOD, surveillance, handling, transportation, and storage).
13. Facilities, support equipment, and training.
14. Safety equipment and safeguards (e.g., interlocks, system redundancy, failsafe design considerations, subsystem protection, fire suppression systems, PPE, and warning labels).
15. Protective clothing, equipment, or devices.
16. Training and certification pertaining to safe operation and maintenance of the system.
17. System phases (e.g., test, manufacture, operations, maintenance, transportation, storage, and disposal).

The PHA is a detailed hazard analysis utilizing structure and rigor. It is desirable to perform the PHA using a specialized worksheet. Although the format of the PHA analysis worksheet is not critical, it is important that, as a minimum, the PHA generate the following information:

System: Subsystem/Function:			Preliminary Hazard Analysis				Analyst: Date:		
No.	Hazard	Causes	Effects	Mode	IMRI	Recommended Action	FMRI	Comments	Status

Figure 2.58 Example PHA worksheet.

- System hazards
- Hazard effects (e.g., actions, outcomes, mishaps)
- HCFs (or potential causal factor areas)
- Mishap risk assessment (before and after DSFs are implemented)
- SCFs and TLMs
- Recommendations for eliminating or mitigating the hazards

Figure 2.58 shows an example columnar format PHA worksheet. This particular worksheet format has proven to be useful and effective in many applications, and it provides all of the information necessary from a PHA.

PHA is an HA technique that is important and essential to system safety. For more detailed information on the PHA technique, see Clifton A. Ericson II, *Hazard Analysis Techniques for System Safety* (2005), chapter 5.

See *Safety Analysis Technique* and *Safety Analysis Type* for additional information.

PRELIMINARY HAZARD LIST (PHL)

PHL analysis is both a system safety analysis Type and Technique for identifying and listing potential hazards and mishaps that may exist in a system. The PHL is performed during conceptual or preliminary design, and is the starting point for all subsequent HAs. Once a hazard is identified in the PHL, the hazard will be used to launch in-depth HAs and evaluations, as more system design details become available. The PHL is a means for management to focus on hazardous areas that may require more resources to eliminate the hazard or control risk to an acceptable level. Every hazard identified on the PHL will be analyzed with more detailed analysis techniques. The primary output from the PHL is a list of hazards and the hazard sources that spawn them. It is also necessary and beneficial to collect and record additional information, such as the prime HCFs (e.g., hardware failure, software error, and human error), the major mishap category for the hazard (e.g., fire, IL, and physical injury), and any safety-critical factors that will be useful for subsequent analysis (e.g., SC function and SC hardware item).

The primary purpose of the PHL is to identify and list potential system hazards. A secondary purpose of the PHL is to identify safety-critical parameters and mishap categories. The PHL analysis is usually performed very early in the design development process and prior to performing any other HA. The PHL is used as a management tool to allocate resources to particularly hazardous areas within the design, and it becomes the foundation for all other subsequent HAs performed on the program. Follow on HAs will evaluate these hazards in greater detail as the design detail progresses. The intent of the PHL is to affect the DFS as early as possible in the development program. The PHL is can be applied to any type of system at the conceptual or preliminary stage of development. The PHL can be performed on a subsystem, a single system, or an integrated set of systems. The PHL is generally based on preliminary design concepts and is usually performed early in the development process, sometimes during the proposal phase or immediately after contract award in order to influence design and mishap risk decisions as the design is formulated and developed.

The PHL technique is similar to a brainstorming session, whereby hazards are postulated and collated in a list. This list is then the starting point for subsequent HAs, which will validate the hazard and begin the process of identifying causal factors, risk, and mitigation methods. Generating a PHL is a prerequisite to performing any other type of HA. The use of this technique is highly recommended. It is the starting point for more detailed hazard analysis and safety tasks, and it is easily performed. Typically, in performing the PHL analysis, the analyst compares the design knowledge and information to hazard checklists. This allows the analyst to visualize or postulate possible hazards. For example, if the analyst discovers that the system design will be using jet fuel, he then compares jet fuel to a hazard checklist. From the hazard checklist it will be obvious that jet fuel is an HE, and that a jet fuel fire/explosion is a potential mishap with many different ignition sources presenting many different hazards. Hazard checklists provide a common source for readily recognizing hazards. Since no single checklist is ever really adequate in itself, it becomes necessary to develop and utilize several different checklists. Utilizing several checklists may generate some repetition, but will also result in improved coverage of HEs. Remember that a checklist should never be considered a complete and final list, but merely a mechanism or catalyst for stimulating hazard recognition.

It is desirable to perform the PHL analysis using a worksheet. The worksheet will help to add rigor to the analysis, record the process and data, and help support justification for the identified hazards. The format of the analysis worksheet is not critical, and typically columnar-type worksheets are utilized.

| Preliminary Hazard List Analysis |||||
| System Hazard Source Type: |||||
No.	Hazard Source Item	Hazard	Hazard Effects	Comments

Figure 2.59 Example PHL worksheet.

The following basic information should be obtained from the PHL analysis worksheet:

- Hazard sources in the system
- Postulated hazards resulting from the hazard sources
- TLM categories stemming from identified hazards
- Recommendations (such as safety requirements/guidelines that can be applied)

An example PHL worksheet for system safety usage is shown in Figure 2.59.

In this PHL worksheet, the second column contains a list of system hazard source items, from which hazards can easily be recognized. For example, by listing all of the system functions, hazards can be postulated by answering the questions "what if the function fails to occur" or "what if the function occurs inadvertently." When using a worksheet, the PHL process provides rigor for focusing on hazards, and it provides an indication of where major system hazards and mishap risk will exist.

PHL is an HA technique that is important and essential to system safety. For more detailed information on the PHL technique, see Clifton A. Ericson II, *Hazard Analysis Techniques for System Safety* (2005), chapter 4.

See *Safety Analysis Technique* and *Safety Analysis Type* for additional information.

PRESCRIPTIVE SAFETY

Prescriptive safety is similar to compliance-based safety. Safety is implemented via safety requirements and design guidance provisions. In the case of prescribed requirements, there is usually no room for different design options, as the prescribed requirement must be implemented as stated. In the case of guidance standards, the guidance is a little more general, allowing for implementation via different design options. Prescriptive-based safety is an approach to safety based on laws, regulations, guidance standards, and so on. In this approach, the contractor or managing authority must follow prescriptive design requirements or guidance and show evidence of compliance. Prescribed design measures must be implemented into the system design. In the case of prescribed requirements, there is usually no room for different design options, as the prescribed requirement must be implemented as stated. In the case of guidance standards, the guidance is a little more general, allowing for implementation via different design options.

This safety approach is effective and useful; however, its major drawback is that it does not ensure that all system potential mishap risk is reduced to the lowest effective value practical. Prescriptive safety provides a known basic level of safety, but it may fall short if further safety features are necessary for a particular system. Prescribed safety requirements only address a known set of hazards. Even for a compliance-based safety program, it is still necessary to perform HA to ensure that all hazards have been identified and the risk mitigated to the lowest level practical.

See *Compliance-Based Safety* and *Risk-Based Safety* for additional related information.

PRIMARY EXPLOSIVE

Primary explosives are sensitive materials, such as lead azide or lead styphnate, which are used to initiate detonation. They are used in primers or detonators, are sensitive to heat, impact, or friction, and undergo a rapid reaction upon initiation.

PRIME CONTRACTOR

Prime contractor is the contractor having responsibility for design control and/or delivery of a system/equipment such as aircraft, engines, ships, tanks, vehicles, guns and missiles, ground communications and electronics systems, and test equipment.

PRINCIPAL FOR SAFETY (PFS)

The PFS is the Government safety manager or lead safety engineer (on DoD programs) who represents the PM in all safety matters. The PFS is tasked with the responsibility of ensuring the SSP is defined in the System Safety Management Plan (SSMP), and that the SSP is implemented and operating effectively. The PFS is the primary safety focal point for the program, and the safety interface to the contractor's SSP. The PFS speaks for the PM in all safety matters; therefore, it is important that the PFS has qualifications commensurate with the size, complexity, and safety-criticality of the system involved. It is the responsibility of the PM to obtain a qualified PFS and approve the appointment. The PFS is responsible for ensuring the SSP is properly implemented and adequately addresses the appropriate safety requirements, guidelines, regulations, and policies.

The PFS is the single POC for SR matters on the government side of a program. The PFS is designated in writing by the PM and has the authority to speak for him on SR matters. The PFS is the technical authority regarding matters of system safety. This position may be referred to as the system safety lead on some programs. The PFS has the specific responsibility for executing an SSP. The PFS may manage one or more safety engineers, depending upon the size of the program. The PFS keeps the PM informed of the status of the SSP; notifies the PM of any impact to mishap risk, schedule, cost, or technical performance affecting system safety tasks; makes recommendations on the content of the contractor's SSP; and coordinates with the procuring contracting officer on all system safety contractual matters.

The PFS is responsible for supporting the contracting effort and ensuring system safety involvement in preparation of the RFP, SOW, and SOOs for the contractors bidding on the SSP. The PFS is responsible for ensuring the proper system safety language and tasks are in the contract. It may be necessary for the PFS to perform a preliminary hazard-risk assessment of the system to judge the SSP size, criticality, and required safety tasks. In preparing for the safety contracting effort, there are many aspects of system safety that the PFS must consider and include in the contract. The PFS should be intimately familiar and knowledgeable with the safety contract.

The PFS is responsible for developing and maintaining the SSMP that documents an SSP specifically tailored for the integrated system under development. It is imperative that the PFS develops a thorough SSMP that will guide the contractor in establishing an effective SSP that meets the government's objectives. For programs involving more than one contractor, the SSMP provides an integrated safety approach. The PFS is responsible for working with the contractor on developing an SSPP and providing final approval for the SSPP. It is the responsibility of the PFS to develop and implement an effective SwS approach for the program. This approach should be documented in the RFP, SSMP, and Software Safety Program Plan (SwSPP), as it will also provide guidance for the contractor's SwSP. The PFS must coordinate with other program disciplines to ensure they are part of the SwS process. The PFS is also responsible for ensuring the appropriate SwS tools are in place, such as the software criticality risk tables and the generic SwS requirements checklist.

The PFS is responsible for planning and designing a thorough hazard identification approach, which should be well defined and described in the RFP, SSMP, and SSPP. Analysis techniques to be used should be thoroughly described so there will be no misunderstandings during the conduct of the analysis. The descriptions should include purpose, methodology, level of detail, format, and resulting output. The PFS is responsible for establishing the SSWG, the SSWG charter, and conducting regularly scheduled SSWG meetings. The PFS is also responsible for ensuring that SSWG meeting minutes are documented and all SSWG action items are resolved on a timely basis. The PFS serves as the chairman, or co-chair, of the SSWG.

The basic tasks and responsibilities of the PFS include, but are not limited to, the following:

1. Defining and executing the SSP
2. Represent the PM in all safety matters
3. Serve as the focal point for the SSP
4. Establish safety policies and guidelines for the program
5. Prepare the contractual safety documents (SOW, SOO, and RFP)
6. Ensure the SSP has adequate resources, including funding, qualified staff, tools, and training
7. Prepare the SSMP and ensure it is properly implemented
8. Prepare the SSWG Charter
9. Convene and chair (or co-chair) regularly scheduled SSWG meetings
10. Maintain meeting minutes file for all meetings and SSWGs
11. Ensure compliance with applicable regulations, policies, guidelines, requirements, and laws
12. Establish the technical hazard identification and risk management approach
13. Establish the HRI matrix for the program
14. Establish and implement the program hazard risk acceptance process (RAP)
15. Schedule and meet SSP tasks and activities
16. Maintain an action item list and ensure all action items are closed on a timely basis
17. Establish, review, and approve system safety design requirements
18. Approve and monitor the contractor SSP and SSPP
19. Review and approve the contractor safety products
20. Maintain the appropriate SSP interfaces
21. Support the required design reviews
22. Ensure SSP audits are periodically performed
23. Initiate contact with and support the Weapons Systems Safety Explosives Review Board (WSESRB)
24. Initiate contact with and support the LSRB
25. Initiate contact with and support the range safety officer
26. Support the flight clearance process for system safety

27. Support the CSI process performed by logistics engineering
28. Serve as a member of the configuration control board to evaluate safety changes
29. Ensure SAR requirements are met
30. Ensure an HTS is established and maintained
31. Ensure the risk for all hazards is accepted and signed for by the appropriate risk acceptance authorities

Overall, the PFS is responsible for the entire SSP from inception through implementation, execution, and final closure. Although the PFS is primarily responsible for the government side of the safety program, he does have indirect responsibility for ensuring the contractor's safety program is adequate and on course.

PRIORITY AND GATE

A Priority AND gate is a logic gate used in FTA where the gate logic states that gate output occurs only if all of the inputs occur together, and the inputs must occur in a specified sequence. For example, input 1 must occur before input 2, for a two input gate. See Figure 2.27 for an example of a priority AND gate.

See *Fault Tree Symbols* for additional related information.

PROBABILISTIC RISK ASSESSMENT (PRA)

PRA is a comprehensive, structured, and logical analysis method for identifying and quantitatively evaluating risk in a complex technological system. The objective of a PRA is to obtain a quantitative risk assessment of a hazard or potential mishap scenario. FTA is one of the primary tools available for conducting a PRA.

PROBABILITY OF FAILURE ON DEMAND (PFD)

PFD is the probability of a system failing to perform its design function upon demand for its operation. It is typically associated with a demand for operation arising from a potentially hazardous condition. This system parameter degrades over time due to equipment failure rates. PFD equals 1 minus safety availability.

PROBABILITY OF LOSS OF AIRCRAFT (PLOA)

PLOA is a metric used primarily in reliability engineering to assess overall system level reliability (or unreliability) of an air vehicle. It represents an estimate of the probability of loss or significant damage to the air vehicle under analysis over a given period of time. While it is usually expressed per hour, it can be calculated for any period of exposure. It should be noted that PLOA is also a metric used by system safety to represent the system level of safety (or un-safety) for loss of an aircraft. Under the reliability aspect, PLOA typically means "loss of function," that is, a failed or lost function affecting air vehicle system reliability. All subsystem functions that keep the air vehicle flying must perform properly under the reliability definition. System reliability is allocated to subsystems using the PLOA metric. Each major subsystem is assigned a probability of loss of function that it must meet in order to meet the desired system reliability when all subsystems are rolled up to a top PLOA number. Under the system safety aspect, PLOA includes many other possible causal factors in addition to loss of function. For example, PLOA can result from events such as fire, collision, human error, and CFIT, just to name a few.

PLOA calculations are commonly generated using FTA tools, which mathematically roll up basic event failure probabilities, through logic gates, to a single top event. In addition to providing a program level metric, the process of generating PLOA helps identify design weakness, SPFs, and redundancy imbalance. PLOA is an effective metric for any stage of design development but provides the most benefit when used to identify problems during initial system architectural studies. The PLOA generation approach and analysis process is often more important than the metric itself. FTA provides the ability to graphically describe functional interdependencies, while mathematically calculating system failure probabilities. FTA is an essential tool for finding SPFs and CCF problems within subsystem architecture.

Figure 2.60 provides an example of the aircraft subsystems that are typically included in PLOA, probability of loss of control (PLOC), and probability of loss of mission (PLOM) analyses.

PROBABILITY OF LOSS OF CONTROL (PLOC)

PLOC is the estimated probability of degraded aircraft control due to failure of any system or part. PLOC includes only those systems and components necessary to maintain aerodynamic stability including speed, pitch, yaw, and roll. Therefore, components such as landing gear would not be included. As well as an indicator of system reliability, PLOC is used to assess the ability to effectively command the air vehicle to a safe termination point should the need arise during an emergency. PLOC is a good SOF metric and can be used to assess airworthiness before flight test.

See *Probability of Loss of Aircraft (PLOA)* for additional related information.

PLOM	PLOA	PLOC	Subsystem	Examples
✓	✓	✓	Processors	Computers, Data Buses, etc.
✓	✓	✓	Software	Aircraft control software
✓	✓	✓	Sensors	Air Data, Transducers, etc.
✓	✓	✓	Communication	VHF, UHF, etc.
✓	✓	✓	Navigation	GPS, INS, radar, etc.
✓	✓	✓	Flight Surface Actuation	Actuators, Servos, Valves, etc.
✓	✓	✓	Electrical Power	Generators, Batteries, Power Busses, Wires, etc.
✓	✓	✓	Hydraulic Power	Pumps, Valves, Filter, Pipes, etc.
✓	✓	✓	Fuel	Tanks, Valves, Lines, etc.
✓	✓	✓	Propulsion	Engine, FADEC, Oil Pumps, etc.
✓	✓	✓	Displays	Aircraft displays or ground operator displays
✓	✓	✓	Data Link	Digital data link to unmanned aircraft
✓	✓		Human	Pilot, Ground Operators for UAVs
✓	✓		Environment	Rain, Ice, Lightning, Temp, Vibration, Fire, EMI, etc.
✓	✓		Operations	Exceeding envelope and margins
✓	✓		Landing Systems	Landing Gear, Brakes, etc.
✓	✓		Erroneous Commands	The inadvertent execution of flight control commands
✓	✓		Flight Termination	Flight terminations systems
✓	✓		Weapon Systems	Delivery, Stores Mgt., etc.
✓			Radar	Non-Navigational
✓			Reduced Redundancy	Loss of one redundant element
✓			Mission Unique	Tail Hook, Refuel, Lasers, etc.
✓			Lighting	Landing, Navigation, Anticollision, etc.

Figure 2.60 PLOA, PLOC, and PLOM breakdown.

PROBABILITY OF LOSS OF MISSION (PLOM)

PLOM is the estimated probability of mission failure due to failure of any system or part. PLOM is the broader term, and is inclusive of PLOA and PLOC.

See *Probability of Loss of Aircraft (PLOA)* for additional related information.

PROCESS

A process is a particular course of action intended to achieve a desired goal or result. Typically, a process becomes a standardized set of steps that are performed in the process. Since a process performs a desired task, the erroneous performance of a process step, or the failure to perform a process step when needed, may result in serious safety consequences. Process-related hazards can result from many different or combined causal factors, such as hardware failures, hardware tolerance errors, system timing errors, software errors, human errors, sneak circuits, and environmental factors. The criticality of the process typically determines the safety vulnerability involved.

PRODUCTION READINESS REVIEW (PRR)

The PRR determines the readiness of the system developers to efficiently produce the required number of systems. It ensures that the production plans; fabrication, assembly, and integration enabling products, and personnel are in place and are ready to begin production.

See *Critical Design Review (CDR)*, *Preliminary Design Review (PDR)*, and *Systems Engineering Technical Reviews (SETR)* for additional information.

PRODUCT SAFETY

Product safety is the process applied to develop and build a safe product. For many types of products, this is achieved by ensuring the product design meets government standards and requirements for that family of products (i.e., compliance-based or prescriptive safety). When the requirements are met, it is assumed the product is safe. A more effective product safety program would be to combine the prescriptive safety requirements with an SSP that implements the system safety process.

PROGRAM

A program is an organized and directed effort that uses resources to achieve desired objectives. For example, a development program is an organization of people and resources for producing a product or system. A computer program is a set of software instructions that achieves a desired result when executed in a computer.

PROGRAMMABLE ELECTRONIC SYSTEM (PES)

A PES is an electronic computer system that does not utilize mechanical parts for decision making. It involves a computer or a PLD that can be programmed to perform the desired system functions. For example, a PES may be a controller unit for a numerical control milling machine. Quite often a PES is used as a safety system in a larger system, such as a safety monitoring device for a chemical process plant.

PES systems can perform a multiplicity of functions, depending on the computing power and speed of the computer and the number of inputs and outputs that the CPU can accept. They can perform these functions rapidly. They allow the system's functions and parameters to be changed rapidly and cheaply, by altering the program. A conventional relay-based electromechanical system capable of performing the same functions would be complex, bulky, and changes to the systems functions would involve physical changes to wiring and components. PES devices are typically certified to functional safety standards, such as:

- IEC 61508, Functional Safety of Electrical/Electronic/Programmable Electronic Safety-Related Systems, 1999.
- IEC 61511, Functional Safety of Safety Instrumented Systems for the Process Industry Sector.

See *Electrical/Electronic/Programmable Electronic System (E/E/PES)* for additional related information.

PROGRAMMABLE LOGIC CONTROLLER (PLC)

A PLC is a digital computer used for automation of electromechanical processes, such as control of machinery on factory assembly lines, amusement rides, or lighting fixtures. Unlike general-purpose computers, the PLC is designed for multiple inputs and output arrangements, extended temperature ranges, immunity to electrical noise, and resistance to vibration and impact. Programs to control machine operation are typically stored in battery-backed or nonvolatile memory. A PLC is an example of a real-time system since output results must be produced in response to input conditions within a bounded time, otherwise unintended operation will result.

A PLC is a computer-based controller that is event driven and can handle logic level signals only: it receives signals and processes them in real time, for example, level and temperature gauges in a chemical plant. A computer-based programmable controller (PC) is program driven: it operates a program of instructions to direct a machine to perform a task and receives feedback. The PC does not usually process feedback signals in real time; that is, it processes them when it gets round to them, which may depend on where it is in the program. However, a PC may be able to operate in real time. In practice, a PLC and PC differ only in complexity and the terms are often used synonymously. PLCs were developed to replace relay logic systems and are often programmed in "ladder logic," which strongly resembles a schematic diagram of relay logic. PLCs can also be programmed in a variety of ways, from ladder logic to more traditional programming languages such as BASIC and C. Another method is State Logic, a very high-level programming language designed to program PLCs based on state transition diagrams.

The functionality of the PLC has evolved over the years to include sequential relay control, motion control, process control, distributed control systems, and networking. The data handling, storage, processing power, and communication capabilities of some modern PLCs are approximately equivalent to desktop computers. A major advantage of PLCs is that it is typically armored for severe conditions, such as dust, moisture, heat, cold. Also, PLCs have the facility for extensive input/output (I/O) arrangements for connecting the PLC to sensors and actuators.

PROGRAMMABLE LOGIC DEVICE (PLD)

A PLD is an electronic component used to build reconfigurable digital circuits. Unlike a logic gate, which has a fixed function, a PLD has an undefined function at the time of manufacture. Before the PLD can be used in a circuit, it must be programmed or reconfigured. PLDs are arrays of logic based on rewritable memory technology. Design changes can be quickly implemented by simply reprogramming the device. PLDs have logic gate capacity in the hundreds of gates range. A PLD is a combination of a logic device and a memory device. The memory is used to store the pattern that was given to the chip during programming. A PLD programming language is used to write the intended device function into code, which is then written to the PLD. PLDs consist of a broad array of devices and systems such as a PLC, PLD, Programmable Array Logic (PAL), Field Programmable Gate Array (FPGA), ASIC, System-on-Chip (SOC), and Complex PLD (CPLD) that can be utilized for digital logic implementation and control.

PROGRAMMATIC ENVIRONMENTAL, SAFETY, AND HEALTH EVALUATION (PESHE)

The DoD 5000 series system acquisition instructions and directives lay out an acquisition strategy for complete integration of safety into an Environment, Safety and Occupational Health (ESOH) program. The PESHE documents ESOH risk and the PMs' strategy to comply with ESOH requirements for each of the ESOH areas. Because the PESHE is a special required program document, it is not intended to supersede or replace other management plans, analyses, or program documents. Rather, the PESHE summarizes the results of other analyses that individually deal with ESOH issues.

An ESOH program is divided into the following technical risk management areas:

1. System safety
2. Occupational health
3. Environmental compliance (environmental safety)
4. HAZMAT management
5. Pollution prevention
6. Explosives safety

The PESHE document is the means for communicating ESOH risk and status to the SSWG, PM, and acquisition executives. The PM uses the PESHE to identify and manage ESOH hazards, and to determine how to best meet ESOH regulatory requirements. The PM shall keep the PESHE updated over the system life cycle. As prescribed by DoD 5000 systems acquisition policies, the following are the minimum required data elements for a PESHE:

- Strategy/plan for integrating ESOH considerations into the systems engineering process
- Identification of ESOH responsibilities
- Approach to identify ESOH risks, to prevent the risks, and to implement controls for managing those ESOH risks where they cannot be avoided
- Identification and status of ESOH risks, including approval authority for residual ESOH risks
- Method for tracking progress in the management and mitigation of ESOH risks
- Method for measuring the effectiveness of ESOH risk controls
- Schedule for completing NEPA/Executive Order (E.O.) 12114 documentation, including document approval by the appropriate decision authority
- Identification of HAZMATs used in the system and the plan for their demilitarization/disposal, as well as the remainder of the system
- Critical issues demanding PM attention
- Identification of open hazards that are high or medium

The PM typically prepares a PESHE document early in the program life cycle and updates it during major system milestones. The PM uses the PESHE to identify and manage ESOH hazards, and to determine how to best meet ESOH regulatory requirements. The PESHE documents HAZMATs used in the system and contains a plan for the system's demilitarization and disposal.

PROGRAMMATIC SAFETY PRECEPT (PSP)

A PSP is a safety precept that is directed specifically at organizational goals, tasks, policy, standards, and/or processes that will help implement safety into the system development process. These precepts offer safety guidance, for the PM, which will effect mishap risk reduction through an effective and well-planned safety program. When the PSPs are closely followed, the culture and tasks are set in place for an efficient and successful SSP. An example PSP might be "The program shall ensure that COTS software is assessed for safety as a component in the system environment and not as an external isolated element."

See *Safety Precept*s for additional related information.

PROGRAM SAFETY ENGINEER (PSE)

The PSE is responsible for the technical tasks relating to the SSP. This primarily involves hazard identification and resolution, risk assessment, hazard tracking and closure, and safety ocumentation. On the government side of a program, the PSE typically reports to the PFS), while on the contractor side of the program, the PSE reports to the Program Safety Manager (PSM).

PSE responsibilities include, but are not limited to:

- Performing HAs
- Performing mishap risk assessments
- Preparing design SSRs for hazard mitigation
- Preparing SARs
- Implementing and maintaining an HTS
- Coordinating system safety matters with interfacing activities
- Performing data collection and recording of system safety documents
- Providing verification that all hazards are successfully mitigated

PROGRAM SAFETY MANAGER (PSM)

The PSM is the person responsible for the SSP on the contractor side of a program. The PSM represents the PM on all safety matters. The PSM is effectively the equivalent to the government PFS, as they share similar roles and responsibilities.

See *Principal for Safety (PFS)* for additional related information.

PROPELLANT

A propellant is a substance or mixture of substances used for propelling projectiles and missiles, or to generate gases for powering auxiliary devices. When ignited, propellants burn at a controlled rate to produce quantities of gas capable of performing work, but in their application they are required not to undergo a deflagration-to-detonation transition (DDT).

PULSED LASER

A pulsed laser is a laser that delivers its energy in the form of a single pulse or a train of pulses when the pulse duration is <0.25 s.

PYROPHORIC

A pyrophoric substance will ignite spontaneously in air. Examples are iron sulfide and many reactive metals including uranium, when powdered or sliced thinly. Pyrophoric materials are often water reactive as well and will ignite when they contact water or humid air. They can be handled safely in atmospheres of argon or (with a few exceptions) nitrogen. Most pyrophoric fires should be extinguished with a Class D fire extinguisher for burning metals.

PYROTECHNIC

A pyrotechnic is a substance or mixture of substances which, when ignited, undergoes an energetic chemical reaction at a controlled rate intended to produce, on demand and in various combinations, specific time delays or quantities of heat, noise, smoke, light, or IR. Examples include illuminants, smoke, delays, decoys, flares, and incendiaries.

QUALIFICATION

Qualification occurs when an item has been demonstrated to function within performance specifications. This is generally achieved via a qualification test that verifies requirements have been met. The test is also intended to uncover deficiencies in design and the method of manufacture. Sometimes the test is under simulated conditions more severe than those expected from handling, storage, and operations. The test is often designed to exceed design safety margins or to introduce unrealistic modes of failure.

See *Design Qualification Test* for additional related information.

QUALIFIED EXPLOSIVE

Qualified explosives are A&E items that possess properties judged to make it safe and suitable, primarily from a safety point of view, for consideration of use in a particular role. The materials/item has been approved for use in munitions development, product improvement, or other programs leading to eventual service application.

QUALITATIVE SAFETY ANALYSIS

In system safety, a qualitative analysis is a nonmathematical process that reviews all factors affecting mishap risk against a predetermined set of parameters. Qualitative analysis involves the use of qualitative criterion in the analysis and provides a qualitative result. Typically, this approach uses categories to separate different parameters, with qualitative definitions that establish the ranges for each category. Engineering judgments are made as to which category something might fit into. This approach has the characteristic of being subjective, but it allows more generalization and is therefore less restricting.

Arbitrary categories have been established in MIL-STD-882 that provides qualitative measures for the most reasonable likelihood of occurrence of a mishap and for the outcome severity of a mishap. For example, if the safety analyst assesses that an event will occur frequently, it is assigned an index level "A," or if it occurs occasionally, it is given an index level "C." This qualitative index value is then used in qualitative risk calculations and assessments.

System safety typically applies the qualitative risk characterization method because for a large system with many hazards, it can become cost-prohibitive to quantitatively model, analyze, and predict the risk of each and every hazard. In addition, low risk hazards do not require the refinement provided by quantitative analysis. It may be necessary to conduct a quantitative analysis only on a select few high consequence hazards. Experience over the years has proven that qualitative methods are very effective, and in most cases provide decision-making capability comparable to quantitative analysis. Qualitative risk characterization provides a very practical and effective approach when cost and time are concerns, and/or when there is very little supporting data available. The key to developing a qualitative risk characterization approach is by carefully defining severity and mishap probability categories.

QUANTITATIVE SAFETY ANALYSIS

In system safety, a quantitative analysis involves the use of numerical or quantitative data in the analysis, and provides a quantitative result. This approach has the characteristic of possibly being more objective and more accurate. It should be noted, however, that quantitative results can be biased by the validity and accuracy of the input numbers. For this reason, quantitative results should *not* be viewed as an exact number, but as an estimate with a range of variability depending upon the goodness quality of the data.

In a quantitative analysis, mathematical theories and models are used to calculate mishap risk factors. It is important to recognize that models are the analyst's viewpoint of a system and not the actual system itself. Do not ever confuse mathematical model results with reality. A probability guarantees nothing; it is an estimate from a model that provides relative information for decision making.

Quantitative risk characterization provides a useful approach when greater accuracy is required for decision making. Occasionally, a numerical design requirement must be met, and the only way to provide evidence that it is satisfied is through quantitative analysis. PRA is a quantitative analysis that estimates the probability factor of mishap risk. For high consequence systems, it is often necessary to conduct a PRA to determine all of the causal factors for a given mishap, and their total probability of causing the mishap to occur.

Scientific theory teaches that when something can be measured (quantitatively), more is known about it, and therefore, numerical results provide more value. This is generally true; however, the strict use of quantitative methods must be tempered by utility and model accuracy. Qualitative judgments can provide useful results while at the same time involving less time and expense. In a risk assessment, precise numerical accuracy is not always necessary. Mishap risks are not easily estimated using probability and statistics when the HCFs are not yet well understood (such as in preliminary design). Qualitative measures provide a useful and valid judgment at much less expense than quantitative measures, and they can be obtained much earlier in the system development life cycle. It makes sense to first evaluate all identified hazards qualitatively, and then, for high-risk hazards, to conduct a quantitative analysis for more precise knowledge.

QUANTITY-DISTANCE (QD)

QD is the quantity of explosives material and the distance of separation that will provide defined types of protection should the explosives detonate. The intent is to prevent sympathetic detonation of nearby explosives and personnel protection. QD requirements are defined in DoD 4145.26-M, the Contractors Safety Manual for Ammunition and Explosives. QD criteria represent physical limits which cannot be breached without incurring explosives mishap risks. QD applies the cardinal principle of explosives safety which is to expose the least amount of people to the least amount of explosives for the least amount of time. QD experts (explosives safety specialists) determine the risk of exposure by examining blast, fragment, and thermal hazards of a given amount of explosive using the NEW and applying accepted protection principles to determine levels of hazard a person is exposed to at a given distance from the explosives. The quantity (amount of NEW) and a given distance from the explosive equals the QD. There are factors that can be used to provide some protection from the effects of the detonation of explosives; however, the effects of a blast wave from an explosion cannot be mitigated. Only distance from the blast will provide protection from the blast wave of an explosion.

RADIATION HAZARD (RADHAZ)

RADHAZ is a term used for EMR hazards. RADHAZ analysis is the computation of distance-areas in which EMR energy hazards to personnel, equipment, and explosives will exist from EMR sources. RADHAZ safety concerns include:

- HERP
- HERO
- HIRF

See *Electromagnetic Radiation (EMR)* for additional related information.

RADIO FREQUENCY IDENTIFICATION (RFID)

RFID is an automatic identification method using devices called RFID tags or transponders. An RFID tag is an object that can be attached to or incorporated into a product, animal, or person for the purpose of identification using radio waves. Chip-based RFID tags contain silicon chips and antennas. Passive tags require no internal power source, whereas active tags require a power source.

RADIO FREQUENCY IMPROVISED EXPLOSIVE DEVICE (RFIED)

An RFIED is an IED that can be initiated via an RF device. It is typically used by terrorists as it is homemade, as opposed to being built by a manufacturer.

RADIO FREQUENCY (RF) RADIATION

RF radiation is EMR in the RF range of the EM spectrum. RF radiation is a primary hazard source in a system, and is therefore of concern to system safety.
See *Electromagnetic Radiation (EMR)* for additional related information.

RADON

Radon is an invisible, radioactive atomic gas that results from the radioactive decay of radium, which may be found in rock formations beneath buildings or in certain building materials themselves. It is colorless, odorless, and tasteless. It is one of the densest substances that remains a gas under normal conditions and is considered to be a health hazard due to its radioactivity. Radon is formed as part of the normal radioactive decay chain of uranium. Radon is responsible for the majority of the mean public exposure to ionizing radiation. It is often the single largest contributor to an individual's background radiation dose, and is the most variable from location to location. Radon gas from natural sources can accumulate in buildings, especially in confined areas such as attics, and basements. It can also be found in some spring waters and hot springs. Epidemiological evidence shows a clear link between breathing high concentrations of radon and incidence of lung cancer; radon is considered a significant contaminant that affects IAQ worldwide.

RANGE SAFETY

During the development or modification of a system, it is often necessary to test the system on a dedicated test range. Since each test range is individual and unique, there are range safety requirements specific to each range. Overall, range safety is the responsibility of the range safety officer (RSO) at the host range. Range safety consists of facility and support equipment safety, as well as the safety built into the system being tested. The RSO is not trying to make the system safe—only the range and its collateral area. The RSO reviews all of the safety data on every system used at the range and examines and evaluates all interfaces between and among the systems being used at the range. Each of the system's SARs is reviewed individually to assure that that system by itself is safe. Then, each of the interfaces will be examined to determine that no hazards are introduced when all these systems are integrated into one test system/range system. It is the responsibility of the PFS or the PSM to make contact with the RSO and plan and coordinate all required activities.

RAMS

RAMS is an acronym meaning a combination of the reliability, availability, maintainability, and safety disciplines.

RANDOM ACCESS MEMORY (RAM)

RAM is a form of computer data storage. RAM takes the form of ICs that allow stored data to be accessed in any order (i.e., at random). The word random thus refers to the fact that any piece of data can be returned in a constant time, regardless of its physical location and whether or not it is related to the previous piece of data. RAM is a volatile type of memory, where the information is lost after the power is switched off. By contrast, storage devices such as magnetic discs and optical discs rely on the physical movement of the recording medium or a reading head. In these devices, the movement takes longer than data transfer, and the retrieval time varies based on the physical location of the next item.

READ-ONLY MEMORY (ROM)

ROM is a form of storage media used in computers and other electronic devices. Because data stored in ROM cannot be modified, it is mainly used as firmware (software that is very closely tied to specific hardware and unlikely to require frequent updates). In its strictest sense, ROM refers only to mask ROM (the oldest type of solid state ROM), which is fabricated with the desired data permanently stored in it, and thus can never be modified. However, more modern types such as EPROM and flash EEPROM can be erased and reprogrammed multiple times; they are still described as "read-only" because the reprogramming process is generally infrequent, comparatively slow, and often does not permit random access writes to individual memory locations.

RED STRIPE

When an aircraft is grounded for safety reasons, the action is referred to as an aircraft Red Stripe in the NAVAIR branch of the U.S. Navy.

REDUNDANCY

In system design, redundancy is the duplication of critical components of a system with the intention of increasing reliability of the system, usually in the case of a backup or fail-safe design. In many safety-critical systems, such as aircraft fly-by-wire and hydraulic systems, some parts of the control system may involve triplex redundancy in order to ensure maximum reliability (and safety). In a triply redundant system, the system has three subcomponents, all three of which must fail before the system fails. Since each one rarely fails, and the subcomponents are expected to fail independently, the probability of all three failing is calculated to be extremely small. Redundancy may also be known by the terms "majority voting systems" or "voting logic." Redundancy does increase design complexity.

There are many forms of redundancy, such as:

- Hardware redundancy
- Information redundancy, such as Error detection and correction methods
- Time redundancy, including transient fault detection methods such as Alternate Logic
- Software redundancy, such as N-version programming

System safety often utilizes redundancy as a safety design feature or mechanism to enhance safety. The probability of two items failing is much smaller than for a single item, thus reducing the potential risk. It should be noted, however, that sometimes redundancy for reliability actually can increase the safety risk. For example, two explosive initiator devices might be used in a system design to make initiation more reliable; however, now the probability of inadvertent initiation becomes twice as likely.

Figure 2.61 shows the redundancy concept using RBDs. In this example, component A has a reliability of 0.9. In order to improve system reliability, a

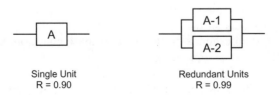

Figure 2.61 Example of redundancy.

redundant design is adopted, where two identical components are implemented in parallel. The redundant design provides a system reliability of 0.99 (using R = 0.9 for each unit). It should be noted that FTA can also be utilized to calculate system reliability.

See *Reliability* and *Reliability Block Diagram (RBD)* for additional related information.

REENGINEERING

Reengineering is the process of examining and altering an existing system to reconstitute it in a new form. This may include reverse engineering, restructuring (transforming a system from one representation to another at the same level of abstraction), re-documenting (analyzing a system and producing user or support documentation), forward engineering (using software products derived from an existing system, together with new requirements, to produce a new system), retargeting (transforming a system to install it on a different target system), and translation (transforming source code from one language to another, or from one version of a language to another).

See *Reverse Engineering* for additional related information.

REGRESSION TESTING

Regression testing is any type of software testing that seeks to uncover software errors by retesting a modified program. The intent of regression testing is to provide a general assurance that no additional errors were introduced in the process of fixing discovered problems. Regression testing is commonly used to efficiently test the system by systematically selecting the appropriate minimum suite of tests needed to adequately cover the affected change. Typically, regression testing involves rerunning previously run tests and checking whether previously fixed faults have reemerged or if new faults have emerged.

When software is modified after undergoing development testing, it would be ideal to repeat all of the original tests to ensure that no new or old errors arise from the changes. This, however, can become costly and time consuming, so the compromise is to perform a limited subset of the original tests to check-out the modifications; this subset of repeated tests is known as regression testing. Experience has shown that as software is fixed, emergence of new errors and/or the reemergence of old errors is quite common. Some common causal factors for these errors include:

- A fix for a problem in one software area inadvertently causes a software bug in another area.

- A fix for a problem resolves the narrow case where it was first observed, but not the general cases that are possible.
- A fix is lost through poor revision control.
- A fix is implemented incorrectly.
- A programmer may be knowledgeable in the software area being fixed, but not in other areas of the software which could be impacted.

In most software development situations, it is considered good practice that when a bug is located and fixed, a test that exposes the bug is recorded and regularly retested after subsequent changes to the program. Although this may be done through manual testing procedures using programming techniques, it is often done using automated testing tools. Such a test suite contains software tools that allow the testing environment to execute all the regression test cases automatically; some projects even set up automated systems to automatically rerun all regression tests at specified intervals and report any failures. As part of the SwS process, the software system safety (SwSS) analyst should ensure that the proposed suite of regression tests is adequate to cover all safety situations and concerns.

REFACTORING

In software engineering, refactoring a source code module refers to modifying the code without changing its external behavior, and is sometimes informally referred to as "cleaning it up." Code refactoring is any change to a computer program which improves its readability or simplifies its structure without changing its results. When refactoring, it is a good idea to have test fixtures in place, which can validate that the refactoring does not change the behavior of the software. Automatic unit testing helps ensure that refactoring does not make the code stop working or change the functionality.

Refactoring neither fixes bugs nor adds new functionality. Rather it is designed to improve the understandability of the code or change its structure and design, and remove dead code, to make it easier for human maintenance in the future. In particular, adding new behavior to a program might be difficult with the program's given structure, so a developer might refactor it first to make it easy, and then add the new behavior. An example of a refactoring is to change a variable name into something more meaningful, such as from "IR" to "interestRate." A more complex refactoring is to change the code within an IF block into a subroutine. The key insight in refactoring is to intentionally clean up code separately from adding new functionality, using a known catalogue of common useful refactoring methods, and then separately testing the code (knowing that any behavioral changes indicate a bug). The process is explicitly intending to improve an existing design without altering its intent or behavior.

In extreme programming and other agile methodologies, refactoring is an integral part of the software development cycle: developers alternate between adding new tests and functionality and refactoring the code to improve its internal consistency and clarity.

RELATIONSHIP

In the most general sense, a relationship is an interaction between the elements of a system and the particular parameters involved.

RELIABILITY

Reliability, for engineering purposes, is defined as "the probability that a device will perform its intended function, without failure, during a specified period of time under stated conditions." Reliability relies heavily on statistics, probability theory, and reliability theory. Reliability theory is the foundation of reliability engineering. The function of reliability engineering is to develop the reliability requirements for the product, establish an adequate reliability program, and perform appropriate analyses and tasks to ensure the product will meet its requirements. These tasks are managed by a reliability engineer following a reliability program plan.

Reliability can be viewed in several ways:

- The capacity of a device or system to perform as designed and as intended
- The idea that something is fit for a purpose with respect to a time period
- The resistance to failure of a device or system
- The ability of a device or system to perform a required function under stated conditions for a specified period of time
- The probability that a functional unit will perform its required function for a specified interval under stated conditions

In reliability theory R is the probability of successful operation, and Q is the probability of unsuccessful operation (i.e., failure). Q is typically represented as P for probability of failure. The following formulas are used extensively in reliability and FTA:

$$R + Q = 1$$
$$R = P_S = e^{-\lambda T}$$
$$Q = P_F = 1 - e^{-\lambda T}$$
$$\lambda = 1/\text{MTBF}$$

where λ is the failure rate, T is the exposure time, and MTBF is the mean time between failures.

System safety and reliability assist each other most of the time: they share common goals and methodologies. Both are dependent upon design architecture and component failure rates for effectiveness. In order to make a system safe and reliable, the design process utilizes two primary factors: design architecture and component failure rates. To be reliable (and safe) a system must not fail within a specified period of time. This is mainly achieved by making the components reliable enough to meet that time period. If the component is not reliable enough then architectural means, such as redundancy, are used to improve the reliability (and safety). The ability of something to "fail well," that is, fail without catastrophic consequences, falls under the purview of system safety rather than reliability.

RELIABILITY BLOCK DIAGRAM (RBD)

An RBD is a reliability model of a product or system. RBDs are used to understand and predict system reliability. RBDs provide a graphical means of evaluating the relationships between different parts of a system and calculating the system reliability from the model. Figure 2.62 shows RBD models for a series system and a parallel system. These basic models are the building blocks for larger more complex systems.

RBDs are valuable to assess relative differences in design alternatives. RBDs are a valuable asset for performing HA and FTA of a system.

RELIABILITY GROWTH

Reliability growth is the improvement in reliability that results when design, material, or part deficiencies are revealed by testing and are eliminated or mitigated through corrective action. Reliability growth can be measured over time, and it can be predicted via various reliability growth models that are available.

REMOTE CONTROL

Remote control is a mode of operation of a UMS wherein the human operator, without benefit of video or other sensory feedback, directly controls

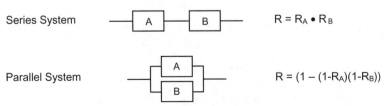

Figure 2.62 RBD models of series and parallel systems.

the actuators of the UMS on a continuous basis, from off the vehicle and via a tethered or radio-linked control device using visual LOS cues. In this mode, the UMS takes no initiative and relies on continuous or nearly continuous input from the user.

REMOTELY GUIDED

A UMS requiring continuous operator input for mission performance is considered remotely guided. The control input may originate from any source outside of the UMS itself. This mode includes remote control and tele-operation.

REMOTELY OPERATED VEHICLE (ROV)

An ROV is a tethered underwater robot. An ROV is sometimes called a remotely operated underwater vehicle to distinguish it from remote control vehicles operating on land or in the air. ROVs are unoccupied, highly maneuverable, and operated by a person aboard a ship. They are linked to the ship by a tether (umbilical cable), a group of cables that carry electrical power, video, and data signals back and forth between the operator and the vehicle. High power applications will often use hydraulics in addition to electrical cabling. Most ROVs are equipped with at least a video camera and lights. Additional equipment is commonly added to expand the vehicle's capabilities, such as sonar, magnetometers, still cameras, manipulators or cutting arms, water samplers, and instruments that measure water clarity, light penetration, and temperature. ROVs are typically used in deepwater industries such as offshore oil and hydrocarbon extraction.

REMOTE TERMINAL UNIT (RTU)

An RTU converts electrical signals from equipment to digital values that can easily be transmitted across computer networks. By converting digital commands to electrical signals and sending these electrical signals out to equipment, the RTU can control equipment, such as opening or closing a switch or a valve, or setting the speed of a pump. The RTU can also convert equipment electrical signals to digital format and transmit them back to the control system for feedback, such as the open/closed status from a switch or a valve, or measurements such as pressure, flow, voltage, or current.

REPAIR

Repair is corrective maintenance action performed on an item as a result of a failure, so as to restore the item back to operational capability. Repair typically improves the item's failure rate.

REPAIRABLE ITEM

A repairable item is a durable item which, when unusable (unserviceable), can be economically restored to a serviceable condition through regular repair procedures.

REPAIR TIME

Repair time is the time spent replacing, repairing, or adjusting all items suspected to have been the cause of the malfunction, except those subsequently shown by interim test of the system not to have been the cause. Repair time generally affects availability.

REQUESTS FOR DEVIATION/WAIVER

A request for deviation/waiver is a request to deviate from the contractual requirements. Each request for deviation/waiver must be evaluated to determine the hazards and the risk of the proposed deviation from or waiver of a requirement, or a deviation from a specified method or process on the existing system. The change in the risk involved in accepting the deviation or waiver must be identified. When the level of safety of the system will be reduced by deviation from or waiver of the requirement, method, or process, the managing authority MA must be so notified.

REQUIREMENT

In engineering, a design requirement is a statement that specifies or describes what the design must do, or what the design must include, in order for an item to meet its objectives. It is an essential condition, state, or form that the design must satisfy. A single requirement is an incremental design definition. A set of design requirements establishes the framework for designing and building a system. For example, one requirement in a series of requirements for a weapon system might be "The Safe and Arm device shall have a window whereby the actual safe/arm state can visually be viewed by the operator." Note that this requirement implies that there must be another design requirement specifying the use of a safe and arm device.

A requirement is a singular documented need of what a particular product or service should be or perform. It is most commonly used in a formal sense in systems engineering or software engineering. It is a statement that identifies a necessary attribute, capability, characteristic, or quality of a system in order for it to have value and utility to a user. An interface requirement is a requirement that specifies the necessary attributes between two interfacing system elements, such as systems, subsystems, and software modules. In the systems engineering approach to system development, sets of requirements are used as inputs into the design stages of product development. Requirements are also an important input into the verification process, since tests should trace back to specific requirements. Requirements show what elements and functions are necessary for the particular project.

There are two basic types of requirements, function and nonfunctional. A functional requirement is a description of what a system must do. This type of requirement specifies something that the delivered system must be able to perform or accomplish. Nonfunctional requirements specify something about the system itself, and how well it performs its functions. Such requirements are often called performance requirements. Examples of such requirements include safety, usability, availability, reliability, supportability, testability, maintainability, and so on. These types of requirements should be defined in a way that is verifiably measurable and unambiguous.

A collection of requirements define the characteristics or features of the desired system. A good list of requirements generally avoids saying how the system should implement the requirements, leaving such decisions to the system designer. Describing how the system should be implemented may be known as implementation bias or "solution engineering." However, implementation constraints on the solution may be validly expressed by the future owner if desired.

Requirements typically fall into the following categories:

- Functional requirements describe the functionality that the system is to execute; for example, formatting some text or modulating a signal. They are also known as capabilities.
- Nonfunctional requirements are the ones that act to constrain the solution. Nonfunctional requirements are known as quality requirements or "ility" requirements (e.g., reliability).
- Constraint requirements are the ones that act to constrain the solution. No matter how the problem is solved, the constraint requirements must be adhered to.

Nonfunctional requirements can be further classified according to whether they are usability requirements, look and feel requirements, humanity requirements, performance requirements, maintainability requirements, operational requirements, safety requirements, reliability requirements, or one of many other types of requirements.

In software engineering only functional requirements can be directly implemented in software. The nonfunctional requirements are controlled by other aspects of the system. For example, in a computer system reliability is related to hardware failure rates, and performance is controlled by CPU and memory. Nonfunctional requirements can in some cases be decomposed into functional requirements for software. For example, a system-level nonfunctional safety requirement can be decomposed into one or more functional requirements. In addition, a nonfunctional requirement may be converted into a process requirement when the requirement is not easily measurable. For example, a system level maintainability requirement may be decomposed into restrictions on software constructs or limits on lines or code. Design requirements have many different aspects, which are summarized in Table 2.16.

Requirements are usually written as a means for communication between the different stakeholders. This means that the requirements should be easy to understand both for normal users and for developers. One common way to document a requirement is stating what the system shall do, for example: "The contractor shall deliver the product no later than xyz date." Use cases are another way to document requirements.

Writing good design requirements is an art and a skill. Table 2.17 lists some typical characteristics of good requirements.

TABLE 2.16 Various Aspects of Requirements

Aspect	Description
Defines the design	Functional—things the product must do, an action that the product must take
	Nonfunctional—properties or qualities that the product must have
	Constraint—a limitation that constrains the design options
Defines the major system elements	Hardware—specifies hardware design
	Software—specifies software design
	HMI—specifies human machine interface design
	Environment—specifies environmental constraints and design
	Interfaces—specifies design of system and subsystem interfaces
Specify user and life-cycle needs	Performance—specifies system performance needs and objectives
	Operations—specifies system operational needs
	Support—specifies system support needs
	Test—specifies system test needs and objectives
Defines system abstraction levels	Physical—defines physical attributes of system
	Functional—defines functional attributes of system
Defines system qualities	Safety—to mitigate hazards and establish system mishap risk level
	Reliability—to establish system reliability level
	Quality—to establish system quality level
Defines source	Prescribed—defined by customer, standards, guidelines, and so on
	Derived—developed to expand higher requirements or solve problems
Defines basic type	Qualitative—qualitative attribute and format
	Quantitative—quantitative objective and format

TABLE 2.17 Characteristics of Good Requirements

Characteristic	Explanation
Unitary (cohesive)	The requirement addresses one and only one thing.
Complete	The requirement is fully stated in one place with no missing information.
Consistent	The requirement does not contradict any other requirement and is fully consistent with all authoritative external documentation.
Nonconjugated (atomic)	The requirement is *atomic*; that is, it does not contain conjunctions. For example, "The postal code field must validate American *and* Canadian postal codes" should be written as two separate requirements.
Traceable	The requirement meets all or part of a business need as stated by stakeholders and authoritatively documented.
Current	The requirement has not been made obsolete by the passage of time.
Feasible	The requirement can be implemented within the constraints of the project.
Unambiguous	The requirement is concisely stated without recourse to technical jargon, acronyms (unless defined elsewhere in the requirements document), or other esoteric verbiage. It expresses objective facts, not subjective opinions. It is subject to one and only one interpretation. Vague subjects, adjectives, prepositions, verbs, and subjective phrases are avoided. Negative statements and compound statements are prohibited.
Mandatory	The requirement represents a stakeholder-defined characteristic the absence of which will result in a deficiency that cannot be ameliorated. An optional requirement is a contradiction in terms.
Verifiable	The implementation of the requirement can be determined through one of four possible methods: inspection, demonstration, test, or analysis.
Concrete	A requirement should be a concrete prerequisite rather than a goal. A goal-oriented requirement is a vague requirement that usually cannot be verified or tested, and usually provides no firm design guidance.
Designer flexible	A requirement should avoid specifying a design implementation, unless specifically intended or necessary. It is better to leave implementation options open to the designer whenever possible.
Realizable/viable	A requirement must be realistically achievable within the constraints of time, cost, and technology.
Nonconflicting/ nonoverlapping	Requirements should not conflict with one another and requirements should not overlap one another. Every requirement should stand on its own merit.

TABLE 2.17 *Continued*

Characteristic	Explanation
Non-duplicated	Requirements should not be duplicated, either partially or fully. This leads to possible conflicts, confusion, and traceability problems.
Must/should clarity	Requirements have to be clear on "must" and "should." Design requirements are generally mandatory; therefore, the term "must" is necessary. Terms such as "may," "should," or "could" are not mandatory properties, and thus do not necessarily require implementation. These nonmandatory terms imply that something would be nice to have, but that it is not positively demanded.
Consistent terminology	Where domain or specialist terms, names, or documents are mentioned in requirements, ensure they are used consistently and correctly. This is often achieved by having a separate section of the requirements specification that contains a glossary, abbreviations, and definitions.

Requirements generally change with time. Once requirements are defined and approved, they should fall under formal change control. For many projects, requirements are altered before the system is complete. This is partly due to the complexity of the system and the fact that users do not know what they want before they see it. Requirement changes should always be evaluated by system safety for possible safety impact on the system design.

Safety design requirements are an intrinsic element of the SSP. The purpose of an SSR is to provide design guidance for intentionally designing safety into a system or product. A SSR is a design requirement that is primarily intended to enhance the safety quality of the system under development. SSRs generally focus on eliminating or mitigating a hazard. Requirements are the lifeblood of a system; however, requirements do not guarantee the system is hazard and risk free; they only assure protection from known hazards. Requirements analysis without HA only achieves half the safety job.

See *Functional Requirement, Nonfunctional Requirement,* and *System Safety Requirement (SSR)* for additional related information.

REQUIREMENTS MANAGEMENT

Requirements are the lifeblood of the system. The components and functions comprising a system are highly interrelated and complex, which means they must be well understood and well defined in order to properly build the system. The requirement and specification process must correctly define the system as a whole, including architectures, functions, interrelationships, and constraints.

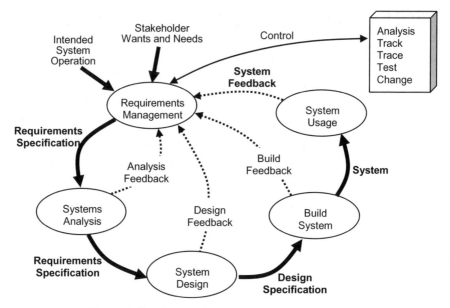

Figure 2.63 The requirements management process.

Figure 2.63 is a diagram illustrating the role of the requirements management in the system development life cycle. It should be noted that requirements management does not stop after the initial requirements are established, it is an ongoing process.

The process of establishing and implementing design requirement is probably one of the most important tasks in the system development process. Requirements management is the process of developing and implementing design requirements for a system or product. This process seeks to minimize design and development problems by using systematic and structured methods to establish and control design requirements. Since it is only logical and feasible to have design requirements prior to the design process, requirements development must begin at an early stage in the development life cycle.

The requirements management process is a gathering and developing process. This process starts by gathering requirements from user needs, wants, and intent. Requirements are developed to provide the best system solution in the form of design requirements. The output from this process is the requirements specification, which is a complete description of the system, including its functions, data, and behavior. It goes without saying that requirements should not be missing. It may be difficult to know when a requirement is missing, but system analysis and traceability analysis helps to identify missing links. Attempting to build a systems analysis model from existing requirements may reveal where requirements are missing. A review of other requirements for implied requirements that cannot be found helps in identifying missing requirements.

Requirements management involves communication between the program team members and stakeholders, and adjustment to requirements changes throughout the course of the project. To prevent one class of requirements from overriding another, constant communication among members of the development team is critical. Requirements management involves the following aspects:

- Requirements Analysis—acquire user needs and translate into design
- Track—an electronic database to record and maintain history of requirements
- Trace—linkage of requirements from inception through test verification
- Test—test all requirements for verification and validation; include evidence
- Change—change management system to control modification of requirements

See *Requirement* for additional related information.

REQUIREMENTS TRACEABILITY

Requirements traceability is part of the requirements management process. Requirements traceability is concerned with documenting the life of a requirement and to provide bidirectional traceability between various associated requirements. It enables users to find the origin of each requirement and track every change which was made to this requirement.

Not only the requirements themselves should be traced but also the requirements relationship with all the artifacts associated with it, such as models, analysis results, test cases, test procedures, test results, and documentation of all kinds. Even people and user groups associated with requirements should be traceable. Traceability should encompass a trace of where requirements are derived from, how they are satisfied, how they are tested, and what impact will result if they are changed. Electronic database traceability tools are available to assist in this task, which can become very large and complex for large systems.

In the SwS domain, requirements traceability is an important element in ensuring SCFs are safely implemented. In addition, software requirements should be traced to hazards in order to ensure the hazards are properly and completely mitigated.

See *Requirements Management* for additional related information.

RESIDUAL RISK

Residual risk is the amount of potential mishap risk presented by a hazard or system, following mitigation effort. This is the risk which is allowed to persist without taking further engineering or management action to eliminate or reduce the risk. The accepted level of residual risk is based on knowledge and decision making when the system safety process is followed. Note that if the decision is made to not perform mitigation, then the residual risk is then the same as the initial risk. If risk mitigation is performed in stages, then there may be an interim residual risk and a final residual risk.

There are effectively two categories of residual risk:

- Hazard residual risk
- System residual risk

Hazard residual risk is the level of risk presented by an individual hazard after risk mitigation has been applied to the hazard. Each and every hazard presents its own unique level of residual risk, and when a proper risk assessment is performed, this value should be fairly accurate. System residual risk is the level of risk presented by a system following the completed effort of an SSP. In essence, this is a total system risk, or a conglomeration of risk from individual hazards. It should be noted that system residual risk may very likely be understated in many cases because not all system hazards may have been identified.

See *Acceptable Risk*, *Hazard*, and *Risk* for additional information.

RESILIENCE ENGINEERING

Resilience is the property of a material to absorb energy when it is deformed elastically and then, upon unloading, to have this energy recovered. When used by psychologists, resilience refers to the ability to recover from trauma or crisis. In engineering, resilience is the property of a system to effectively adapt and cope with unexpected events with breaking down. The opposite of resilience is brittleness, referring to systems that break down when boundary conditions or underlying assumptions are challenged by new events. Examining a system's resilience means studying how the system in question performs when it is pushed near the boundaries of how it has been designed to operate. The principle ideas in resilience engineering stem from the recognition that failure does not always stem from malfunctions or poor design. Instead, many adverse events stem from the network of interactions and adaptations that are often necessary for complex systems to be useful in the "real world."

System resilience is the ability of organizational, hardware, and software systems to mitigate the severity and likelihood of failures or losses, to adapt to changing conditions, and to respond appropriately after the fact. Safety culture is a key element in system resilience. The goal of system resilience is to always function and never fail, but if failure should occur, then recovery is the next goal. System resilience focuses on a combination of fault tolerance with robustness. Resilience is the persistence of performance when facing changes. Qualities of a resilient system include:

- Anticipation—knowing what to expect
- Focus—knowing what to look for
- Feedback—knowing what to respond to
- Response—knowing what to do

RETINAL HAZARD REGION

The retinal hazard region refers to laser safety and is the optical radiation with wavelengths between 0.4 mm and 1.4 mm, where the principal hazard is usually to the retina.

REVERSE ENGINEERING

Reverse engineering is the process of analyzing an existing system and producing a semi-duplicate version from scratch, without knowing exactly how the original system was designed. The input and output of the two systems is identical, but the internal workings may be different. Software designers use this technique to copy an existing software module because they do not have access to the original and/or because they do not want to violate copyrights.

REWORK

Rework is corrective maintenance action performed where an article is reprocessed to conform to its original specifications or drawings.
 See *Repair* for additional related information.

RISK

The concept and reality of risk has been around for some time. There are many different types of risk, such as safety risk, hazard risk, mishap risk, schedule risk, cost risk, investment risk, product risk, and sports risk. Risk also involves many contending factors, such as perceived risk, real risk, individual risk, group risk, societal risk, high risk takers, low risk takers, and risk aversion. On the surface, risk appears to be a very simple concept; however, risk can easily become very complex due to all the types, factors, possibilities, and considerations involved. Risk and risk management are not just safety concepts. Risk analysis and risk management are used in many different fields, such as finance, project management, and health care, just to name a few. Risk is not about the present, it is about the future. Risk deals with uncertainty and outcomes.

Risk is a vector value combining event likelihood with event outcome. Risk is a metric expressing the expected value of a future event based on the parameters creating the potential event. It expresses the likelihood of a potential gain/loss from a given decision. Risk involves three parameters: (a) a potential future event, (b) the likelihood of the event occurring, and (c) the potential consequences from the event when it occurs. Each of these aspects involves an element of uncertainty. Risk is defined as the product of the event likelihood and the event outcome, where the outcome can be either a positive or negative consequence depending on the event. Risk outcome is the final expected result of the future event, given that it occurs. Risk outcome can be a threat or an opportunity; however, in system safety, it is treated as a threat of loss, damage, death, injury, or any combination of these outcomes. Risk is a way of quantifying uncertainty and danger.

Risk is an intangible quality; it does not have physical or material substance (a mishap does, but not risk). It is a future value concept with some quantifiable metrics, likelihood and severity, which characterize the future event. Risk can be thought of as the net present value of a future event. In system safety, risk is a measure of the future event, where the event is an expected mishap. Risk likelihood can be characterized in terms of probability, frequency, or qualitative criteria, while risk severity can be characterized in terms of death, injury, damage, dollar loss, and so on. Future safety events can only be identified as a hazard, which means that safety risk is the metric characterizing the amount of danger presented by a hazard. Recognizing that a hazard is the precursor (or blueprint) to a mishap, safety risk is the common denominator between the hazard and a mishap, and also the measure of the relative threat presented by a hazard.

Safety risk is sometimes stated in two different ways by individuals with different backgrounds, which can lead to confusion. The two types of safety risk terms are:

- **Hazard risk**—a safety metric characterizing the amount of danger presented by a hazard, where the likelihood of a hazard occurring and transforming into a mishap is combined with the expected severity of the mishap predicted by the hazard.
- **Mishap risk**—a safety metric characterizing the amount of danger presented by a potential mishap, where the likelihood of the mishap's occurrence is combined with the resulting severity of the mishap. Mishap risk likelihood defines the likelihood of the mishap occurring, while mishap risk severity defines the expected final consequences and loss outcome expected from the mishap event. The mishap likelihood and severity can only be computed from the information contained in the hazard description.

It should be noted from these definitions that hazard risk and mishap risk are really the same entity, just viewed from two different perspectives. HA looks

Figure 2.64 Hazard/mishap risk.

into the future and predicts a potential mishap from a hazard perspective. A potential mishap can only be recognized and understood in terms of a hazard; therefore, mishap risk can only be reached by first determining the hazard and then evaluating its risk in terms of a mishap from the HCFs. Since a hazard merely predefines a potential mishap, the risk has to be the same for both of them. Figure 2.64 depicts the hazard–mishap risk relationship.

Some basic principles regarding safety risk include the following:

- Risk is a metric of the likelihood and consequence of a potential future event.
- Risk exposure is a component of the likelihood measure.
- When a hazard exists, there is always risk associated with it.
- Hazard risk varies based on the hazard components involved.
- Risk is in effect regardless if it is known or unknown.
- Risk can usually be eliminated or reduced through DSFs.
- Hazards and their associated risk must be identified before risk can be assessed.
- Risk is automatically accepted, unless some action is taken to eliminate or reduce it.
- Risk is automatically accepted if a hazard is unrecognized (unidentified).

Risk is a measure used to characterize the uncertainty associated with future potential events, in order that decisions regarding these events can be made today. Risk decisions can result in either negative or positive outcomes. Risk management is a tool used to make decisions in the present that will help to produce a desired outcome in the future. Risk is a measurement that rates the overall safety significance (or danger) of a hazard. This significance rating allows decision makers to determine what action to take. Risk is about choices, and risk metrics provides data to assist in making choices. Some of the possible options include the choice to:

- Ignore the risk or investigate it and understand it
- Analyze the risk thoroughly or superficially
- Take appropriate action to change risk to an acceptable level
- Communicate the risk or cover it up
- Live with high risk or lower the risk

Measurement provides a mechanism for understanding an entity, and it also provides a means for evaluating changes to that entity. As Lord Kelvin stated "Anything that exists, exists in some quantity and can therefore be measured." Even though a hazard is a potential future event, its current value can be measured using the parameters of risk—likelihood and severity. Risk likelihood is the measure of the future event occurring, whereas risk severity is the measure for the amount of undesired consequence resulting from the future event when it occurs.

See *Hazard* and *Mishap* for additional related information.

RISK ACCEPTANCE

Risk is the potential danger presented by a hazard. Acceptable risk is the amount of potential mishap risk (i.e., danger) presented by an identified hazard that is allowed to persist without further risk reduction action. In system safety, risk acceptance is the formal process of accepting the risk presented by an identified hazard, thereby acknowledging its existence and the actions taken to control it.

The definition of safety is "freedom from unacceptable mishap risk." This definition implies that in order to know if a system is safe, all the hazards and their attached risk must be known, and then the risk must be communicated and controlled until it is deemed acceptable. This requires an established risk management process, which includes a risk acceptance step.

Hazards and risk will always be with us; as a result, we are forced to make risk acceptance decisions every day. Some risk decisions are simple, such as deciding to cross a busy street, while others are more complicated, such as selecting a new car based on its unique safety features or choosing to implement a tri-redundant flight control system for aircraft safety. Sometimes, however, we are not even aware that we are participants in a risk acceptance situation, such as when a trucking company transports HAZMATs through our city streets, which could have a safety impact on our lives should a mishap occur.

Risk acceptance is not a simple clear-cut process. One thing is certain: hazards and risk exist regardless of our perception, knowledge, or awareness of their presence. Hazards and risk do not care if we know about them or try to do anything about them. If you recognize a hazard and have control over the situation, you can apply your own risk criteria and judgment. This means that although you have some independence, you are a risk manager, by understanding risk and knowing when it is not acceptable. If someone else has control over the situation, then you are accepting their hazard recognition, risk criteria, risk knowledge, and risk judgment. This means you are dependent upon trust, and thereby defer all risk responsibility and knowledge. If no one has control over the situation, or it is ignored, then everyone accepts the risk by default (unknowingly).

Making the decision of risk acceptability is a difficult, yet a necessary responsibility of the system MA, system developer, or system user. This decision is made with full knowledge that it is the user who is exposed to the risk, and the decision is often negotiated or tempered with competing factors such as cost, schedule, and operational effectiveness. We are all risk managers; however, we must sometimes acquiesce in our risk decision making and let others decide for us, assuming they have more information and knowledge. Ignorance of a hazard (and its risk) will not prevent the hazard from causing us harm, so we must hope that the risk acceptors are competent and ethical.

On the surface, risk acceptance sounds like a simple concept ... the approval of the risk presented by an identified hazard. This sounds like a quick and easy process, but in reality, it can sometimes become quite complicated. The term implies that an accepted risk is a low risk; however, that is not always the case. There are some important questions and dilemmas attached to the concept of risk acceptance, such as:

1. Who specifically is the risk applicable to?
2. What if the hazard is not recognized (unknown)?
3. Who exactly is accepting the risk and by what authority?
4. What specifically is the risk acceptance criterion used and is it rational?
5. Is everyone exposed to the risk aware of the risk acceptance decision?
6. Is the accepted risk level embraced by all involved?
7. Does allowing a higher level authority accept higher risks provide an ethical level of safety or does it merely dilute it?

The acceptance of risk presented by identified hazards is an established process set forth in MIL-STD-882 (series) which requires that all identified hazards be eliminated or reduced to an acceptable level of risk. In addition, the Office of the Secretary of Defense (OSD) policy clearly specifies the requirement for system safety and a formal RAP. This OSD policy also establishes four risk levels and who can accept the risk for each level on DoD programs. These risk acceptance levels are as follows:

- High risk—accepted by the Component Acquisition Executive (CAE)
- Serious risk—accepted by the Program Executive Officer (PEO)
- Medium risk—accepted by the program manger
- Low risk—accepted by the program manger

The RAP should be a formal procedure that is described and documented in the SSPP. The RAP can be tailored to meet individual and program unique needs. There are seven basic steps in the RAP; these steps include the following:

1. Identify the hazard
2. Perform a risk assessment on the hazard

3. Obtain technical concurrence and validation of the hazard and risk assessment
4. Develop a risk mitigation plan (RMP)
5. Implement the RMP
6. Verify the risk mitigation results
7. Obtain formal risk acceptance (this includes communication)

Risk acceptance is the culmination of a risk management process that results in a decision that the potential mishap risk presented by a hazard is known, understood, and acceptable. Risk acceptance involves a decision-making process involving risk analysis and risk mitigation, in conjunction with program trade-offs involving factors such as cost, schedule, design complexity, and effectiveness. The risk acceptance decision may involve a group effort; however, there is usually a specific decision authority responsible for the final decision, which requires a signature. When the decision authority decides to accept the risk, the decision must be coordinated with all affected organizations and then documented so that in future years everyone will know and understand the elements of the decision and why it was made.

Accepted risk does not necessarily equate to the lowest risk possible, it may be a high risk (labeled as unacceptable) that must be accepted for various program reasons or mission needs. The system user is consciously exposed to this risk. Risk acceptance documentation also provides necessary data if the decision must be revisited.

Two significant questions in the RAP are how should hazard risk be characterized for acceptance judgment and what acceptance criteria should be used. The risk acceptance method selected must address the concern of complexity versus utility. If the judgment criteria method is too complex it will not be used effectively. Figure 2.65 shows three different example approaches. The Hazard Risk Index (HRI) approach provided as an example in MIL-STD-882 (or some variation) is the most commonly used approach.

Method 1 shows an example of quantitative risk characterization by probability and severity cost. In this example, the mishap probability and cost of a hazard must be determined. The hazard risk point is then plotted on the graph and is determined as acceptable or unacceptable based on where it falls in the predefined acceptance region. In this method, mishap cost represents a method for measuring mishap severity. The major concern with this method is that there is little gradation in the acceptance level.

Method 2 shows an example of combined quantitative and qualitative risk characterization. In this example, all hazards are plotted by mishap risk on one axis. The hazards are then determined as acceptable or unacceptable based on where they fall in the predefined acceptance regions. The major concern with this method is how the risk metric is characterized (e.g., cost, deaths).

Method 3 shows a risk acceptance approach using the Hazard Risk Index (HRI) method suggested in MIL-STD-882. This method provides for a good

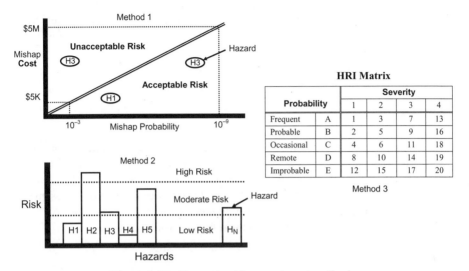

Figure 2.65 Example risk acceptance methods.

characterization of risk, which can be estimated qualitatively or quantitatively. It also provides a relatively simple methodology that is cost-effective to perform.

See *Hazard Risk Index (HRI)* for additional related information.

RISK ACCEPTANCE AUTHORITY (RAA)

This is the person, or persons, that have the authority to accept potential mishap risk. Typically, the mishap risk for a hazard falls into one of several different possible levels based on the risk matrix, and a person of higher rank or authority is required to accept the corresponding level of risk. DoDINST 5000.2 states that high, serious, medium, and low risk must be accepted by the CAE, PEO, PM, and PM, respectively. In theory, the risk level should be pushed down, through design measures, to the lowest level in order that the risk can be accepted by the lowest level decision authority. In reality, this is not always possible due to various system constraints, thus requiring a decision authority hierarchy.

See *Risk Acceptance* for additional related information.

RISK ACCEPTANCE PROCESS (RAP)

This is the formal process for obtaining a risk acceptance signature from the Risk Acceptance Authority (RAA) for a hazard. The process involves analysis, risk assessment, risk mitigation, coordination, and documentation. The RAP should be a formal procedure that is described and documented in the SSPP. The RAP can be tailored to meet the individual and unique needs of a program.

See *Risk Acceptance* for additional related information.

RISK ANALYSIS

Risk analysis is activity of examining each identified risk to refine the description of the risk, isolate the causal factors, and determine the effects. It refines each risk in terms of its likelihood and consequence. It also involves establishing measures to control or mitigate the risk. It is essentially the risk assessment and risk control stages of the risk management process.

See *Risk Management* for additional related information.

RISK-BASED SAFETY

Risk-based safety is a nonprescriptive approach to safety. In this approach hazards are identified, the risk presented by the hazards is determined, and the risk mitigated to the lowest level practical through the incorporation of design safety methods.

Risk-based safety is a hazard identification and risk management approach to implementing safety. In this approach, system hazards are identified and assessed for potential mishap risk. The hazards presenting unacceptable mishap risk are then controlled through DSFs to reduce the risk to an acceptable level. Risk-based safety is more adept at ensuring that nothing is overlooked in regard to implementing a safe system design.

Prescriptive design safety guidance provides a mechanism for implementing a safe system design at the beginning of a program before the entire set of necessary HAs and risk assessments can be completed. An effective SSP should be prescriptive (compliance) based and risk based; however, risk based is the most important aspect because many hazards are unique to a particular system.

See *Compliance Based Safety* and *Prescriptive Safety* for additional information.

RISK COMPENSATION

Risk compensation is a human behavioral effect whereby individual people may tend to adjust their behavior in response to perceived changes in risk. Individuals will tend to behave in a more cautious manner if their perception of risk or danger increases. Another way of stating this is that individuals will behave less cautiously in situations where they feel "safer" or more protected.

RISK MANAGEMENT

Risk management is an important tool for making critical decisions and, in some cases, meeting regulatory requirements. In system safety, risk management is a process involving the identification of hazards and their causes, determining the consequences of the hazards, calculating the probability of their occurrence, and determining whether the risk is acceptable or if corrective actions are needed to make the risk acceptable. Mishap risk management is a key element of the system safety process.

Risk management is somewhat of a balancing act, where potential mishap risk and the cost of controlling it is weighed against typical program constraints, such as cost, schedule, and user needs. The system safety process actually has risk management built into it; system safety balances the constraints of potential mishap risk management against the constraints of program management. Overall, the risk management process involves three major actions: (1) risk assessment, (2) risk control, and (3) risk communication. Step 1 involves risk identification and assessment, step 2 involves risk reduction, and step 3 involves communicating both the risks and risk controls to the appropriate authority and end user for acceptance. These risk management steps are depicted in Figure 2.66.

Risk Assessment Segment. In order to manage risk, it must first be identified, measured, and evaluated. This is called the risk assessment step. Individuals perform personal risk assessments every day as part of normal routine. In the field of system safety, risk assessment involves identifying and assessing risks that can lead to mishaps. Potential mishap risk can only be identified by first identifying system hazards that create the risk. By assessing risks, priorities can set for the allocation resources to mitigate hazards and risk, and thereby minimize losses. Mishap risk assessments involve identifying the consequences of a UE and

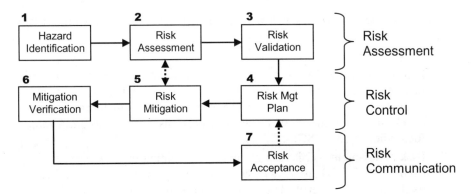

Figure 2.66 Risk management model.

its likelihood of occurrence. For example, the risks associated with a ship running aground would be greater if the cargo is hazardous (increased consequences) or if the crew is inexperienced (higher probability). Note that a more hazardous cargo does not make the ship more likely to run aground, but the risks associated with its grounding are greater.

Risk Control Segment. Once hazards and risks are known, it is imperative to take steps to reduce the risk when necessary. Risk control is the process of prioritizing risks, determining which risks must be mitigated, and then implementing DSFs to the reduce risk. An example would be designing an automobile with air bags to reduce the risk of personnel injury should an accident occur.

Risk management controls typically target the reduction of hazard severity by reducing the severity of the consequences of a hazard. Risk controls also target reducing the likelihood of a hazard becoming a mishap. Rick control measures such as double hulls and life-saving equipment are intended to reduce the consequences of a ship collision mishap. Equipping a ship with the advanced redundant navigational systems and increased training to improve crew competency are methods to reduce the likelihood of a ship collision mishap. The keys to successful risk controls are that they be effective, specific to the hazards that create the risk, and that they are relatively easy to implement. This will increase the likelihood of successfully mitigating the hazard and the risk.

Risk Communication Segment. After hazards and their associated risk have been mitigated to a level that appears acceptable to program management, each hazard risk must be formally accepted by the appropriate decision authority. Risk acceptance criteria should be preestablished and documented in the SSPP for the program. In addition, the risk acceptance authorities should be identified in the SSPP. Obtaining concurrence and written acceptance by the appropriate authority involves intensive risk communication on many different levels.

In the system safety process, risk management involves the following critical steps:

1. **Hazard Identification**—This step involves identifying system hazards through as many means as possible. The primary method for identifying hazards is via HA, utilizing information from system hazard sources, checklists, experience, lessons learned, past mishaps, and so on. A major focus should be on SCFs.
2. **Risk Assessment**—This step involves determining the risk presented by each identified hazard. Analysis must be performed to assess hazard likelihood and severity. This requires knowing and understanding the HCFs and hazard effects.

3. **Risk Validation**—This step involves validating the results of the risk assessment to ensure correctness and concurrence by technical authorities in the technical areas involved.
4. **Risk Management Plan**—This step involves selecting those hazards with unacceptable risk levels and planning corrective action to reduce the risk. Determining unacceptable hazards must be based on established risk selection criteria. An effective procedure is to prioritize each hazard by its relative risk, and then apply mitigation resources on higher risk hazards.
5. **Risk Mitigation**—This step involves establishing design safety methods needed to mitigate risk to an unacceptable level. This involves translating safety design features into design requirements to support the system development process. Mitigation methods should be selected using the SOOP. Mitigation methods should also be approved by all program stakeholders.
6. **Mitigation Verification**—This step involves ensuring that the risk mitigation methods have been implemented and that they are effective. This involves requirements traceability of safety requirements. It also involves testing safety requirements and verifying test result evidence for completeness and success.
7. **Risk Acceptance**—This step involves communicating the risk to the proper program authorities and having them formally accept the risk for each identified hazard. If the risk acceptance authority believes the risk can be further reduced or an alternative method should be used, then the hazard will be sent back to step 4.

When developing hazard mitigation controls, the following general risk reduction rules apply:

- If any side of the hazard triangle is eliminated through risk mitigation techniques, the hazard and its associated risk are also eliminated.
- If none of the hazard triangle sides can be eliminated, the hazard likelihood and/or severity must be controlled.
- When a hazard is identified, it is very difficult to reduce the hazard severity, it is much easier to reduce the hazard likelihood. For this reason, the risk severity category almost always remains the same, even after mitigation.

Another aspect of risk control is risk sensitivity. Sensitivity addresses the ability of the risk management approaches to affect the risk in the most effective places. For maximum impact, risk management control efforts should be directed toward those items driving risk, that is, those which are most sensitive to intervention. For example, suppose the failure of a specific resistor, and a specific diode, cause a hazard. Sensitivity analysis would establish that redesign of the resistor would provide an 80% reduction in the risk, while the diode would only provide a 20% reduction. Another sensitivity example is the risk related to weather. While risk related to weather hazards may be reduced through better forecasting, planning, and preparation, the weather itself is not sensitive to control. Thus, addressing those factors with higher sensitivity, such as forecasting, preparation, and avoidance will maximize the effectiveness of risk management controls.

RISK MANAGEMENT PLANNING

Risk management planning is the activity of developing and documenting an organized and comprehensive strategy for identifying program risks. It includes establishing methods for mitigating risk and for tracking risk.

RISK MITIGATION

Risk mitigation is the strategy and methods employed to reduce potential mishap risk presented by a hazard. It involves establishing and implementing DSFs to reduce the likelihood or severity presented by a hazard. Risk reduction is most effectively achieved when the SOOP is followed. The resulting risk following risk mitigation is known as residual risk.

See *Design Safety Measures*, *Risk*, and *Safety Order of Precedence (SOOP)* for additional information.

RISK MITIGATION PLAN (RMP)

The RMP is a plan devised to mitigate or reduce the risk presented by a specific hazard. It involves establishing DSFs to reduce the likelihood or severity presented by a hazard after evaluating all options. It also involves obtaining concurrence from the appropriate organizations and disciplines involved in technical areas affected by the hazard. The RMP should identify the cost, schedule, materials, tasks, and responsibilities involved. The RMP may be a formal or informal plan. The RMP should include the specifics of what should be done, when it should be accomplished, who is responsible, and the funding required to implement the RMP. It is typically a formal plan when it must be presented to a risk acceptance authority for the approval of future development, maintenance, or repair work.

See *Risk Management* for additional related information.

RISK MITIGATION PLAN (RMP) IMPLEMENTATION

The activity of executing the RMP to ensure successful risk mitigation occurs. It determines what planning, budget, and requirements and contractual changes are needed, provides a coordination vehicle with management and other stakeholders, directs the program teams to execute the defined and approved RMP, performs the risk required reporting, and documents the change history.

RISK PRIORITY NUMBER (RPN)

An RPN is the risk ranking index for reliability, where RPN = (probability of occurrence) × (severity ranking) × (detection ranking). This number typically appears on the reliability oriented FMECA. It should be noted that the RPN is not the same as a Hazard Risk Index which is derived from probability of occurrence and severity of a hazard. The RPN provides a risk ranking order of components for reliability improvement.

See *Hazard Risk Index (HRI)* for additional related information.

RISK TRACKING

Risk tracking is the activity of systematically tracking, monitoring, and evaluating the performance of risk mitigation actions against established mitigation plans. It feeds information back into the other risk management activities of identification, analysis, mitigation planning, and mitigation plan implementation. For tracking the risk presented by hazards, the system safety HTS is the vehicle used for the risk tracking process.

See *Hazard Tracking System (HTS)* for additional related information.

RUNAWAY VEHICLE

A runaway vehicle is a vehicle experiencing unintended acceleration, braking failure, operator control failure, or any combination of these effects.

SAFE

Safe is the condition of being protected from danger, mishaps, or other undesirable consequences. This can take the form of being protected from something that can cause health, physical, social, financial, political, emotional, occupational, psychological, or economical losses. It can include protection of people, possessions, systems, animals, or the environment. Safe is typically defined as relative freedom from danger or the risk of harm; secure from danger or loss. Safe is a state that is secure from the possibility of death, injury, or loss. A person is considered safe when there is little danger of harm threatening them. A system is considered safe when it presents low mishap risk (to users, bystanders, environment, etc.). Safe can be regarded as a state ... a state of low mishap risk (i.e., low danger); a state where the threat of harm or danger is nonexistent or minimal.

SAFE SEPARATION

Safe separation is a term used to reflect the safe distance between two objects under parameters and conditions. For example, the safe separation distance for an artillery round is the distance from the gun barrel exit to the point of detonation where the detonation fragments will not affect friendly forces or the firing forces.

Some example applications of the safe separation term include:

- Separation distance between aircraft in flight
- The ability (and safe distance) of a missile or bomb to be released without hitting the carrier aircraft
- The distance an artillery round must travel before its explosive can be safely initiated (so it will not injure friendly forces)

SAFE SOFTWARE

Safe software is software that executes within a system context and environment with an acceptable level of potential mishap risk. This means the software will not cause any system hazards or prevent system design safety mechanisms from performing correctly, or the likelihood of causing these conditions is within acceptable bounds.

See *Software Safety (SwS)* for additional related information.

SAFETY

Safety is freedom from unacceptable mishap risk. Safety is the condition of being protected against physical harm or loss. Safety as defined in MIL-STD-882D is "freedom from those conditions that can cause death, injury, occupational illness, damage to or loss of equipment or property, or damage to the environment." Safety is the state of being safe. Safety is also recognized as a discipline; the activity performed to intentionally making something safe. As a discipline, it is the process of making certain that adverse effects will not be caused by some event under defined conditions and that acceptable mishap risk is achieved for a system or product. Safety is the reduction of potential mishap risk to an acceptable level.

Since 100% freedom from mishaps is not possible, safety is effectively "freedom from conditions of unacceptable mishap risk." Safety is the "condition" of being protected against physical harm or loss (i.e., mishap). It should be noted that safety itself is not a device (as some dictionaries state); it is a state of being safe or an activity working toward creating a safe state. A "safety device" is a special device or mechanism used to create safe conditions or a safe design.

The term safety is often used in various casual ways, which can sometimes be confusing. For example, "the designers are working on aircraft safety" implies the designers are establishing the condition for a safe state in the aircraft design. Another example, "aircraft safety is developing a redundant design" implies a branch of safety, "aircraft safety" that is endeavoring to develop safe system conditions.

Safety as a discipline can be broken down into many different specialty categories, such as system safety, explosives safety, personnel safety, fire safety, occupational health safety, laser safety, industrial safety, range safety, aircraft safety, and nuclear safety.

Safety is often viewed from different perspectives, such as:

- Normative safety—refers to products or designs that meet applicable design standards.
- Substantive safety—refers to the real-world historical safety of a product or system as favorable, whether or not standards are met.
- Perceived safety—refers to the subjective level of comfort of users. For example, auto safety is perceived as safer than flying, yet statistically commercial flying is safer.
- Achieved safety—refers to the actual level of safety achieved for a system or product.
- Functional safety—refers to the safety of functional equipment.
- System safety—refers to the predicted safety (risk) of an entire product or system.

SAFETY ANALYSIS TECHNIQUE

HAs are performed to identify hazards and their associated potential mishap risk. There are two categories or divisions of HAs: *types* and *techniques*. The HA type defines an analysis category or class, whereas the HA technique defines a unique analysis methodology (e.g., FTA). The analysis type establishes timing, depth of detail, and system coverage of the analysis. The analysis technique refers to a specific and unique analysis methodology that provides specific results and is performed according to an established set of rules or guidelines. System safety is built upon seven basic types, while there are well over 100 different techniques.

In general, analysis type defines the "what and when" to analyze for safety, while the analysis technique defines the specific "how to" perform the analysis.

An HA technique defines a specific and unique analysis methodology that provides a specific procedure, along with a specific type of expected results. The overarching distinctions of an analysis technique include the following:

- Establishes "how to" perform the analysis methodology
- Establishes the analysis rules, guidelines, and graphics
- Establishes the level of detail of the information required for the analysis
- Establishes the technical expertise required
- Provides the information needed to satisfy the intent of a particular analysis type

HA techniques can have many different inherent attributes, which makes their utility different. The appropriate technique to use can often be determined from the inherent attributes of the methodology itself. The following is a list of the most significant attributes for an HA methodology:

- Qualitative or quantitative
- Level of analysis detail
- Data required for the analysis
- Program timing (when performed in the development life cycle)
- Time required to perform the analysis
- Inductive or deductive approach
- Complexity of the analysis
- Difficulty of the analysis
- Technical expertise required to perform the analysis
- Tools required to support the analysis
- Cost of the analysis
- Subjectivity of the method

The System Safety Analysis Handbook, published by the System Safety Society, contains a list of over 100 different analysis techniques. An example of some typical HA techniques include, but are not limited to, the following:

- FTA
- SCA
- BA
- ETA
- FHA
- Markov Analysis (MA)
- BPA
- HAZOP analysis

- CCFA
- MORT
- CCA
- PHA
- SSHA
- SHA
- O&SHA
- HHA
- SRCA
- THA

See *Safety Analysis Type* for additional related information.

SAFETY ANALYSIS TYPE

HAs are performed to identify hazards and their associated potential mishap risk. There are two categories or divisions of HAs: *types* and *techniques*. The HA type defines an analysis category or class, whereas the HA technique defines a unique analysis methodology. The analysis type establishes timing, depth of detail, and system coverage of the analysis. The analysis technique refers to a specific and unique analysis methodology that provides specific results and is performed according to an established set of rules or guidelines. System safety is built upon seven basic types as originally established in MIL-STD-882.

The overarching distinctions of the analysis type include:

- Establishes where, when, and what to analyze
- Establishes a specific analysis task at specific time in program life cycle
- Establishes what is desired from the analysis
- Provides a specific design focus (e.g., preliminary, detailed)

The analysis type defines the "what and when" to analyze for safety, while the analysis technique defines the specific "how to" analyze. An HA type defines the analysis purpose, timing, scope, level of detail, and system coverage; it does not specify how to perform the analysis. HA type describes the scope, coverage, detail, and life-cycle phase timing of the particular HA. Each type of analysis is intended to provide a time- or phase-dependent analysis that readily identifies hazards for a particular design phase in the system development life cycle. Since more detailed design and operation information is available as the development program progresses, so in turn more detailed information is available for a particular type of hazard analysis. The depth of detail for the analysis type increases as the level of design detail progresses.

Each of these analysis types define a point in time when the analysis should begin, the level of detail of the analysis, the type of information available, and the analysis output. The goals of each analysis type can be achieved by various analyses techniques. The analyst needs to carefully select the appropriate techniques to achieve the goals of each of the analysis types.

The concept of seven HA types was intentionally developed by early system safety practitioners and has been proven successful for over 45 years for both military and commercial applications. The HA types were defined and refined in MIL-STD-882. One confusing area of system safety is that there are also seven HA techniques with the same name as the types; however, their purpose is to satisfy each type. The following seven HA types provide the analysis baseline for the system safety discipline:

1. PHL
2. PHA
3. SSHA
4. SHA
5. O&SHA
6. HHA
7. SRCA

An important principle about HA is that one particular HA type does not necessarily identify all the hazards within a system; identification of hazards may take more than one analysis type, hence the seven types. A corollary to this principle is that one particular HA type does not necessarily identify all of the HCFs; more than one analysis type may be required. After performing all seven of the HA types, all hazards and causal factors should have been identified, assuming an adequate analysis program was conducted. Additional hazards that were overlooked may be discovered during the test program.

See *Safety Analysis Technique* for additional related information.

SAFETY AND ARMING (S&A) DEVICE

The purpose of the S&A is to reliably initiate an explosive device, but only after all safety criteria have been met. Typically, an S&A physically separates an explosive from the initiator. When the S&A is in the safe position an inadvertent initiation signal cannot reach the explosive due to the physical barrier. When the S&A is in the arm position an initiation signal will reach and detonate the explosives; at this point the initiation signal should be intentional and not inadvertent. The S&A device is also sometimes referred to as a safe & arm device.

See *Fuze* for additional related information.

SAFETY ASSESSMENT REPORT (SAR)

The SAR is a comprehensive report that provides a risk assessment of the system and evidence of the SSP effectiveness. The SAR is a snapshot of the potential mishap risk a system design presents at a particular point in time in the program. The SAR is a living or evolving document updated at each program milestone to reflect the current mishap risk status. This provides decision makers with objective information on the safety status of the system design. The SAR is built upon safety analyses and test results; it summarizes all safety activities at a specific point in time. The thoroughness of the SAR depends upon the thoroughness of the HAs and testing conducted up to that point in time. The SAR matures in levels of detail as the system design progresses and matures. It should be noted that an SAR is not an HA, but a summary of risk presented by all of the hazards identified by all the various HAs. Where the SSP is a process over time, the SAR is a snapshot in time.

The SAR is an important safety product because it provides an overview of the system level of risk at that point in time. It can show where the system has been effectively designed for safety and it can also point out where more safety focus is needed. Development of the SAR is an iterative process during the life of the program and should be prepared for each major program milestone to assist in program decision making. Total system risk will be underestimated if anything is omitted from the assessment, such as hazards, software, COTS items, or human factors.

The SAR is an evolving document updated at each program milestone to reflect the current mishap risk status. This provides the decision makers with objective information on the safety status of the system design. The SAR is built upon completed safety analyses and testing, and is a summary of all safety activities at a specific point in time. The thoroughness of the SAR depends upon the thoroughness of the HAs and testing conducted through that point in time. The SAR matures in levels of detail as the system development progresses and matures. The SAR is also a reflection of the success (or failure) of the SSP.

The SAR does not utilize an analysis worksheet since it is a summary document of safety analyses, risk assessments, test results, and safety studies already performed. As a minimum, the following basic information is required from the SAR document:

- Introduction (purpose, scope)
- Brief system description (including hardware, software and COTS/NDI)
- SSP description
- Hazard and TLM assessment summary
- Risk assessment summary
- A discussion of the results of tests conducted to validate safety requirements

	Severity							
Likelihood	I Catastrophic		II Critical		III Marginal		IV Negligible	
(A) Frequent	1	(0)	3	(0)	7	(0)	13	(0)
(B) Probable	2	(0)	5	(0)	9	(0)	16	(0)
(C) Occasional	4	(0)	6	(1)	11	(0)	18	(2)
(D) Remote	8	(0)	10	(10)	14	(7)	19	(0)
(E) Improbable	12	(4)	15	(0)	17	(0)	20	(0)
Total		4		11		7		2

Figure 2.67 Risk assessment summary.

- HAZMATs summary
- List of DSFs (e.g., triple redundant flight control systems)
- Operational limitations
- Safety claim and evidence summary
- Conclusions and recommendations
- Appendix (detailed HAs or reference to source document)

When summarizing the system risk in the SAR, a typical approach is to use the Hazard Risk Index (HRI) matrix with the cells containing the number of identified hazards that fall within each cell. This is demonstrated in Figure 2.67 through a fictitious example of a program risk assessment summary. Below the severity categories, the left-hand side of the cell contains the assigned risk level and the right side contains the number of hazards the program has in that risk cell (in parentheses).

The suggested risk levels from MIL-STD-882D are as follows: high risk = 1–5, serious risk = 6–9, medium risk = 10–17, and low risk = 18–20. Based on this information, the risk assessment conclusions that can be derived from this example are:

1. A total of 24 system hazards have been identified (4 + 11 + 7 + 2)
2. A total of 0 hazards have a risk level of high
3. A total of 1 hazards have a risk level of serious
4. A total of 21 hazards have a risk level of medium (4 + 10 + 7)
5. A total of 2 hazards have a risk level of low
6. On a government program, the PM can accept the residual risk for the 23 medium and low risk hazards without further design action
7. On a government program, serious and high risk must be further reduced or go to the PEO to obtain acceptance

The SAR was formally instituted and promulgated by MIL-STD-882 and is typically a CDRL item. The SAR format is specified in DID DI-SAFT-80102A, "Safety Assessment Report." Remember, the SAR is a summary assessment document and is not a regurgitation of existing documentation or analyses.

See *Top-Level Mishaps (TLMs)* for additional related information.

SAFETY AUDIT

A safety audit is an independent examination of an SSP to assess compliance with specifications, standards, contractual agreements, or other criteria. The purpose is to conduct an independent review and examination of system records and activities in order to determine the adequacy and effectiveness of the work performed, to ensure compliance with established policy and operational procedures, and to recommend any necessary changes.

The safety audit can be either a formal or informal review of a program, to determine if objectives and requirements have been met. An audit also involves identifying deficiencies, problems, and issues. The following are typical categories of items resulting from an audit:

1. Compliance—A compliance is the complete satisfaction of an objective or requirement.
2. Finding—A finding is the identification of a failure to show compliance to one or more of the objectives or requirements. A finding may involve an error, deficiency, or other inadequacy. A finding might also be the identification of the nonperformance of a required task or activity.
3. Observation—An observation is the identification of a potential improvement. An observation is not a compliance issue and does not need to be addressed before approval.
4. Issue—An issue is a concern; it may not be specific to compliance or process improvement but may be a safety, system, program management, organizational, or other concern that is detected during the audit review.
5. Action—An action is an assignment to an organization or person, with a date for completion, to correct a finding identified during the audit.

In system safety an audit is typically performed on an SSP. The purpose of the audit is to determine if the SSP is on track and if all contractual requirements are being satisfied and that mishap risk is being properly identified, assessed, controlled, and accepted. A safety audit is typically conducted according to a preplanned schedule that is established in the SSPP. A safety audit typically consists of a review of the system developer, contractor, or subcontractor's documentation, hardware, and/or software to verify that it complies with project requirements and contractual requirements. A safety audit should be performed by someone not working on the program under review, that is, someone independent from the program. An audit should be well documented in order to provide an audit trail. An audit trail is a chronological record of system activities or audit evidence that is sufficient to enable the reconstruction of the final results by an independent person or group of people.

Some of the aspects to be considered when performing an SSP audit include the following:

1. A system safety organization exists for the purpose of conducting an SSP.
2. The SSP has an adequate SSMP and SSPP.
3. The SSP has a lead safety person (PFS, safety manager, safety lead).
4. The SSP staff have the appropriate qualifications and experience.
5. The SSP is meeting the requirements of the SOW, SSPP, and SSMP.
6. The SSP is meeting the requirements of the applicable safety standards.
7. The SSP has adequate resources and funding.
8. HAs are complete and thorough.
9. Risk assessments appear complete and thorough.
10. Identified hazards are being appropriately eliminated and/or mitigated.
11. Hazard risk is being accepted by the proper acceptance authorities.
12. SSWG meetings are being conducted per plan and minutes are documented.
13. Hazards are being tracked in a formal HTS.
14. Hazards are being closed on a scheduled and timely basis.
15. Contractor safety CDRLs are received on schedule and are appropriate.

See *Audit* for additional related information.

SAFETY BARRIER DIAGRAM

A safety barrier diagram is a graphical depiction of the progression of unwanted events as they traverse through a path of safety barriers toward a UE. The diagram considers failures, events, and conditions necessary to defeat a series of barriers intended to prevent a mishap. A barrier diagram represents possible accident scenarios, each having different outcomes. ETs, FTs, and cause–consequence diagrams are often used as barrier diagrams. A barrier diagram essentially models a safety function or barrier function as it is designed to inhibit an unwanted function or energy source.

A barrier function is a function planned and designed to prevent, control, or mitigate the propagation of failures, condition, or event into a UE or mishap. A safety barrier can be a series of elements that implement a barrier function, each element consisting of a technical system or human action.

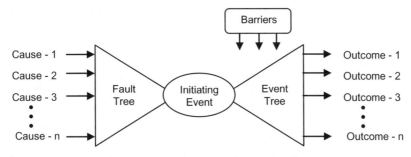

Figure 2.68 Bow-tie analysis of barriers.

An advantage of safety barrier diagrams is that by focusing on the safeguards that are deliberately inserted into the system to prevent or mitigate accidents, safety barrier diagrams directly show the issues that are of primary concern. Safety barriers should be directly related to the event sequence or accident scenario, and it should not be confused with organizational factors that affect the performance of the barrier.

Figure 2.68 shows a bow-tie analysis model (also referred to as X-tree analysis) of barriers. This analysis technique is a combination of FTA and ETA. The analysis begins with identification of the IE of concern in the center. An FTA is performed to identify the causal factors and probability of this event. Then an ETA is performed on all the barriers associated with the IE, and the possibilities of each barrier function failing. The various different failure combinations provide the various outcomes possible, along with the probability of each outcome.

See *Barrier*, *Barrier Analysis (BA)*, and *Barrier Function* for additional related information.

SAFETY CASE

In general, a safety case is a special risk assessment making the case, or justification, that the potential mishap risk presented by a product or system is acceptable. In the United States, this could be achieved by using the SAR format. However, in the United Kingdom, the term safety case is slightly more formal and involved. The concept of a safety case has been well established in the United Kingdom since the enactment of the Health and Safety at Work Act of 1974. Safety cases are required in a number of industries, such as aerospace, chemical, nuclear, offshore, and railway. Different industries have specific requirements regarding safety cases. In many cases, the safety case is a legal obligation for a commitment to safety, and demonstration that the commitment has been achieved. Safety standards such as the U.K. Defence Standard 00–56 require that safety case development be treated as an evolutionary activity that is integrated with the rest of the design and safety life cycle. At each stage of the evolution of the safety case, the safety argument is expressed in terms of what is known about the system being developed. At the early stages of project development the safety argument is limited to presenting high-level objectives; as design and safety knowledge increases during the project, these objectives (and the corresponding arguments) can be expressed in increasingly tangible and specific terms.

A safety case is a formal documented body of evidence that provides a convincing and valid argument that a system is adequately safe for a given application in a given environment. The safety case documents the safety requirements for a system, the evidence that the requirements have been met, and the argument linking the evidence to the requirements. Elements of the safety case include safety claims, evidence, arguments, and inferences. Claims are simply propositions about properties of the system supported by evidence. Evidence may either be factual findings from prior research or scientific literature or sub-claims that are supported by lower-level arguments. The safety argument is the set of inferences between claims and evidence that leads from the evidence forming the basis of the argument to the top-level claim, which is typically that the system is safe to operate in its intended environment. Producing a safety case does not require a specific process; any methodology is acceptable as long as it provides a compelling argument that the system meets its safety requirements. System safety is argued through satisfaction of the requirements, which are then broken down further into more specific goals that can be satisfied directly by evidence

A safety case is a well-reasoned argument, supported by evidence that a system is acceptably safe to operate in a particular context. Maturing a safety case in step with design maturity has proven to be an effective means of identifying and addressing safety concerns during a system's life cycle. It is common in the United Kingdom that developers of safety-critical systems are required to produce a corresponding safety case communicating an argument, supported by evidence that a system is acceptably safe to operate.

Figure 2.69 shows the basic elements of a safety case and the relationship between these elements. The safety case should consist of a structured argument supported by a body of evidence. The quantity and quality of the evidence depends on the systems risks, complexity, and the familiarity of the circumstances involved. It should be noted that in many safety assessments, the safety case argument is often neglected or stated very weakly, thus the intent of this formal approach is to avoid that situation.

The three basic elements of the safety case are can be described as:

- Claims—statements or requirements about a safety property of the system, subsystem, or component.
- Evidence—facts, assumptions, or sub-claims derived from lower-level sub-arguments; used as the basis for the safety argument.
- Arguments—the logical and rational link between safety claims and the evidence supporting the claims; logical information linking the evidence to the claim, where inference is typically the mechanism used.

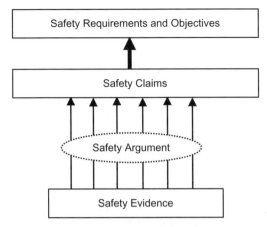

Figure 2.69 Elements of the safety case.

The safety argument is that which communicates the relationship between the evidence and the objectives. Both argument and evidence are crucial elements of the safety case that must be well integrated together. Argument without supporting evidence is unfounded, and is therefore unconvincing. Evidence without argument is unexplained, making it unclear that safety objectives have been satisfied. There are three types of argument that can be used in the safety case:

- Deterministic—relying upon axioms, logic, and proof
- Probabilistic—relying upon probabilities and statistical analysis
- Qualitative—relying upon adherence to standards, design codes, and so on

Although it is possible to communicate clear arguments in a textual narrative report, many arguments expressed in this manner are often poorly expressed and are difficult to comprehend. The biggest problem with the use of free text is in ensuring that all stakeholders involved share the same understanding of the argument. Without a clear and shared understanding of the argument, safety case management is often an inefficient and ill-defined activity. To alleviate this problem a structured technique, the Goal Structuring Notation (GSN) methodology, has been developed to address the problems of clearly expressing and presenting safety arguments.

GSN is a graphical argument notation that explicitly represents the individual elements of any safety argument (requirements, claims, evidence, and context) and the relationships that exist between these elements. GSN shows how individual requirements are supported by specific claims, how claims are supported by evidence, and the assumed context that is defined for the argument. The principal purpose of a goal structure is to show how goals (claims about the system) are successively broken down into subgoals until a point is reached where claims can be supported by direct reference to available

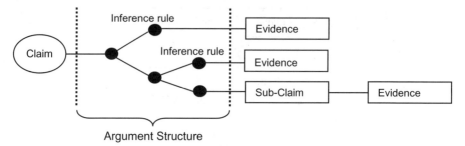

Figure 2.70 Basic model for GSN.

evidence (solutions). As part of this decomposition, the GSN helps to make clear the argument strategies adopted (such as a quantitative or qualitative approach), the rationale for the approach (such as assumptions and justifications), and the context in which goals are stated (such as the system scope or the assumed operational role). Figure 2.70 shows a general model for the GSN process. Actual GSN diagrams follow a formal set of symbols, terms, and definitions that are strictly followed.

SAFETY-CRITICAL

SC is a designation given to any item that will lead to a catastrophic or critical mishap (i.e., loss of life and/or major damage) should the item fail to function, malfunction, or fail to operate properly. SC items can be any system level of hardware, software, or human tasks involving an event, operation, process, or procedure. SC items can be functions, requirements, paths, tasks, procedures, components, or component tolerances. SC is a high consequence or high-risk subset of the SR designation and represents items presenting high risk in some manner.

SC is a special safety designation given to any system element whose failure to operate, or incorrect operation, will directly lead to a mishap that could result in death, serious injury, major system loss, or severe environmental damage. A system element can consist of a system, subsystem, component, item, function, process, procedure, and so on. The purpose is to indicate that anything labeled as SC will always present a major system risk factor and therefore must be given a higher level of safety rigor and continued safety vigilance. The potential mishap risk associated with an SC element must be reduced to an acceptable level; however, this does not reduce or change the SC designation. SC is a term applied to any condition, event, operation, process, or item whose proper recognition, control, performance, or tolerance is essential to keep the risk of a Class A or Class B accident, as defined by DoDINST 6055.7, as low as reasonably practicable during system operation and support.

It is generally the task of system safety to identify SC elements of the system, which is primarily accomplished through HA. It should be noted that when the risk presented by a SC item is mitigated to an acceptable level, the item still remains designated as SC, in order to maintain a vigilant focus on it.

SAFETY-CRITICAL FUNCTION (SCF)

SCF is a special safety designation given to any function (hardware, software, or procedure) whose failure to operate, or incorrect operation, will directly lead to a mishap with catastrophic or critical outcome (as opposed to marginal or negligible outcome), such as death, serious injury, major system loss, or severe environmental damage. The purpose is to indicate that this function will always present a major system risk factor and therefore must be given a higher level of safety rigor and continued safety vigilance. The potential mishap risk associated with an SCF must be reduced to an acceptable level; however, this does not reduce or change the SC designation.

It is generally the task of system safety to identify SCFs of the system, which is primarily achieved through HA—FHA in particular. Ensuring that the SCFs of a system present acceptable mishap risk is a large part of the SSP effort.

SAFETY-CRITICAL FUNCTION (SCF) THREAD

An SCF thread is the linked set of system elements (items, components, functions, software modules, etc.) comprising an SCF. These elements are necessary for the successful performance of the function. It also includes the input and output elements to the function. In order to make the SCF safe, each element in the SCF thread must be safe.

Figure 2.71 shows an example SCF thread for the brake function in an automobile system. It was determined from FHA of this hypothetical system that the brake function was safety-critical, because if the braking function failed when needed it could lead to a loss of life mishap. All of the systems (or subsystems) in the brake function are identified in this thread and they in turn become safety-critical. Further HA can now be performed on these critical components and a risk assessment performed. Derived design safety requirements can be generated for each element as necessary to mitigate unacceptable risk. This thread can be used to understand the HCFs and their interrelationships within the thread.

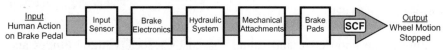

Figure 2.71 SCF thread for brake function.

SAFETY-CRITICAL ITEM (SCI)

An SCI is any item that meets the criteria of being safety-critical. It is a special safety designation given to any item (hardware or software) whose failure to operate, or incorrect operation, will directly lead to a mishap with catastrophic or critical outcome (as opposed to marginal or negligible outcome), such as death, serious injury, major system loss, or severe environmental damage. Typically, any item in an SCF thread is considered to be an SCI.

An SCI is a hardware or SI that has been determined, through system safety analysis, to potentially contribute to a catastrophic or critical hazard, or that may be implemented to mitigate a catastrophic or critical hazard. An SCI is essentially the same as a CSI except that systems required to identify CSIs have additional statutory and regulatory requirements that the contractor must meet in supplying those CSIs to the government. SCI and CSI lists are typically developed from HA and potential mishap risk assessment.

See *Critical Safety Item (CSI)* and *Safety-Critical* for additional related information.

SAFETY-CRITICAL OPERATION

A safety-critical operation is any operation whose failure to operate, or incorrect operation, will directly lead to a mishap with catastrophic or critical outcome (as opposed to marginal or negligible outcome), such as death, serious injury, major system loss, or severe environmental damage. The purpose is to indicate that this operation will always present a major system risk factor and therefore must be given a higher level of safety rigor and continued safety vigilance. The potential mishap risk associated with a safety-critical operation must be reduced to an acceptable level; however, this does not reduce or change the safety-critical designation. It is generally the task of system safety to identify safety-critical operations of the system, which is achieved primarily through HA.

SAFETY-CRITICAL REQUIREMENT

A safety-critical requirement is a design requirement that is necessary to mitigate, or assist in the mitigation, of a hazard that has been designated as SC. It is also a requirement that is involved in the implementation of a SC function, operation, task, and so on.

SAFETY CULTURE

Culture is the knowledge and values shared by a society; the attitudes and behavior that are characteristic of a particular social group or organization.

Safety culture is the organizational atmosphere, attitude, environment, and conditions surrounding the personnel work environment that reflects the organizations attitude toward safety. The safety attitude that permeates a work environment tends to have positive or negative influence on performance in regard to safety. If the system culture is a strong advocate for safety, then the safety culture will have a positive influence, and personnel will strive toward better safety practices. If the system culture is negative toward safety, then personnel do not perceive overall concern for safety, and they become lackadaisical toward safety, in both system design and operation. It is imperative that a positive safety culture be established and backed by management, during both design and operational phases.

The safety culture of an organization is an important factor in implementing and maintaining safety in a development, operational, or manufacturing program. It is an important factor in an organization developing a safe system or product because the attitude will directly influence how much actual effort and integrity is put into the effort necessary to develop a safe system or product design.

See *Culture* for additional related information.

SAFETY DEVICE

A safety device is a special and intentional feature in the design of a system or product employed specifically for the purpose of eliminating or mitigating the risk presented by an identified hazard. A safety device may not be necessary for system function, but it is necessary for safety (i.e., risk reduction). A safety device can be any device, technique, method, or procedure incorporated into the design to specifically eliminate or reduce the risk factors comprising a hazard. In general, safety devices often tend to be static interveners intended to serve as hazard countermeasures. Examples include physical guards, barricades, revetments around explosive storage facilities, guardrails, machine guards, safety eyewear, hearing protection, guards, and barricades.

See *Design Safety Feature (DSF)* and *Safety Order of Preference (SOOP)* for additional related information.

SAFETY FACTOR (SF)

SF is a term describing the cushion between expected load and actual design strength in a component or product design. With an SF the component should work successively with the expected load, and it should still operate successively with an unexpected heavier than anticipated load. The SF provides a cushion (safety margin) for unexpected conditions. The SF allows for uncertainty in the design process, such as calculations, strength of materials, duty and quality, and actual operational conditions.

An SF is typically the ratio of the safety design to the required design. The required design is what the item is required to be able to withstand for expected conditions, whereas the safety design includes an SF that provides a measure of how much more the actual design can tolerate (before failing). For example, a beam in a structure may be required to carry a design load of 3000 lb force. For safety and reliability purposes, the engineer selects a beam that will be able to handle 9000 lb force, providing a 3 to 1 design SF. In this example, the selected safety design provides a cushion of 6000 lb of unanticipated loading.

Even if each component of a larger product has the same SF, the product as a whole does not necessarily have that same SF. If one part is stressed beyond maximum force, the distribution might be changed throughout the entire product, and its ability to function could be affected. Determining SF is a balancing act between cost reduction and safety.

Figure 2.72 shows a load diagram. The expected load is the required design and the load capacity is the safety design. The gray area where the two load curves intersect is the area where stress is applied and wearout failure can be expected.

Care must be taken when applying the SF because a device does not always fail at exactly the specified load limit. As shown in Figure 2.72, a certain quantity of devices will fail below the specified load limit and a certain quantity will fail above. Therefore, if the load-carrying device is safety-critical, the failure confidence level must be taken into account. The safety margin must be very high to provide an adequately safe margin, or the device quality must be very good in order to narrow the failure distribution. SFs can be misleading and have been known to imply greater safety than is the case. A high design SF, well over the required factor, generally results in excessive weight and cost, which sometimes implies over-engineering. Appropriate factors of safety are based on several considerations, such as the accuracy of load, strength, wear estimates, and the environment to which the item will be exposed in service;

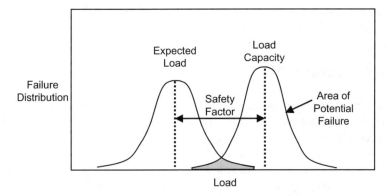

Figure 2.72 Load versus failure distribution.

the consequences of failure, the component failure distribution, and the cost of over-engineering the component to achieve that factor of safety. For example, components whose failure could result in substantial financial loss, serious injury, or death should use a high SF, perhaps four or greater. Noncritical components can generally have a lower safety, perhaps two or higher, for example.

SFs are most generally used in structural applications. However, they can also be used in electronics applications. For example, in a critical application, a diode may be selected using an SF in order to provide a safety margin in current capacity in case unexpected over current occurs in the circuit. This would protect the diode from failing (up to the cushion end point) in a critical application under unexpected stress.

It should be noted that the terms SF, factor of safety, margin of safety, and safety margin are all essentially synonymous.

SAFETY FEATURE

A safety feature is a special and intentional feature in the design of a system or product employed specifically for the purpose of eliminating or mitigating the risk presented by an identified hazard. A safety feature may not be necessary for system function, but it is necessary for safety (i.e., risk reduction). A safety feature can be any device, technique, method, or procedure incorporated into the design to specifically eliminate or reduce the risk factors comprising a hazard.

See *Design Safety Feature (DSF)* and *Safety Order of Preference (SOOP)* for additional related information.

SAFETY INSTRUMENTED FUNCTION (SIF)

An SIF is a function that is safety-critical in process system and is monitored by a SIS. SIFs are determined and implemented in a SIS as part of an overall risk reduction strategy which is intended to reduce the likelihood of identified hazardous events involving a catastrophic release. The SIS safe state is a state of the process operation where a mishap cannot occur. The safe state should be achieved within one-half of the process safety time. Most SIFs are focused on preventing catastrophic mishaps.

See *Safety Instrumented System (SIS)* for additional related information.

SAFETY INSTRUMENTED SYSTEM (SIS)

An SIS is an independent system used to provide safe control functions for processes, for example, an ESD system function or a fire detection and extinguishing system function. SISs typically are composed of sensors, logic solvers, and final control elements. A SIS is a control safety system that is typically implemented in industrial processes, such as those of a factory or an oil refinery. The SIS performs specified functions to achieve or maintain a safe state of the process when dangerous process conditions occur. SISs are separate and independent from regular control systems but are composed of similar elements, including sensors, logic solvers, actuators, and support systems.

The specified functions that are to be made safe are referred to as SIFs. SIFs are determined and implemented as part of an overall risk reduction strategy which is intended to reduce the likelihood of identified hazardous events involving a catastrophic release. The SIS safe state is a state of the process operation where a mishap cannot occur. The safe state should be achieved within one-half of the process safety time. Most SIFs are focused on preventing catastrophic mishaps.

The correct operation of a SIS requires a series of equipment to function properly. It must have sensors capable of detecting abnormal operating conditions, such as high flow, low level, or incorrect valve positioning. A logic solver is required to receive the sensor input signals, make appropriate decisions based on the nature of the signals, and change its outputs according to user-defined logic. The logic solver may use electrical, electronic, or programmable electronic equipment, such as relays, trip amplifiers, or PLCs. The logic solver outputs results in the final system devices taking action on the process (e.g., closing a valve) to bring it to a safe state. Support systems, such as power, instrument air, and communications, are generally required for SIS operation. The support systems should be designed to provide the required safety integrity and reliability, and they should be independent of the primary process systems. The SIS concept is shown in Figure 2.73.

International standard IEC 61511 provides guidance to end users on the application of SISs in the process industries. This standard is based on IEC 61508, a generic standard for design, construction, and operation of E/E/PESs. Other industries also have standards that are based on IEC 61508, such as IEC 62061 (for machinery systems), IEC 62425 (for railway signaling systems), IEC 61513 (for nuclear systems), and ISO 26262 (for road vehicles).

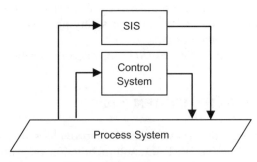

Figure 2.73 SIS concept.

What an SIS shall do (the functional requirements) and how well it must perform (the safety integrity requirements) may be determined from HAs, LOPA, risk assessments, and so on. During SIS design, construction, installation, and operation, it is necessary to verify that these requirements are met. The functional requirements may be verified by design reviews, such as FMECA and various types of testing, for example, factory acceptance testing, site acceptance testing, and standard functional testing.

SISs are most often used in process facilities, such as refineries, chemical plants, and nuclear power plants, to provide safety protection. Example protection functions include:

- High fuel gas pressure initiates action to close the main fuel gas valve.
- High reactor temperature initiates action to open cooling media valve.
- High distillation column pressure initiates action to open a pressure vent valve.

Standards involving SIS systems include:

- IEC 61508, Functional Safety of Electrical/Electronic/Programmable Electronic Safety-Related Systems, 1999.
- IEC 61511, Functional Safety of Safety Instrumented Systems for the Process Industry Sector.
- ISA 84.01–2003, Functional Safety: Safety Instrumented Systems for the Process Industry Sector.

SAFETY INTEGRITY LEVEL (SIL)

SIL is defined as the relative level of risk reduction provided by a safety function, or a specified target level of risk reduction. SIL is a measurement of performance required for an SIF. Safety integrity is the probability of an SR system satisfactorily performing the required safety functions under all the stated conditions and the stated time period.

The IEC's standard IEC 61508, Functional Safety of Electrical/Electronic/ Programmable Electronic Safety-Related Systems, 1999, defines SIL using requirements grouped into two broad categories: hardware safety integrity and systematic safety integrity. A device or system must meet the requirements for both categories to achieve a given SIL. Four SIL levels are defined, with SIL4 being the most dependable and SIL1 the least dependable. A SIL is determined based on a number of quantitative factors in combination with qualitative factors such as development process and safety life-cycle management. The SIL requirements for hardware safety integrity are based on a probabilistic analysis of the device. To achieve a given SIL, the device must have less than the specified probability of dangerous failure and have greater than the specified safe failure fraction.

TABLE 2.18 SILs Defined in IEC61508

SIL	PFD	
1	0.1–0.01	$\geq 10^{-2}$ to 10^{-1}
2	0.01–0.001	$\geq 10^{-3}$ to 10^{-2}
3	0.001–0.0001	$\geq 10^{-4}$ to 10^{-3}
4	0.0001–0.00001	$\geq 10^{-5}$ to 10^{-4}

The SIL requirements for systematic safety integrity define a set of techniques and measures required to prevent systematic failures from being designed into the device or system. These requirements can either be met by establishing a rigorous development process or by establishing that the device has sufficient operating history to argue that it has been proven in use. An SIL is a statistical representation of the reliability of the SIS when a process demand occurs. SILs are correlated to the PFD, which is equivalent to the unavailability of a system at the time of a process demand. The actual targets required vary depending on the likelihood of a demand, the complexity of the device, and types of redundancy used. SILs defined in IEC61508 are shown in Table 2.18.

The amount of risk reduction means the residual risk presented by the device. The SIL table provides a class of safety integrity. An SIL 1 device is not as reliable in providing risk reduction as a SIL 2 device; SIL 3 is better and SIL 4 is best. An SIL is the safety classification or grading of a system. The more LOP, the greater the RRF, where RRF = 1/PFD. The particular SIL that is required drives the device architecture to achieve that level.

The term SIL is also defined and used in DEF (AUST) 5679, The Procurement of Computer-Based Safety Critical Systems, Australian Defence Standard, 1999.

SAFETY INTERLOCK

A safety interlock is a design safety arrangement whereby the operation of one control or mechanism allows, or prevents, the operation of another function. The safety interlock is a special safety device used in a system design to increase the level of safety of a specific function; it is a DSF. The primary purpose of an interlock is to provide a mechanism to make or break an SR function, based upon a set of predetermined safety criteria. It should be noted that interlocks are absolutely not necessary for the operational functionality of a system. An interlock is a device added to the design in order to achieve the needed safety required of the system, not for the operational effectiveness of the system. Interlocks control state transitions in an attempt to prevent the system from entering an unsafe state or to assist in exiting from an unsafe state.

See *Interlock* for additional related information.

SAFETY LATCH

A safety latch is a mechanical device designed to slow direct entry to a controlled area, such as a safety latch on the door to a laser laboratory.

SAFETY MANAGEMENT SYSTEM (SMS)

A major contributing factor in operational accidents is the failure to adequately manage known risk, due to the lack of a systematic process, including leadership and accountability. SMS is a process for managing risk that ties all elements of the organization together laterally and vertically and ensures appropriate allocation of resources to safety issues. Note that SMS is not the same as system safety. System safety impacts the design of a system or product, whereas SMS impacts the operations of an organization, and is very similar to ORM.

SMS is a coordinated, comprehensive set of processes designed to direct and control resources to optimally manage the safety of an operational aspect of an organization. SMS takes unrelated processes and builds them into one coherent structure to achieve a higher level of safety performance, making safety management an integral part of overall risk management. SMS is based on leadership and accountability. It requires proactive hazard identification, risk management, information control, auditing, and training. It also includes incident and accident investigation and analysis.

SMS facilitates the proactive identification of hazards and maximizes the development of a better safety culture, as well as modify attitudes and actions of personnel in order to make a safer workplace. SMS helps organizations avoid wasting financial and human resources and management's time being focused on minor or irrelevant issues. SMS lets managers identify hazards, assess risk, and build a business case to justify controls that will reduce risk to acceptable levels.

There are typically 11 fundamental attributes that will assist in ensuring the SMS is effective for any organization. The core attributes of a SMS are:

1. SMS management plan
2. Safety promotion
3. Document and data information management
4. Hazard identification and risk management
5. Occurrence and hazard reporting
6. Occurrence investigation and analysis
7. Safety assurance oversight programs
8. Safety management training requirements
9. Management of changes

10. Emergency preparedness and response
11. Performance measurement and continuous improvement

It is important to view an SMS as a business/operations tool for owners and managers. The risk management process within the SMS includes the need to determine the cost of implementing versus not implementing control measures. In an SMS, policies and procedures are the ways organizations express and achieve their desired level of safety. Policies characterize the nature and performance of an organization, and procedures define how to execute policies. Policy is information that establishes a basic requirement for how the organization functions (what you want to do). It should be short and to the point. Customers should also know what the organization's policies are so they can base their expectations on them. Policies guide the development of procedures. Procedures define the actual methods that the organization uses to apply their policies (how you do what you want done).

SMS is a proactive, integrated approach to safety management. SMS is part of an overall management process that the organization has adopted in order to ensure that the goals of the organization can be accomplished. It embraces the principle that the identification and management of risk increases the likelihood of accomplishing the mission. Hazards can be identified and dealt with systematically through the hazard reporting program that facilitates continuing improvement and professionalism. Auditing and monitoring processes ensure that aircrafts are operated in such a way as to minimize the risks inherent in flight operations. For an effective SMS, an organization must have a culture of open reporting of all safety hazards in which management will not initiate disciplinary action against any personnel, who in good faith, due to unintentional conduct, discloses a hazard or safety incident.

In order to develop an effective SMS, management must embrace the following SMS safety principles and establish policies to implement these principles:

- Always operate in the safest manner practicable
- Never take unnecessary risks
- Safe does not mean risk free
- Everyone is responsible for the identification and management of risk
- Familiarity and prolonged exposure without a mishap leads to a loss of appreciation of risk
- An absence of accidents does not necessarily equate to safety
- Development of a safety culture
- Continued pursuit of an accident-free workplace and operations
- Support for safety training and awareness programs
- Conducting regular audits of safety policies, procedures, and practices

- Monitoring industry activity to ensure best safety practices are incorporated into the organization
- Providing the necessary resources to support the SMS policy
- All levels of management are accountable for safety performance, starting with the owner/chief executive officer (CEO)

SMS supports the principle that all personnel have the duty to comply with approved standards, which include organization policy, procedures, aircraft manufacturer's operating procedures, and limitations and government regulations. Research shows that once you start deviating from the rules, you are almost twice as likely to commit an error with serious consequences. Breaking the rules usually does not always result in an accident; however, it always results in greater risk for the operation. Violating the rules breaks the principle of "never take unnecessary risks." Undesired outcomes are a function of behavior; therefore, management must be committed to identifying deviations from standards and taking immediate corrective action. Corrective action must be consistent and fair, and can include actions such as counseling, training, discipline, grounding, or dismissal.

SAFETY MARGIN

Safety margin of safety is a term describing the cushion between expected load and actual design strength in a component or product design. A safety margin is typically the ratio of the safety design to the required design. The required design is what the item is required to be able to withstand for expected conditions, whereas the safety margin includes a safety margin that provides a measure of how much more the actual design can tolerate. The terms SF, factor of safety, margin of safety, and safety margin are synonymous.

See *Safety Factor (SF)* for additional related information.

SAFETY MEASURE

A safety measure is a special and intentional feature in the design of a system or product employed specifically for the purpose of eliminating or mitigating the risk presented by an identified hazard. A safety measure may not be necessary for system function, but it is necessary for safety (i.e., risk reduction). A safety measure can be any device, technique, method, or procedure incorporated into the design to specifically eliminate or reduce the risk factors comprising a hazard.

See *Design Safety Feature (DSF)* and *Safety Order of Preference (SOOP)* for additional related information.

SAFETY MECHANISM

A safety mechanism is a special and intentional feature in the design of a system or product employed specifically for the purpose of eliminating or mitigating the risk presented by an identified hazard. A safety mechanism may not be necessary for system function, but it is necessary for safety (i.e., risk reduction). A safety mechanism can be any device, technique, method, or procedure incorporated into the design to specifically eliminate or reduce the risk factors comprising a hazard.

See *Design Safety Feature (DSF)* and *Safety Order of Preference (SOOP)* for additional related information.

SAFETY OF FLIGHT (SOF)

SOF determines the property of a particular air system configuration to safely attain, sustain, and terminate flight within prescribed and accepted limits for injury/death to personnel and damage to equipment, property, and/or environment. The intent of assessing SOF is to show that appropriate risk management has been completed and the level of risk (hazards to the system, personnel, property, equipment, and environment) has been appropriately identified and accepted by the MA. Ascertaining SOF is part of the airworthiness and flight clearance processes.

See *Airworthiness* and *Flight Clearance* for additional information.

SAFETY ORDER OF PRECEDENCE (SOOP)

Following the identification of a hazard and its associated risk, the safety objective is to either eliminate the hazard or reduce the hazard risk. It is preferable to eliminate the hazard when possible; however, it is often not possible to eliminate a hazard because the basic hazard components involved are necessary for desired system function. When unable to eliminate an identified hazard, system safety must mitigate or reduce the mishap risk presented by the hazard via the implementation of a DSF. It has been found through experience that DSFs fall into several different categories, with some categories being more effective than others. Mishap risk is most effectively reduced when the SOOP is followed, where the methods at the top of the hierarchical list are the most effective, and the least effective and desirable methods are at the bottom. The SOOP is a preferred approach to achieve optimum risk reduction.

The following is the SOOP in preferred order:

1. **Eliminate the hazard through design alternatives.** Select a design alternative that removes the hazard altogether. This is most often done by removing the HS component of the hazard.

2. **Reduce risk through DSFs.** When unable to eliminate a hazard, reduce the risk by reducing the risk likelihood and/or the risk severity. The preferred order of methods for achieving this is as follows:
    - **Incorporate modified design selections.** Adopt an alternative design selection that reduces the HE component of a hazard. For example, reduce the amount of voltage required, reduce the amount of HAZMATs used, change to a less dangerous chemical, and use a less volatile fuel.
    - **Incorporate DSFs.** Utilize special DSFs, such as redundancy, interlocks, SISs, backup systems, fault detection, fault tolerance, partitions, and design diversity.
    - **Incorporate safety devices.** Use protective safety methods or devices. In general, safety devices are static interveners. Examples include physical barriers; machine guards; barricades; revetments around explosives storage facilities; guardrails; toe boards; safety eyewear; hearing protectors
    - **Provide warning devices.** Use detection and warning systems to alert personnel to the particular hazard.
    - **Develop procedures and training.** Incorporate special procedures and special training to guide personnel in safe operations. Procedures may include the use of PPE. For hazards assigned catastrophic or critical mishap severity, avoid using this step as the only risk reduction method.

SAFETY PRECEPT

Safety precepts form basic truths, laws, or presumptions that can be used as a basis for safety reasoning in developing system design and operational requirements. Good safety precepts traverse a narrow path; their intent is to influence and guide critical safety design without being overly prescriptive in a manner which might constrain the design or the design options. In essence, safety precepts articulate a desirable fundamental safeguard.

A safety precept is worded as a nonspecific and unrestricted safety objective that provides a focus for addressing potential safety issues that present notable mishap risk. Precepts are intentionally general and not prescriptive in nature; they provide a goal, which may be achieved via numerous possible options. Precepts provide a focus and objective as opposed to a detailed solution. The need for a safety precept may result from the desire to mitigate certain hazards, hazard types, TLMs, or special safety concerns.

The Three Laws of Robotics developed by Isaac Asimov in his books on robots (circa 1939) provide a good example of safety precepts. These precepts merely state what the end goal must be, but not how to specifically accomplish the end goal. The detailed implementation could be different for different robotic systems, or it could change as technology improves. The Three Laws of Robotics are:

1. A robot may not injure a human being or, through inaction, allow a human being to come to harm.
2. A robot must obey orders given it by human beings except where such orders would conflict with the First Law.
3. A robot must protect its own existence as long as such protection does not conflict with the First or Second Law.

Safety precepts should meet the following criteria, as a minimum, in order to be valuable and effective for system designers:

1. Safety precepts must provide value in the guidance they provide. If they do not help the safety or the design organizations, then they are of no worth to the program.
2. Safety precepts must provide general safety guidance or safety objectives. The goal is to provide a general focus and direction on a particular safety issue, not a detailed design requirement.
3. Safety precepts must provide a framework for safety reasoning. Precepts should stimulate thinking on both primary and secondary aspect of the safety issue.
4. Safety precepts should allow for tailoring or options in the specific implementation, since different applications may have some unique variances.
5. A safety precept should give system developers and users' confidence that the particular safety concern will be relatively safe if the precept is followed.

In order to completely facilitate safety guidance and direction, safety precepts have been subdivided into three categories, whereby each category is aimed at a specific area of program influence. These categories include the following:

PSP

A PSP is a safety precept that is directed specifically at organizational goals, tasks, policy, standards, and/or processes that will help implement safety into the system development process. These precepts offer safety guidance, for the PM, which will effect mishap risk reduction through an effective and well-planned safety program. When the PSPs are closely followed, the culture and tasks are set in place for an efficient and successful SSP. An example PSP might be "The program shall ensure that COTS software is assessed for safety as a component in the system environment and not as an external element."

OSP

An OSP is a safety precept that that is directed at the operational phase. These precepts are operational safety rules that must be adhered to

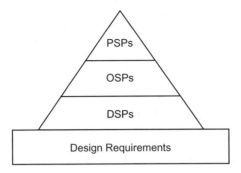

Figure 2.74 Safety precepts pyramid.

during system operation. One aspect of OSPs is that they may directly spawn the need for DSPs that will help realize the OSP via design methods. Although the development PM has no influence during system operation, the OSPs will provide guidance for many design features that will need to be implemented in order to meet the OSPs. An example OSP might be "The system shall be considered unsafe until a safe state can be verified."

DSP

A DSP is a safety precept that is directed specifically at system design. These precepts are general design objectives intended to guide and facilitate the design of more detailed solutions, without dictating the specifics within the precept. General design direction allows for the selection of specific solutions that are focused on the particular application, along with the technologies available. An example DSP might be "The system shall be designed to only perform valid commands issued from a valid authorized source."

Safety precepts are building block in the system safety process. The three types of safety precepts can be viewed as forming a pyramid as depicted in Figure 2.74. The design safety requirements process begins with PSPs at the top of the pyramid. At the next level are the OSPs and then the DSPs. Once the three levels of safety precepts have been established, a baseline has been created for developing and guiding the detailed design requirements.

SAFETY-RELATED

Safety-related is a special safety designation given to any item that will lead to a hazard or unsafe condition should the item fail to function, malfunction, or perform properly. An SR item can be any system level of hardware, software, or human task involving an event, operation, process, function, procedure, or tolerance measurement.

Safety-related is used in two different ways:

1. Safety-related is used as a general overarching term applied to any item that is related to safety. In this usage it is inclusive of than safety-critical and safety-significant items (SSIs).
2. Safety-related is also used as a specific term referring to anything less than safety-critical in severity, typically in situations where they are being used together.

Typically, if an item is safety-critical, it is then referred to as being safety-critical, as opposed to safety-related. Items less than safety-critical in severity are referred to as safety-related. When safety-critical and SR items are being compared together, safety-related refers to items whose failure to operate, or incorrect operation, will contribute to a mishap that could result in injury, minor system loss, or minor environmental damage (the same as safety-significant), whereas safety-critical hazards refer to hazards with catastrophic or critical severity.

SAFETY REQUIREMENTS/CRITERIA ANALYSIS (SRCA)

The SRCA is an analysis for evaluating SSRs for completeness and thoroughness. SRCA has a two-fold purpose:

1. To ensure that every identified hazard has at least one corresponding safety requirement
2. To verify that all safety requirements are implemented and successful

The SRCA is essentially a traceability analysis to ensure that there are no holes or gaps (i.e., no hazard has been left unmitigated) in the safety requirements and that all identified hazards have adequate and proven design mitigation coverage. The SRCA applies to hardware, software, firmware design requirements.

SRCA is applicable to analysis of all types of systems, facilities, and software where hazards and safety requirements are involved during development. SRCA is particularly useful when used in an SwSP. The SRCA technique, when applied to a given system by experienced safety personnel, is very thorough in providing an accurate traceability of safety design requirement verification and safety test requirement validation.

The SRCA process consists of comparing the SSRs to design requirements and identified hazards. In this way, any missing safety requirements will be identified. In addition, SSRs are traced into the test requirements to ensure that all SSRs are tested. The idea behind this thought process is that a matrix worksheet is used to correlate safety requirements with design requirements, test requirements, and identified hazards. If a hazard does not have a corresponding safety requirement, then there is an obvious gap in the safety requirements. If a safety requirement is not included in the design requirements, then there is a gap in the design requirements. If a safety requirement is missing from the test requirements, then that requirement cannot be verified and validated. If an SSR cannot be shown to have passed testing, then the associated hazard cannot be closed.

The SRCA is a detailed correlation analysis, utilizing structure and rigor to provide traceability for all SSRs. The SRCA begins by acquiring the system hazards, SSRs, design requirements, and test requirements. A traceability matrix is then constructed that correlates the hazards, SSRs, design requirements, and test requirements together. The completed traceability matrix ensures that every hazard has a corresponding safety requirement and that every safety requirement has a corresponding design and test requirement.

The SRCA consists of two separate correlation traceability analyses: (1) an SSRs correlation and (2) a guideline compliance correlation. The guideline correlation only applies to systems where guidelines exist and are applied to the system design. For example, generic SwS guideline requirements form are generally applied to the design of software, and the guideline requirements from MIL-STD-1316 are applied to fuze system designs. Figure 2.75 exemplifies this process.

The SRCA is a detailed analysis utilizing structure and rigor. It is desirable to perform the SRCA using worksheets. Although the format of the analysis worksheet is not critical, typically, matrix or columnar-type worksheets are used to help maintain focus and structure in the analysis. Software packages are available to aid the analyst in preparing these worksheets. Figure 2.76 provides an example SRCA worksheet for the traceability of SSRs.

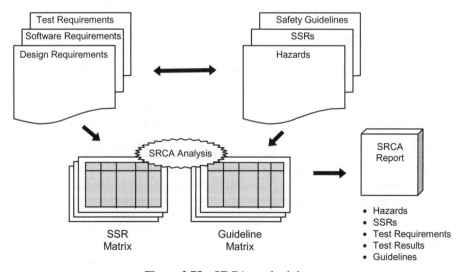

Figure 2.75 SRCA methodology.

368 SYSTEM SAFETY TERMS AND CONCEPTS

System: Subsystem:			SSR Traceability Matrix				SRCA			
SSR No.	System Safety Requirement (SSR)	SC	HAR No.	TLM No.	Design Req. No.	Test Req. No.	Test			
							M	C	R	
							M=Method C=Coverage R=Result			

Figure 2.76 Example SRCA requirements correlation matrix worksheet.

SRCA is a safety analysis technique that is important and essential to system safety. For more detailed information on the SRCA technique, see Clifton A. Ericson II, *Hazard Analysis Techniques for System Safety* (2005), chapter 10.

SAFETY-SIGNIFICANT ITEM (SSI)

An SSI is any item that is safety-related but is not safety-critical. An SSI is any function, subsystem, or component, the failure of which (including degraded functioning or functioning out of time or out of sequence) could result in a hazard or contribute to a hazard with marginal or negligible severity.

SAFETY VULNERABILITY

Safety vulnerability is the susceptibility of a system to hazards and mishap risk. Safety vulnerability is not equal among systems; some systems have more vulnerability than others. Safety vulnerability results from many different driving factors, such as hazardous system components, system size, system complexity, and system application.

Understanding the safety vulnerabilities within a system is important because it drives the SSP tasks and cost. It also delineates the relative system risk and criticality that can be expected from the system. A system's safety vulnerability is determined through HA and risk assessment.

SAFING

Safing is the act of disarming an armed device; moving the device from arm to safe.

SAMPLING

Sampling is that part of statistical practice concerned with the selection of an unbiased or random subset of item observations within a population of items, intended to yield some knowledge about the population of concern, especially for the purposes of making predictions based on statistical inference. Sampling involves taking a representative portion of a material, product, or group to test, measure, or observe. Sampling typically provides information to determine if a product meets design requirements, quality requirements, and/or regulatory requirements. Sampling of individuals provides information about a population on factors such as preferences, satisfaction, personal attributes, and so on. Sampling may involve testing or observing the measurable parameters of a product, testing a one-shot device such as an explosive item, taking surveys using questionnaires, and so on.

Analysts generally cannot test or survey the entire population of an item because the cost is too high or the item will be destroyed in the test. The advantages of sampling are that the cost is lower, data collection is faster, and fewer items are damaged or destroyed by certain tests.

In software development, sampling involves selecting a representative set of software life-cycle data for inspection or analysis. The purpose is to determine the compliance of all software life-cycle data developed up to that point in time in the project. Sampling is the primary means of assessing the compliance of the software processes and data. Examples of sampling may include the following:

- Inspecting the traceability from system requirements to software requirements to software design to source code to object code to test cases and procedures to test results.
- Reviewing analyses used to determine system safety concerns and requirements.
- Examining the structural coverage of source code modules.
- Examining software quality assurance (SQA) records and CM records.

SECONDARY FAILURE

A secondary failure is the failure of a component due to an external factor or force on the component, causing the component to fail by exceeding its design parameters. For example, "diode fails due to excessive RF/EMI energy in the system." In this example, excessive electromagnetic energy on the component causes early failure of the diode component. A secondary failure is typically the result of out-of-tolerance operational or environmental conditions. A secondary failure is a component failure that is directly caused by a separate independent event, which is the true root cause. A secondary failure involves a cause–effect relationship, which is also a dependency relationship.

Some example causes of secondary failure include:

- Temperature
- RF energy
- Water
- Electrical voltage or current

It should be noted that this term is primarily used in FTA. Conditional probability should be used on secondary failures because it is dependency situation; however, in FTA, the conditional aspect is often ignored, and the failure event is simply treated as an independent failure and assigned an appropriate failure that considers both the component failure and the external event failure rate. The mathematical error produced by this approach is typically minimal.

See *Dependent Failure* for additional related information.

SECTION

A section is a structurally integrated set of components and integrating hardware that forms a subdivision of a subsystem, module, and so on. A section forms a testable level of assembly, such as components/units mounted into a structural mounting tray or panel-like assembly, or components that are stacked.

SHALL

With regard to design requirements, the term "shall" indicates a mandatory action in the requirement. It is typically used in a requirement when the application of the requirement or procedure is mandatory. Requirements using "shall" statements are considered to be binding and to require formal verification.

See *Should* and *Will* for additional related information.

SHARP EDGES

Sharp edges on mechanical parts or equipment can cause personnel injury and/or equipment damage. To prevent mishaps, sharp edges should be rounded to a safe radius. MIL-STD-1472E Human Engineering, Design Criteria Standard, 1998, states the following:

> 5.13.5.4 Edge Rounding
> Where applicable, all exposed edges and corners shall be rounded to a radius not less than 0.75 mm (0.03 in.). Sharp edges and corners that can present a personal safety hazard or cause equipment damage during usage shall be suitably protected or rounded to a radius not less than 13 mm (0.05 in.).

SHOCK HAZARD

Electrical current/voltage in a system is a basic hazard source for many types of electrical-related hazards, all of which should be identified in a system safety HA. Electric shock is the sudden pain or convulsion which results from the passage of an electric current through the body. Minor electrical shocks may cause mishaps due to involuntary reactions. Major electrical shocks may cause death due to burns or paralysis of the heart or lungs. An electric shock results from the passage of direct or alternating electrical current through the body or a body part.

See *Electrical Shock* for additional related information.

SHOP REPLACEABLE UNIT (SRU)

An SRU indicates the operations or maintenance level at which a system element can be repaired or replaced. An SRU is typically a subsystem or assembly that has to be sent back to the depot or manufacturer for repair. It is usually a system element that is at a lower level in the system hierarchy, such as a circuit board for an aircraft flight control system.

See *System Hierarchy* for additional related information.

SHOULD

With regard to design requirements, the term "should" indicates standard policy, and deviation is discouraged in the requirement. It is typically used in a requirement when the application of the requirement or procedure is (strongly) recommended. Requirements defined using "should" statements are objectives and are optional for formal verification.

See *Shall* and *Will* for additional related information.

SINGLE POINT FAILURE (SPF)

An SPF is the failure of a single item or component that would, in turn, directly lead to the occurrence of a specified UE. The intent is to identify component safety significance and criticality in order to understand where potential design safety weaknesses exist, so they can be eliminated or mitigated. Typically, the UE is a critical-safety condition, such as loss of life, loss of the system, loss of mission, and environmental damage. The particular UE depends upon the particular system and its objectives.

It should be noted that from a very broad perspective, a system could have hundreds or even thousands of SPFs in the design that could potentially cause anything from minor to catastrophic outcomes. The design concern is which of these SPFs will cause a specified UE adversely affecting safety (or reliability) objectives. Therefore, a design statement that all SPFs will be eliminated is unreasonable; only those SPFs adversely impacting safety or reliability goals should be addressed.

Relatively speaking, every component in a system can fail and is therefore a potential SPF. So, what is the significance and value of the SPF concept? SPF is a concept and tool for design safety and reliability. The intent of the SPF concept is to identify only those single failures that will directly cause a *specific* UE to occur. SPF does not apply to every possible single failure, but only those of predetermined safety or reliability relevance. The term could be applied to different safety and reliability-related UEs in a particular system, such as aircraft crash and loss of flight controls.

Intuitively, an SPF is not a good thing; loss of a desirable function due to one act does not seem acceptable. However, SPFs are not necessarily a bad thing either. Goodness and badness is a function of criticality and failure probability. The human body only has a single heart, possibly because it is less exposed, and it seems to have a low failure rate when properly treated. Think of the timing, weight, and plumbing issues that could be involved with a redundant heart design; it appears that a single, more reliable unit is the best design in this case.

In general, prohibiting SPFs for all SR UEs seems reasonable due to their significance. However, there is a real danger in blindly specifying that no SPF shall result in a catastrophic or severe hazard without taking into deliberation system unique factors and constraints. Consider the following:

- The failure rate for a certain single component may be smaller than the combined failure rate for two redundant items, thus a redundant design could actually be less safe.
- The cost, weight, and/or size of two redundant items may be prohibitive for the particular system, and acceptable risk safety can be achieved through the reliability of a single item.

In trying to establish a good safety policy, many system development programs will mandate that no catastrophic or critical hazards can exist in the system design due to an SPF. This is quite often done with good intentions by establishing a program safety precept with this stipulation. It must be noted, however, that trying to mitigate hazards by a parts count policy is a very tricky and risky business. SPF risk mitigation should be accomplished primarily by probability control, as opposed to a parts count rule.

The objective of identifying SPFs is to prevent SPFs that could directly result in a *specified UE* of significant safety consequence. These UEs usually have significant safety consequence attached to them. It should be noted that loss of life is the typical ultimate outcome of an UE, and therefore should not really be in the SPF definition. For example, a properly stated SPF prohibition might be something like "no SPF shall result in loss of the three redundant

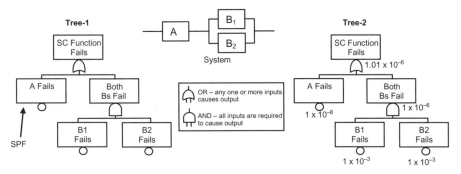

Figure 2.77 FTA of single point failure (SPF).

power supplies and power distribution network in the aircraft." An inadequate SPF prohibition would be "no SPF shall cause loss of life." The latter example is far too broad to provide adequate design guidance.

To determine which is better from a safety standpoint, consider the example system shown in Figure 2.77, which is a hypothetical SCF where successful safe operation requires successful operation of components A and B. There is only one component A, but there are two redundant components of type B, only one of which must function for system success. Figure 2.77 contains an FTA of this system for the UE "SC function fails." This FT clearly depicts that component A is an SPF for this system. There is only one component A, and its failure directly causes the UE. If component B1 fails, there is still a backup type B component B2; thus, B1 and B2 are redundant and prevent vulnerability due to an SPF.

The most obvious conclusion from Tree-1 is that component A should be designed out of the system or be made redundant. The SPF sticks out like a sore thumb and looks very unsafe. However, consider Tree-2 where the failure probability values have been added for each failure. The added information provided in Tree-2 presents a slightly different story. This FT shows that the combined probability of both B1 and B2 failing is $P = 1 \times 10^{-6}$, whereas the probability of the SPF event A is $P = 1 \times 10^{-8}$. Note that the probability driver is the redundant B components due to their each having a lower probability than the SPF. This means that the SPF is actually safer than the redundant design. Rather than spending money to fix the SPF, it might be wiser to consider spending money to improve the redundant design. SPF resolution and implementation should be based on risk likelihood for system safety and a reliability level for system reliability.

SITUATION AWARENESS (SA)

SA is the perception of environmental elements within a volume of time and space, the comprehension of their meaning, and the projection of their status in the near future. It is also a field of study concerned with perception of the environment critical to decision makers in complex, dynamic areas from aviation, power plant operations, and military operations, to more ordinary, but complex tasks, such as driving an automobile or a motorcycle.

SA involves being aware of what is happening around you to understand how information, events, and your own actions will impact your goals and objectives, both now and in the near future. Lacking SA or having inadequate SA has been identified as one of the primary factors in accidents/mishaps attributed to human error. Having complete, accurate, and up-to-the-minute SA is essential where technological and situational complexities on the human decision maker are a concern. Essentially, SA refers to knowing exactly where you are and what the conditions and circumstances are surrounding you. For a UMS, SA involves knowing where the UMS is exactly located and what the surrounding environment and conditions are.

SNEAK CIRCUIT

A sneak circuit is an unexpected (and unintended) path or logic flow within a system that, under certain conditions, can initiate an undesired function or inhibit a desired function. Sneak circuits are not the result of hardware failure but are latent conditions that are inadvertently designed into the system and cause it to malfunction under certain conditions.

SNEAK CIRCUIT ANALYSIS (SCA)

SCA is a safety analysis technique for identifying a special class of hazards known as sneak circuits or sneak paths. SCA is accomplished by examining electrical circuits and searching out unintended electrical paths which, without component failure, can result in:

- Undesired operations
- Desired operations but at inappropriate times
- The prevention of desired operations

A sneak circuit is a latent path or condition in an electrical system that inhibits a desired condition or initiates an unintended or unwanted action. This condition is not caused by component failures, but has been inadvertently designed into the electrical system to occur as normal operation. Sneak circuits often exist because subsystem designers lack the overall system visibility required to electrically interface all subsystems properly. When design modifications are implemented, sneak circuits frequently occur because changes are rarely submitted to the rigorous testing that the original design undergoes. Some sneak circuits are evidenced as "glitches" or spurious operational modes and can be manifested in mature, thoroughly tested systems after long use. Sometimes, sneaks are the real cause of problems thought to be the result of EMI or grounding "bugs." SCA can be applied to both hardware and software design; however, software SCA has proven to be less effective and less cost-effective.

Although a very powerful analysis tool, the benefits of an SCA are not as cost-effective to the system safety analyst as HA other tools. Other safety analysis techniques, such as SSHA and FTA, are more cost-effective for the identification of hazards and root causes. SCA is highly specialized, and only assists in a certain niche of potential safety concerns dealing with timing and sneak paths. The technique is not recommended for everyday safety analysis usage, and should be used when required for special design or safety-critical concerns. Specific reasons for performing an SCA include:

1. The system is safety-critical or high consequence and requires significant analysis coverage to provide safety assurance (e.g., safe and arm devices, fuzes, guidance systems, launch commands, and fire control system).
2. When an independent design analysis is desired.
3. The cause of unresolved problems (e.g., accidents, test anomalies) cannot be found via other analysis techniques.

The purpose of SCA is to identify sneak paths in electrical circuits which result in unintended operation or inhibited operation of a system function. There are several ways by which this can be achieved, such as systematic inspection of detailed circuit diagrams, manually drawing simplified diagrams for manual examination, or by using the automated topgraph-clue method developed by Boeing. The topograph-clue method is more structured and rigorous, and it was the genesis for the SCA concept. The theory behind the automated topgraph-clue method of SCA is conceptually that electrical circuit diagrams are transformed into network trees through the use of special computer programs. The network trees are then reduced down into topographs. The topographs are evaluated in conjunction with clue lists to identify sneak circuits. Although the concept appears to be very simple, there is much more complexity and manual labor involved in the process.

SCA is a powerful analysis technique for identifying design flaws that can result in hazard. Some of the unique characteristics of SCA include the following:

- SCA is somewhat of a proprietary technique. Only corporations that have researched and developed the tool have the clues that are necessary for identifying the sneak paths. The clues are not available in any public domain form. Therefore, the cost of SCA can be considerable.
- Entering the data for the SCA is a time-consuming process. Therefore, it is usually only done once during the program development, and this is usually with detailed design information. Consequently, any identified design changes can be more costly than if identified earlier in the program development life cycle.

- SCA does not identify all system hazards, only those dealing with sneak paths.
- SCA only considers normal component operation; it does not consider component failures.
- SCA requires an experienced analyst to actually recognize the sneak paths from the clues and topological diagrams.

SCA is an HA technique that is important and essential to system safety. For more detailed information on the SCA technique, see Clifton A. Ericson II, *Hazard Analysis Techniques for System Safety* (2005), chapter 16.

SOCIETAL RISK

Societal risk is the risk presented to society as whole from a potential event, hazard, or mishap. It involves the relationship between frequency and the number of people suffering from a specified level of harm in a given population from the realization of specified hazard.

SOFTWARE CAPABILITY MATURITY MODEL (CMM)

Software CMM is the application of the CMM to the development of software. There are five CMM levels, as shown in Figure 2.78, and at each higher level, the organization is able to develop better software on a cost-effective basis.

See *Capability Maturity Model (CMM)* for additional related information.

SOFTWARE CHANGE CONTROL BOARD (SCCB)

During the system development process, a CCB is a group delegated to make decisions regarding whether or not proposed changes to a project should be

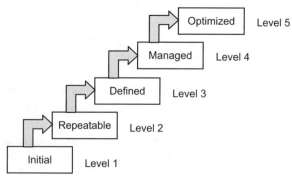

Figure 2.78 CMM levels.

implemented. The CCB is constituted of project stakeholders or their representatives. The authority of the CCB may vary from project to project, but decisions reached by the CCB are often accepted as final and binding. The CCB is part of the configuration control process and is formally documented as part of the configuration control plan. The SCCB is identical to the CCB except for its focus solely on software; on some projects, it is a part of the overall CCB.

Typically, system safety is a member of the SCCB in order to evaluate all changes and proposed changes for safety impact. A proposed change may make an existing safe design unsafe by compromising existing safety features, or it may introduce new hazards into the design. The SCCB works with ECPs for each change. ECPs should have a safety box on the form to ensure safety assessment of the ECP. ECPs should only be evaluated for safety by an inexperienced system safety analyst who understands hazards, mishaps, and risk.

See *Configuration Control* and *Engineering Change Proposal (ECP)* for additional related information.

SOFTWARE CRITICALITY INDEX (SCI)

The Software Criticality Index (SCI) refers to the index number obtained from the Software Criticality Level (SCL) matrix. It is essentially the SCL number, an index number (1 through 5) derived from the SCL matrix. This term is intended to replace the older term Software Hazard Risk Index (SHRI), primarily because the word "risk" in SHRI is misleading for the application.

See *Software Criticality Level (SCL)* for additional related information.

SOFTWARE CRITICALITY LEVEL (SCL)

The SCL is an indicator of the degree of safety importance of a software module. It is an arbitrary index number indicating how significant the safety impact will be should the software module fail or work incorrectly. In a sense, it is an implied risk indicator, suggesting that the software module contains the potential for a level of risk concomitant with the SCL. There are five SCLs: high, serious, medium, low, and not safety. The more critical the SCL assigned to the software module, the greater the implied potential risk significance (however, the actual risk is unknown). The more serious the SCL, the more carefully and rigorously the software module should be treated during development to make it safe.

When establishing and working with SCLs the following principles regarding the SCL concept should be kept in mind:

- The SCL is an index number ranking the relative safety importance of a software module. This ranking implies that in order to make the software safe, greater development rigor must be applied to each successive criticality level.

- Once assigned, the SCL level never changes, unless the basic software architecture is changed. The application of design risk mitigation measures does not change the SCL because it is not a risk measure.
- A low index number (i.e., 1) from the software criticality matrix does not mean that a design is unacceptable from a risk or safety standpoint. Rather, it indicates that a more significant level of effort is necessary for the requirements definition, design, implementation, and test of the software and its interactions with the system.
- The SCL does not identify risk acceptance levels or authorities; it establishes the safety importance of a software module and the LOR that is required to assure the software module presents acceptable risk.

The overall SCL concept is shown in Figure 2.79. It is based on several predefined tables and a predefined matrix. It should be noted that the criteria established in the tables and matrix can be tailored to meet the needs of a project. Example criteria can be found in the industry standards provided below; however, they are primarily for guidance only.

The following steps are involved in determining the SCL of a software module:

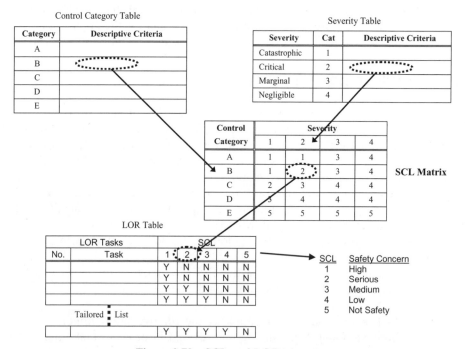

Figure 2.79 SCL and LOR concept.

- Step 1: Identify control category that is applicable to the software module. This is done by comparing the software module's function to the control category criteria in the table until a suitable match is found.
- Step 2: Identify the severity category of the mishap which the software would pertain to. This is done by anticipating the mishap outcome that might occur if the software module failed or operated incorrectly and comparing the outcome to the severity category criteria in the table until a suitable match is found.
- Step 3: Using the classifications from steps 1 and 2, determine the SCL from the SCL matrix. This is done by finding the SCL matrix cell that results from the intersection of the control category and severity categories selected in steps 1 and 2.
- Step 4: Based on the SCL, determine the LOR tasks that are required for the software module. The LOR table lists the specific tasks that must be performed for each particular SCL level.

The SCL is classified into five levels based on a set of criteria for each level. The derived SCL is based on the contribution of the software to potential failure conditions as determined by the system safety analysis process. The impact of failure, both loss of function and malfunction, is addressed when making this determination. The SCL establishes the safety-criticality of the software, which in turn drives the amount of effort (i.e., LOR) required to provide evidence that the software is considered safe. There are various architectural strategies, which during the evolution of the system design, may result in the SCLs being revised.

Note that the term "SCL" goes by different names, such as SL, DAL, or SHRI, depending on the guidance documentation used. There are several industry standards that apply the SCL concept, except they use different terms and tables. Figure 2.90 demonstrates the general SCL concept. To find the exact criteria for each table, refer to any one of the following industry standards:

1. MIL-STD-882C, System Safety Program Requirements, January 1993. This document uses the term SHRI.
2. RTCA/DO-178B, Software Considerations in Airborne Systems and Equipment Certification, 1992. This document uses the term SL.
3. SAE/ARP-4754, Certification Considerations for Highly-Integrated or Complex Aircraft Systems, Aerospace Recommended Practice, 1996. This document uses the term DAL.
4. RTCA/DO-254, Design Assurance Guidance for Airborne Electronic Hardware, 2000. This document uses the term DAL.
5. DoD Joint Software Systems Safety Engineering Handbook, September 2009 draft. This document uses the term SCL.

Table 2.19 contains the software Control Category criteria from MIL-STD-882C, which demonstrates the type of information that is utilized. Other industry standards use different criteria; in addition, the criteria can be tailored.

Table 2.20 provides some example LOR tasks that can be applied. This is just a limited example to demonstrate the breadth and depth of LOR tasks. The program LOR list should be developed carefully and be tailored for each individual program. It is recommended that tasks be selected from current industry SwS standards.

The SwS development assurance and integrity process is based on a stringent set of LOR tasks, which when successfully completed increase the confidence that:

- Software-related hazard mitigations are mapped to SSRs
- SR functions are identified in hardware, software, and firmware
- SR functions are mapped to the design SSRs
- SR requirements are traced from derivation source to SSR, code, test requirement, and test completion
- The SR functions perform as intended with no safety consequence
- The SR functions do not possess any functional capability that is not intended

TABLE 2.19 Software Control Categories (CC) from MIL-STD-882C

CC	Definition
I	Software exercises autonomous control over potentially hazardous hardware systems, subsystems, or components without the possibility of intervention to preclude the occurrence of a hazard. Failure of the software or a failure to prevent an event leads directly to a hazard's occurrence.
IIa	Software exercises control over potentially hazardous hardware systems, subsystems, or components allowing time for intervention by independent safety systems to mitigate the hazard. However, these systems by themselves are not considered adequate.
IIb	Software item displays information requiring immediate operator action to mitigate a hazard. Software failures will allow or fail to prevent the hazard's occurrence.
IIIa	Software item issues commands over potentially hazardous hardware systems, subsystems, or components requiring human action to complete the control function. There are several, redundant, independent safety measures for each hazardous event.
IIIb	Software generates information of a safety-critical nature used to make safety-critical decisions. There are several, redundant, independent safety measures for each hazardous event.
IV	Software does not control safety-critical hardware systems, subsystems, or components, and does not provide safety-critical information.

TABLE 2.20 Example LOR Task Table

Level of Rigor Tasks	SCL I	II	III	IV	V
Requirements					
High-level requirements are verifiable.	Y	Y	Y	Y	N
Low-level requirements are verifiable.	Y	Y	N	N	N
Design					
Trace of safety-critical requirements to design.	Y	Y	N	N	N
Trace of safety-critical requirements to code.	Y	N	N	N	N
Process implementation					
Software development plan exists	Y	Y	N	N	N
Software coding standards are documented and applied.	Y	N	N	N	N
Test					
Modified condition/decision test coverage is achieved.	Y	Y	N	N	N
Decision structure test coverage is achieved.	Y	N	N	N	N
100% regression testing.	Y	Y	N	N	N
Partial regression testing.	Y	N	N	N	N
Functional test of safety-critical threads.	Y	Y	N	N	N
Support					
Proposed changes are evaluated for system safety.	Y	Y	Y	N	N
Analysis					
FHA					

See *Software Safety (SwS)* and *Software Safety Process* for additional related information.

SOFTWARE DEVELOPMENT FILE (SDF)

An SDF is a repository for a collection of material pertinent to the development or support of software. Contents typically include design considerations and constraints, design documentation and data, schedule, and status information, test requirements, test cases, test procedures, and test results.

SOFTWARE DEVELOPMENT LIBRARY (SDL)

An SDL is a controlled collection of software, documentation, and associated tools and procedures used to facilitate the orderly development and subsequent support of software. The SDL includes the development configuration as part of its contents.

SOFTWARE HAZARD RISK INDEX (SHRI)

The SHRI is identical to the SCL. The SHRI was first introduced in MIL-STD-882C to indicate through an index numbering scheme the level of safety criticality of a software module. Since the term had the word "risk" in it, many casual users thought that SHRI was a measure of software hazard risk, similar to the HRI concept for hardware hazards. Thus, the term SHRI has evolved to SCL so that it is no longer confused with an actual risk value.

See *Software Criticality Level (SCL)* for additional related information.

SOFTWARE LEVEL (SL)

The term SL comes from RTCA/DO-178B, Software Considerations in Airborne Systems and Equipment Certification, 1992. In this document, software is classified into five levels based on a set of criteria for each level. The derived SL is based on the contribution of the software to potential failure conditions as determined by the SSA process. The SL implies that the failure condition category of the software drives the amount of effort (level of rigor [LOR]) required to show compliance with certification requirements.

The SL definitions are as follows:

Level A: Software whose anomalous behavior, as shown by the SSA process, would cause or contribute to a failure of system function resulting in a catastrophic failure condition for the aircraft.

Level B: Software whose anomalous behavior, as shown by the SSA process, would cause or contribute to a failure of system function resulting in a hazardous/severe-major failure condition for the aircraft.

Level C: Software whose anomalous behavior, as shown by the SSA process, would cause or contribute to a failure of system function resulting in a major failure condition for the aircraft.

Level D: Software whose anomalous behavior, as shown by the SSA process, would cause or contribute to a failure of system function resulting in a minor failure condition for the aircraft.

Level E: Software whose anomalous behavior, as shown by the SSA process, would cause or contribute to a failure of system function with no effect on aircraft operational capability or pilot workload. Once software has been confirmed as level E by the certification authority, no further guidelines apply.

The SL is an index number ranking the safety-criticality of a software module. This ranking implies that in order to make the software safe, greater development rigor must be applied to each successively critical level.

The SL determination is initially derived from the SSA process, which determines the SL appropriate to the software components of a particular system without regard to system design. The impact of failure, both loss of function and malfunction, is addressed when making this determination. There are various architectural strategies, which during the evolution of the system design, may result in the SL(s) being revised.

The failure condition categories are defined as follows:

1. Catastrophic: Failure conditions which would prevent continued safe flight and landing.
2. Hazardous/severe-major: Failure conditions which would reduce the capability of the aircraft or the ability of the crew to cope with adverse operating conditions to the extent that there would be
   - A large reduction in safety margins or functional capabilities,
   - Physical distress or higher workload such that the flight crew could not be relied on to perform their tasks accurately or completely, or
   - Adverse effects on occupants including serious or potentially fatal injuries to a small number of those occupants.
3. Major: Failure conditions which would reduce the capability of the aircraft or the ability of the crew to cope with adverse operating conditions to the extent that there would be, for example, a significant reduction in safety margins or functional capabilities, a significant increase in crew workload or in conditions impairing crew efficiency, or discomfort to occupants, possibly including injuries.
4. Minor: Failure conditions which would not significantly reduce aircraft safety, and which would involve crew actions that are well within their capabilities. Minor failure conditions may include, for example, a slight reduction in safety margins or functional capabilities, a slight increase in crew workload, such as routine flight plan changes, or some inconvenience to occupants.
5. No effect: Failure conditions which do not affect the operational capability of the aircraft or increase crew workload.

The SLs from DO-178B are summarized in Table 2.21.

TABLE 2.21 Software Levels from DO-178B

Level	Safety Impact
Level A	Software whose failure would cause or contribute to a catastrophic failure of the aircraft
Level B	Software whose failure would cause or contribute to a hazardous/severe failure condition
Level D	Software whose failure would cause or contribute to a major failure condition
Level C	Software whose failure would cause or contribute to a minor failure condition
Level E	Software whose failure would have no effect on the aircraft or on pilot workload

The SL of an item may be reduced if the system architecture:

- Provides multiple independent implementations of a function (redundancy)
- Isolates potential faults in part of the system (partitioning)
- Provides for active (automated) monitoring of the item
- Provides for human recognition or mitigation of failure conditions

The level to which a particular system must be certified is selected by a process of failure analysis and input from the device manufacturers and the certifying authority (FAA or Joint Aviation Authority [JAA]), with the final decision made by the certifying authority. Note that different software components do not need to be certified specifically at each designated level. Certification at any level automatically covers the lower-level requirement but, obviously, the converse is not true. Software certified at Level A can be used in any avionics application. It should be noted that following the advent of RTCA/DO-254 and SAE/ARP-4754, the term SL is now often referred to as DAL in order to be in alignment with the terms used in these documents.

See *Design Assurance Level (DAL)* and *Development Assurance Level (DAL)* for additional related information.

SOFTWARE PROBLEM REPORT (SPR)

SPRs are reports of problems discovered in the software during software development and testing. Each SPR must be evaluated to determine the potential safety implications. If safety impacts are identified, the PM shall be notified of any decrease in the level of safety of the system. SPRs are also known as software trouble reports (STRs).

SOFTWARE REUSE

Software reuse is the act of using already developed software (typically COTS or NDI) in a software application that is under development. Software reuse involves using previously developed software because it is already built and it performs a function that is required by the new system. The reused software may be generic COTS software, such as a computer operating system, or it may be a specialized software module from another project, such as the Missile X guidance software module being used on the new Missile Z system for missile guidance and control. Reused software could also be a mathematical routine from a library. Software reuse can present many problems for safety, particular when used in SCFs or applications. For purposes of SwS, it is recommended that reused software be evaluated for safety impact via HA and system testing.

See *Commercial Off-the-Shelf (COTS)* for additional related information.

SOFTWARE SAFETY (SwS)

SwS is the process of developing safe software, where safe software is software that executes within a system context and environment with an acceptable level of potential mishap risk. This means the software will not cause any system mishaps or prevent system design safety mechanisms from performing correctly, or the likelihood of causing these conditions is within acceptable bounds. The SwS process is the intentional and planned application of management and engineering principles, criteria, and techniques for the purpose of developing software safe for use in a specific system.

SwS is a special aspect, and subset, of system safety; it is also sometimes referred to as SwSS. The scope and coverage of SwS includes computer software, firmware, and programmable logic arrays. SwS is primarily concerned with application software developed as part of a system development program. However, due to the permeating nature of software, SwS must also consider operating systems, compilers, software tools, and reused software, including any form of COTS software that is utilized in the system. In the case of SwS, actual hazard risk cannot be calculated; thus, acceptable risk is nebulous and is based on a diverse SwS process.

SwS is built upon the following basic elements, each of which involves an extensive course of action:

- SwS Program Plan (SwSPP)
- SwS program (SwSP)
- SwS process (which includes analysis, development rigor, and test)

Although SwS generally applies the system safety process that was established for hardware safety, software must be treated slightly differently due to the complexity, special attributes, and unique nature of software. For example, when software is involved in a hazard, it is not possible to obtain the failure rate of potential software causal factors (as is done with hardware), and therefore, a risk likelihood value cannot be computed for a risk assessment. In addition, when a software-related hazard is identified, it is often difficult or impossible to determine if specific causal factors actually exist in the complex and abstract code modules. Software HA is not sufficient for SwS; extensive testing must also be performed to determine if identified hazards can occur, or if previously unidentified hazards exist. The development of safe software involves a mix of HA, design safety requirements scrutiny, significant testing, and using rigorous software development methods and tools.

The use of software in a system presents a paradoxical situation. On the one hand, software provides many benefits to the user, including increased flexibility and speed, greater accuracy, and enhanced system control. On the other hand, that same software can create unforeseen hazards that are not always well understood or easily recognizable. Software definitely increases the potential mishap safety risk of a system and requires significantly more effort to ensure safety. Technical advancements have made the digital computer both less expensive and more powerful in capability. The resultant effect is that computers now dominate the control of system functions, and the software that operates these computers has become a major system element that presents potential mishap risk.

Software is becoming so pervasive that it not only impacts the safety of military weapon systems with significant adverse consequences, but it also impacts the safety of everyday life with its incorporation into products and systems such as microwaves, cell phones, traffic lights, banking, home security, air travel, rail travel, and automobiles. Software can, and already has, caused mishaps with automobiles, medical equipment, spacecraft, trains, aircraft, and weapon systems. Therefore, SwS is an important factor in system design and development which cannot be ignored.

Software has a unique nature that can make it more difficult to apply system safety techniques than for hardware. Some of the unique software features that make it more difficult to apply safety techniques include the following:

- Software does not degrade or wear out over time, as does hardware (no failure rate).
- Software has functional failure modes, whereas hardware physical failure modes.
- Hardware failure modes are random, whereas software failure modes are deterministic.
- Software can be modified easier than hardware, but changes are more expensive due to testing.
- Software is more conceptual (abstract model), whereas hardware is more visual (physical model).
- Software contains more distinct paths than hardware (an impediment to testing).
- Hardware utilizes standard high-reliability parts (e.g., HiRel resistor), whereas software has no equivalent.
- Hardware repair restores to original condition, whereas software repair creates new (unknown) baseline.
- Software has a greater complexity level than hardware.
- Hardware alone can create hazards, whereas software cannot (software must be combined with hardware).
- Software can involve or generate more system states than hardware.
- Software can contain errors that may not be apparent or encountered for many years.
- Software can be (unintentionally) modified during operation by hardware faults.
- Hardware has known failure modes with statistically predictable failure rates, whereas software failure mode probabilities do not exist.

- Software can experience code rot (assumptions go out of date), bit rot (memory deterioration), memory leaks (memory shortage), and so on.
- A good application software package can be adversely affected by the computer operating system software, software development tools, or the compiler.

The following are some SwS paradigms that have been established from the unique characteristics of software. These paradigms are useful when applying the SwS process and when identifying software-related hazards:

- Software by itself is not hazardous; it is only hazardous in a system when performing system functions involving hardware.
- Hardware causes the damage in a mishap (i.e., explosives, radiation equipment, flight controls, and chemicals); however, software can be an initiating factor.
- When identifying software-related hazards, look for hardware–software relationships since software can contribute to hazards only via hardware.
- SwS requires multiple perspectives: system, hardware interfaces, and functions.
- Software code has no concrete failure modes as does hardware; software does have functional failure modes
- Hardware faults can induce software functional failures (modifies or fails software intent).
- Software can have errors and still function (this is a safety concern).
- Not all software errors are safety-related.
- Software hazard risk is difficult to quantify (cannot quantify software errors or functional failures).
- Software always works exactly as coded, but complexity makes comprehension difficult (i.e., sometimes software does more than intended or expected).

It should be noted that SwS is not the same as software reliability or SQA, and it cannot be achieved solely through these processes. The SwS methodology involves an independent stand-alone process that must be integrated with the software development process. The SwS process should be established at the start of a project and be documented in the SwSPP. The SwS program implements the SwSPP and carries out the SwS process.

See *Software Safety Process* and *Software Safety Program Plan (SwSPP)* for additional related information.

SOFTWARE SAFETY (SwS) PROCESS

In theory, the SwS process is a subset of the system safety process and is similar in methodology. However, due to the unique characteristics and nature of software (see *Software Safety [SwS]*), the SwS process deviates slightly and takes a more diverse approach. Whereas system safety is risk based, SwS is assurance based (also sometimes referred to as integrity based). Hardware safety is primarily focused on mitigating hazard risk to an acceptable level. In the case of SwS, actual hazard risk cannot be calculated; thus, acceptable risk is nebulous and is based on a diverse SwS assurance process. The SwS assurance process focuses on functional safety (design) assurance and software development assurance.

The unique nature of software causes the existence of two enigmas associated with SwS, which in turn cause the need for a more diverse approach in order to ensure adequate safety of software. The two stumbling blocks in SwS are:

1. Software functional failures (and hazards) can be postulated, but they cannot be definitively proven by specific identifiable causal factors.
2. Failure rates cannot be determined for software functional failures; therefore, hazard risk cannot be calculated for software-related risk assessments.

Because of these two SwS dilemmas, it is evident that software presents potential mishap risk that is unknown and cannot be precisely determined. Therefore, the most pragmatic way to ensure that software is safe is by applying a bilateral safety approach consisting of (1) software functional coverage and (2) software development coverage. The software functional coverage scheme focuses on the functional design to provide hazard identification and mitigation assurance and SCF identification and assurance. The software development coverage scheme utilizes the software development process to assist in the forced focus on specific development tasks that ensure higher quality software that is presumably safer. The theory is that if the software is developed to a specified set of rigorous requirements, analyses, tests, and development procedures, the resulting product will present acceptable safety risk. When all the appropriate hazard mitigation tasks and software development tasks are successfully performed, the overall software mishap risk is "judged" to be acceptable. This bilateral approach is a strategy intended to broadly cover all aspects of software that can impact safety. This SwS scheme provides a "level of assurance" that the software has received complete safety coverage and the risk presented by the software is deemed acceptable. SwS assurance requires visibility of both the product and the process.

Figure 2.80 shows the overall approach to SwS. It should be noted that these SwS tasks do not provide a quantitative estimate of the potential mishap risk associated with the software. What this approach does provide is a level of confidence that the software can be considered as being safe. Software-related hazards can be accepted for risk based on the conclusions drawn from the safety case, where the safety case is built upon the results of both the functional and developmental completion evidence.

SOFTWARE SAFETY (SwS) PROCESS

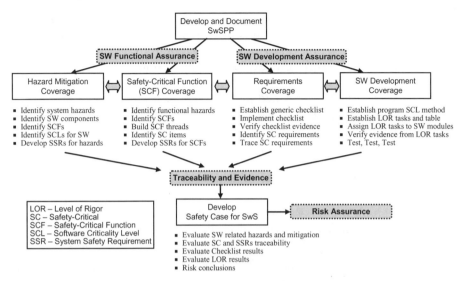

Figure 2.80 Overview of the SwS process.

All of the elements in this four-pronged approach are interrelated and are dependent upon each other. Also, the diagram does not imply any sort of order or sequence; the necessary steps should be performed as makes sense for the project. The four aspects are summarized as follows:

1. Perform SHA to identify hazards. Identify both hardware and software causal factors. The software causal factors lead to design mitigation requirements in the form of SSRs. This step also leads to the identification of SCLs, which will impact the LOR tasks performed by the software development effort.
2. Perform FHA to identify SCFs. This also leads to the identification of safety-critical code modules and design SSRs. SC requirements are tagged in the requirements tracking system for close scrutiny and testing. This branch also supports establishing SCLs for the software modules.
3. Identify the appropriate design safety requirements and apply them to the software. These requirements will include baseline, derived, and generic requirements. Baseline safety requirements stem from the contractual functional design requirements. Derived safety requirements are SSRs established to mitigate hazards and potential safety issues. Generic SwS requirements are industry-known guidelines and requirements, such as those from Standardization Agreement (STANAG) 4404. The generics are very basic and general software requirements that have been found to be useful in helping to assure safe software design. The software design group implements the generics and completes a checklist that provides evidence of completion.

4. Establish the LOR tasks required for each SCL. Perform the LOR tasks on the software modules and document the evidence of successful completion (see *Software Criticality Level [SCL]*). This effort is done primarily by the software development organization.

To finally accept the risk provided by the software, a safety case is developed which provides assurance that the software is considered to be acceptably safe. The safety case for SwS is largely based on the following evidence, as a minimum:

- SR hazards have appropriate mitigation methods, which are in the form of SSRs.
- Hazard mitigation SSRs have been successfully tested.
- SCFs have been documented, and the appropriate SSRs have been established to protect these functions against adverse behavior.
- Software design requirements have been reviewed, and those that are safety-critical have been tagged as safety-critical.
- SCF SSRs and safety-critical SSRs have been successfully tested.
- All LOR tasks have been successfully performed.
- All levels of requirements can be traced to their roots (and hazards) and each is fully tested.

Note that the term SCL goes by different names, such as SHRI, SL, DAL, or SIL, depending on the guidance documentation used. These alternate terms and similar SwS processes are defined in MIL-STD-882C, the DoD Joint Software Systems Safety Engineering Handbook and RTCA/DO-178B.

See *Software Criticality Level (SCL)*, *Software Safety (SwS)*, and *Software Safety Program Plan (SwSPP)* for additional related information.

SOFTWARE SAFETY PROGRAM (SwSP)

An SwSP is the combined set of people and tasks that implement the SwS process on a program or project; it is also sometimes referred to as a software system safety program (SwSSP). Optimally, the SwSP is an integral aspect of the overall system safety (SSP) and the methodology is documented as a subset of the SSPP. This approach provides an integrated and effective method for the identification and control of software contributions to system level hazards, and it also minimizes any impact to the overall program cost and schedule. Detailed SwS analyses and test verification activities provide evidence that safety risks associated with the use of safety-critical software are mitigated or low risk. Software's contribution to system level hazards, and hazard mitigation, must be assessed within a structured and disciplined SwSP.

The objective of the SwSP is to ensure that system design meets applicable safety requirements, and that all hazards associated with the system are identified and eliminated, or controlled in a manner consistent with program objectives, constraints, and risk. The SwSP also provides management visibility of safety risks inherent in the design and planned operations, and defines the process required for management to formally reduce and accept the system safety risks. The SwSP conducts the intentional and planned application of management and engineering principles, criteria, and techniques for the purpose of developing safe software for a system.

An SwSP involves an organization that performs the necessary system safety tasks and activities to implement the system safety process, along with providing the necessary evidence of safety achievement. The scope of an SwSP includes hardware, software, firmware, and HSI for all system life-cycle phases. It should be noted that in addition to the core elements, an SwSP also includes many program support tasks that are necessary to implement safety. The support tasks may vary depending on the type of system. The SwSP and organization can be tailored to a program's size, complexity, and safety-criticality. Typically, the system safety manager or lead is responsible for the SwSP. This includes ensuring all of the SwSP tasks are performed, program milestones are met, and the appropriate artifacts are produced that support certification that the software is safe for use.

See *Software Safety (SwS)* and *Software Safety Process* for additional related information.

SOFTWARE SAFETY PROGRAM PLAN (SwSPP)

An SwSPP is the plan for implementing the SwS process on a program or project; it is also sometimes referred to as a software system safety program plan (SwSSPP). It is very similar in nature to the SSPP. While the SwSP consists of the actual people, activities, and products involved in implementing the SwS process, the SwSPP is the document that formally defines the tasks, products, interfaces, and milestones that will be required of the safety organization. The SwSPP also defines the scope of the safety program and decision criteria to be used by the program. The SwSPP should cover the essential core components of an SwSP, plus all of the relevant support activities that will be required.

The SwSPP is the management tool used to implement and manage an effective SwSP; it is a blueprint for the overall SwS process. It is a formal document that describes the system safety organization and the required management and engineering tasks and activities to be conducted by the SwSP. The SwSPP establishes the objectives, responsibilities, and artifacts required for an effective SwSP. The SwSPP should be inclusive of the entire system life cycle. In addition, the SwSPP should cover all SwS aspects relating to system design, system operational concept, software design, and HSI.

Optimally, the SwSPP is an integral part of the SSPP, and they are combined together in the same document. This approach provides an integrated and effective method for the identification and control of software contributions to system level hazards, and it also minimizes any impact to the overall program cost and schedule. Detailed SwS analyses and test verification activities provide evidence that safety risks associated with the use of safety-critical software are mitigated or low risk. Software's contribution to system level hazards and hazard mitigation must be assessed within a structured and disciplined SSP/SwSP. It is the responsibility of the safety manager or safety lead to develop and implement an effective SwS approach for the project and to develop the SwSPP. The safety manager must coordinate with other program disciplines to ensure they are participants in the SwS process. The safety manager is also responsible for ensuring the appropriate SwS tools are in place, such as the software criticality risk tables, the LOR tasks, and the generic SwS requirements checklist.

Although the SwSPP format is not critical, it is essential that the SwSPP contain the necessary content. The basic program elements for an SwSPP include the following, as a minimum:

1. Introduction
2. System description (short overview)
3. Software architecture and code overview
4. List of software modules and their functions
5. Safety requirements, guidelines, and criteria
6. Organization—SSP, SwSP, and overall program
7. Roles, responsibilities, and interfaces
8. Schedules—SwSP and overall program
9. SwS approach
10. SCL tables
11. Software LOR table
12. Generic SwS code checklist
13. SwSP products
14. Evaluation of changes and problem reports
15. COTS/NDI safety
16. SwS safety case or SAR
17. Review boards
18. SSWG support

See *Software Safety (SwS)*, *Software Safety Process*, and *Software Safety Program (SwSP)* for additional related information.

SPACE

Space denotes applications peculiar to spacecraft and other systems designed for operation near or beyond the upper reaches of the Earth's atmosphere environment.

See *Aerospace* and *Airborne* for additional related information.

SPIRAL DEVELOPMENT

In the spiral development process a desired capability is identified, but the end-state requirements are not known at program initiation. Requirements are refined through experimentation, demonstration, risk management, and continuous user feedback. Development progresses in incremental spirals of preliminary design, detailed design, and test, but the requirements for future increments depend on user feedback and technology maturation.

In this methodology, the user is provided the best possible capability within each increment, and continuous user feedback is important. The requirements for future increments are dependent on the feedback from users and technology maturation. It is an iterative process designed to assess the viability of technologies while simultaneously refining user requirements. Spiral development complements an evolutionary approach by continuing in parallel with the acquisition process to speed the identification and development of the technologies necessary for follow-on increments. Each incremental spiral provides the best possible capability. Spiral development is a form of evolutionary acquisition.

See *Engineering Development Model* and *System Life-Cycle Model* for additional related information.

SPECIFICATION

A specification is a collection of requirements which, when taken together, constitute the criteria that define the functions and attributes of a system or product. A design requirement is an identifiable element of a specification that can be validated and against which an implementation can be verified.

SPECIFICATION CHANGE NOTICE (SCN)

An SCN is a proposed change to the specifications of the already solidified and controlled design. SCNs are part of the CM process and the CCB. The SSP should review each and every proposed and actual SCN for safety impact. Each SCN must be evaluated to determine the potential effect on safety-critical components or subsystems; HAs and mishap risk assessments may be affected. The CCB should be notified when the level of safety of the system will be reduced by an SCN.

STATE

Modes and states are terms used to divide a system into segments which can explain different physical and functional configurations of the system that occur during the operation of the system. Although the two terms are often confused with one another and are sometimes used interchangeably, there are some clear definitions that make the two terms distinct and separate. The purpose of modes and states is to simplify and clarify system design and architecture for a more complete understanding of the system design and operation. A mode is a functional capability, whereas a state is a condition that characterizes the behavior of the functional capability.

A system state is the physical configuration of system hardware and software. Hardware and/or software are configured in a certain manner in order to perform a particular mode. A system state characterizes the particular physical configuration at any point in time. States identify conditions in which a system or subsystem can exist. A system or subsystem may be in only one state at a time. There is a connection between modes and states in that for every operational mode there is one or more states the system can be in.

A system mode is the functional configuration of the system, which established the system manner of operation. A mode is a set of functional capabilities that allow the operational system to accomplish tasks or activities. Modes tend to establish operational segments within the system mission. A system can have primary modes and sub-modes of operation. The system can only be in one mode at any one time. Correctly defining the mode established all of the modal constraints, which impact both design and operation. For example, a stopwatch is a system that typically has modes such as on, off, timing, and reset. The "on" mode may go into the initialization state, which requires the "initialization" software package.

Modes and states are important to system safety because some modes and states are safety-critical and should not be performed erroneously or inadvertently. Evaluating system modes and states is an important factor during HA.

See *Mode* for additional related information.

STAGED PHOTOGRAPHS

Staged photographs are those constructed to gain a better understanding of the sequence of events surrounding a mishap. Staged photographs may include but are not limited to photos of mishap sites with personnel pointing to various objects, a series of photographs showing similar personal actions which may have led to a mishap, equipment which is highlighted or specifically identified for safety investigators, and so on. Photographs of the actual mishap site, a broken piece of equipment, injured or deceased personnel are not considered staged photographs unless the photos have been marked by safety investigation personnel.

STATEMENT OF OBJECTIVES (SOO)

The SOO is that portion of a contract which establishes a broad description of the government's required performance objectives. The SOO identifies the broad, basic, top-level objectives of the acquisition and is used as a focusing tool for both the government and contract bidders. A SOO may be provided in lieu of a SOW.

STATEMENT OF WORK (SOW)

The SOW is that portion of a contract which establishes and defines all non-specification requirements for contractor's efforts either directly or with the use of specific cited documents. The initial SOW is generally a detailed list or description of what the government wants in a product. The final SOW is a contractor's response detailing how the contractor will answer the government solicitation; it provides a detailed list and description of tasks and activities that will be performed during the acquisition development process.

STERILIZATION

Sterilization is a design feature which permanently prevents a fuze from functioning.

STORE

Any device intended for internal or external carriage, mounted on aircraft suspension and release equipment, and which may or may not be intended to be separated in flight from the aircraft. Stores include missiles, rockets, bombs, nuclear weapons, mines, fuel and spray tanks, torpedoes, detachable fuel and spray tanks, dispensers, pods, targets, chaff and flares including external dispensing equipment, and suspension equipment (racks, pylons). Note that individual rockets, gun rounds, and submunitions are not considered to be stores.

Stores are classified in two categories as follows:

- Carriage store—An article of suspension and release equipment that is mounted on an aircraft on a nonpermanent basis. Pylons are not considered carriage stores.

- Mission store—A device which supports a specific mission. This term excludes suspension and release equipment and carriage stores. Examples of mission stores include, but are not limited to, the following:
  - Missiles
  - Bombs
  - Nuclear weapons
  - Rocket pods, dispensers capable of ejecting multiple submunitions, guns, and gun pods
  - Torpedoes
  - Pyrotechnic devices
  - Sonobuoys
  - Flares, chaff dispensers
  - Drones
  - Pods (laser designator, electronic countermeasures, store control, data link, reconnaissance)
  - Fuel and spray tanks
  - Target and cargo drop containers

STRESS TESTING

Stress testing is a form of testing that is used to determine the stability of a given hardware or SI. It involves testing beyond normal operational capacity, often to a breaking point, in order to observe the results. Stress testing may have a more specific meaning in certain industries, such as fatigue testing for materials. In software testing, "system stress tests" refers to tests that put a greater emphasis on robustness, availability, and error handling under a heavy load, rather than on what would be considered correct behavior under normal circumstances. In particular, the goals of such tests may be to ensure the software does not crash in conditions of insufficient computational resources (such as memory or disk space), unusually high concurrency, or denial of service attacks.

SUBASSEMBLY

A subassembly is an integrated set of components and/or parts that comprise a well-defined portion of an assembly, for example, a video display with its related integrated circuitry. A subassembly would be reflected as a specific level in a system hierarchy.

See *System Hierarchy* for additional related information.

SUBJECT MATTER EXPERT (SME)

A person, whether military or civilian, who through knowledge, skill, experience, training, or education, possesses scientific, technical, or other specialized knowledge that may assist to understand or to determine a particular fact in issue. Such an expert may provide information by way of facts, opinions, or otherwise. An SME is the technical authority on a subject. An experienced system safety engineer is typically considered as an SME.

SUBSTANTIAL DAMAGE

Substantial damage refers to damage sufficient to create a Class A mishap as defined by DoDINST 6055.7, or mishap of severity Category I (Catastrophic) as defined by MIL-STD-882.

SUBSYSTEM

A subsystem is a subset of a system; a smaller system that is part of a larger system. A subsystem is a coherent and somewhat independent component of a larger system. A subsystem can include all of the same basic components that a system does, such as hardware, software, components, personnel, processes, and procedures. Subsystems perform a specific function that contributes to accomplishing the system objective. Figure 2.81 displays the general aspects of a subsystem.

See *System* for additional related information.

SUBSYSTEM HAZARD ANALYSIS (SSHA)

The SSHA technique is a safety analysis tool for identifying hazards, their associated causal factors, effects, level of risk, and mitigating design measures.

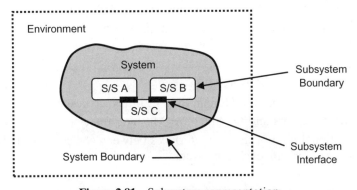

Figure 2.81 Subsystem representation.

The SSHA is performed when detailed design information is available as it provides a methodology for analyzing in greater depth the causal factors for hazards previously identified by earlier analyses such as the PHA. The SSHA helps derive detailed SSRs for incorporating design safety methods into the system design.

The purpose of the SSHA is to expand upon the analysis of previously identified hazards, and to identify new hazards, from detailed design information. The SSHA provides for the identification of detailed causal factors of known and newly identified hazards and, in turn, provides for the identification of detailed SSRs for design. The SSHA provides a safety focus from a detailed subsystem viewpoint, through analysis of the subsystem structure and components. The SSHA helps verify subsystem compliance with safety requirements contained in subsystem specifications.

The SSHA is applicable to the analysis of all types of systems and subsystems, and is typically performed at the detailed component level of a subsystem. The SSHA is usually performed during detailed design development and helps to guide the detailed DFS. The technique provides sufficient thoroughness to identify hazards and detailed HCFs when applied to a given system/subsystem by experienced safety personnel. An understanding of HA theory, as well as knowledge of system safety concepts, is essential. Experience with, and/or a good working knowledge of the particular type of system and subsystem is necessary in order to identify and analyze all hazards. The methodology is uncomplicated and easily learned. Standard SSHA forms and instructions have been developed that are included as part of this chapter.

The SSHA is an in-depth and detailed analysis of hazards previously identified by the PHA. The SSHA also identifies new hazards. It requires detailed design information and a good understanding of the system design and operation. As a minimum, when performing the SSHA, consideration should be given to:

1. Performance of the subsystem hardware
2. Performance degradation of the subsystem hardware
3. Inadvertent functioning of the subsystem hardware
4. Functional failure of the subsystem hardware
5. CMFs
6. Timing errors
7. Design errors or defects
8. Human error and the Human System Interface design
9. Software errors and the Software–Machine Interface
10. Functional relationships or interfaces between components and equipment comprising each subsystem

It is desirable to perform the SSHA analysis using a worksheet. The worksheet helps to add rigor to the analysis, record the process and data, and help support

System:			Subsystem Hazard Analysis				Analyst:		
Subsystem:							Date:		
No.	Hazard	Causes	Effects	Mode	Initial HRI	Recommended Action	Final HRI	Comments	Status

Figure 2.82 Example SSHA worksheet.

justification for the identified hazards and safety recommendations. The format of the analysis worksheet is not critical, and typically columnar-type worksheets are utilized. An example SSHA worksheet using the columnar format is shown in Figure 2.82. This particular worksheet format has proven to be useful and effective in many applied situations and it provides all of the information required from an SSHA.

The following are some basic guidelines that should be followed when performing the SSHA:

- Remember that the objective of the SSHA is to identify detailed subsystem causes of identified hazards, plus previously undiscovered hazards. It refines risk estimates and mitigation methods.
- Isolate the subsystem and only look within that subsystem for hazards. The effect of an SSHA hazard only goes to the subsystem boundary. The SHA identifies hazards at the SSHA interface and includes interface boundary causal factors.
- Start the SSHA by populating the SSHA worksheet with hazards identified from the PHA. Evaluate the subsystem components to identify the specific causal factors to these hazards. In effect, the PHA functional hazards and energy source hazards are transferred to the SSHA subsystem responsible for those areas.
- Identify new hazards and their causal factors by evaluating the subsystem hardware components and software modules. Use analysis aids to help recognize and identify new hazards, such as TLMs, hazard checklists, lesson learned, mishap investigations, and hazards from similar systems.
- Most hazards will be inherent-type hazards (contact with high voltage, excessive weight, fire, etc.). Some hazards may contribute to system hazards (e.g., inadvertent missile launch), but generally, several subsystems will be required for this type of system hazard (thus the need for SHA).
- Consider erroneous input to subsystem as the cause of a subsystem hazard (command fault).

- The PHA and SSHA hazards establish the TLMs. The TLMs are used in the SHA for hazard identification. Continue to establish TLMs and SCFs as the SSHA progresses, and utilize in the analysis.
- A hazard write-up in the SSHA worksheet should be clear and understandable with as much information as necessary to understand the hazard.
- The SSHA hazard column does not have to contain all three elements of a hazard: HS, IMs, and TTO. The combined columns of the SSHA worksheet can contain all three components of a hazard. For example, it is acceptable to place the HE in the Hazard section, the IMs in the Cause section, and the outcome in the Effect section. The Hazard, Causes, and Effects columns should together completely describe the hazard. These columns should provide the three sides of the hazard triangle.
- The SSHA does not evaluate system functions, only functions that reside entirely within the subsystem. Functions tend to cross subsystem boundaries, and are therefore evaluated in the SHA.

The SSHA is an HA technique that is important and essential to system safety. For more detailed information on the SSHA technique, see Clifton A. Ericson II, *Hazard Analysis Techniques for System Safety* (2005), chapter 6.

SUITABILITY

Suitability is a systems engineering metric of the degree to which a system is appropriate for its intended use with respect to nonoperational factors such as MMI, training, safety, documentation, producibility, testability, transportability, maintainability, manpower availability, supportability, and disposability. The level of suitability determines whether the system is the right one to fill the customers' needs and requirements. Suitability measures can be used as performance requirements, design constraints, and so on.

SUPERVISORY CONTROL AND DATA ACQUISITION (SCADA)

SCADA refers to an industrial control system for monitoring and controlling a process. The process can be industrial, infrastructure, or facility-based. Industrial processes include those of manufacturing, production, power generation, fabrication, and refining, and may run in continuous, batch, repetitive, or discrete modes. Infrastructure processes may be public or private, and include water treatment and distribution, wastewater collection and treatment, oil and gas pipelines, electrical power transmission and distribution, civil defense siren systems, and large communication systems. Facility processes occur both in public facilities and private ones, including buildings, airports, ships, and space stations. They monitor and control heating, ventilation and air conditioning (HVAC), access, and energy consumption.

A SCADA System typically consists of the following subsystems:

- An HMI Interface or HMI which presents process data to a human operator, and through which the human operator monitors and controls the process.
- A supervisory (computer) system, gathering (acquiring) data on the process and sending commands (control) to the process.
- Remote terminal units (RTUs) connecting to sensors in the process, converting sensor signals to digital data and sending digital data to the supervisory system.
- PLCs for field devices because they are more economical, versatile, flexible, and configurable than special-purpose RTUs.
- Communication infrastructure connecting the supervisory system to the RTUs.

A SCADA system is usually a centralized system that monitors and controls entire sites, or complexes of systems spread out over large areas, anything between an industrial plant and a country. Most control actions are performed automatically by RTUs or by PLCs. Host control functions are usually restricted to basic overriding or supervisory-level intervention. For example, a PLC may control the flow of cooling water through part of an industrial process, but the SCADA system may allow operators to change the set points for the flow and enable alarm conditions, such as loss of flow and high temperature, to be displayed and recorded. Feedback control loops pass through the RTUs or PLCs, while the SCADA system monitors the overall performance of the loops. Security and safety are significant issues with SCADA systems because of the increasing use of open architectures and communications between SCADA systems and office networks using the Internet.

SURVIVABILITY

Survivability is the capability of a system and crew to avoid or withstand a man-made hostile environment without suffering an abortive impairment of its ability to accomplish its designated mission. Survivability consists of susceptibility, vulnerability, and recoverability.

SYNERGISM

Synergism is the working together of two things to produce an effect greater than the sum of their individual effects. Synergism is a condition whereby the combined effect of several items is greater than the sum of the effects of individual items. Synergism is one of the emergent properties of a system.

SYNERGY

Synergy is the process by which a system generates emergent properties resulting in the condition in which a system may be considered more than the sum of its parts, and equal to the sum of its parts plus their relationships. This resulting condition can be said to be one of synergy.

SYMPATHETIC DETONATION

Sympathetic detonation is the detonation of a munition or an explosive charge induced by the detonation of another like munition or explosive charge.

SYSTEM

A system is an integrated composite of components that provides function and capability to satisfy a stated need or objective. A system is a holistic unit that is greater than the sum of its parts. Systems have structure, function, behavior, characteristics, and interconnectivity. Systems vary in size, purpose, type, and complexity. Modern-day systems are typically composed of people, products, processes, and environments that together generate great complexity and capability. A system is typically viewed as an integrated composite of people, products, and processes that provide a capability to satisfy a stated need or objective.

In a very general sense, a system is any group of interrelated, interacting, and interdependent parts that form a complex and unified whole that has a specific function or purpose. The key is that if all the parts are not interrelated and interdependent, it would not be a system but merely a collection of parts. Figure 2.83 displays the general aspects of a system.

A system is typically composed of any combination of the following elements:

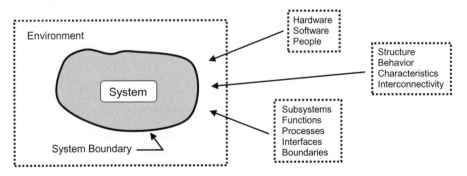

Figure 2.83 System representation.

- Subsystems (sub-subsystems, units, assemblies, components)
- Hardware (electrical, hydraulic, structures, explosives, fuel)
- Software (program, segment, unit, module, logic, algorithms)
- People (operators, testers, maintainers, procedures, tasks)
- Processes (course of action, timing, material combining)
- Procedures (instructions, tasks, manuals, warning notes)
- Interfaces (hardware, software, documentation, communications)
- Functions (modes, phases, tasks, and objectives)
- Facilities (building, location, storage, transportation)
- Boundaries (physical, theoretical, limitations)
- Environment (weather, external equipment, temperature, vibration)

Due to the size and complexity of a system, it can often be difficult for one person to fully understand all aspects and nuances of the system design. The use of a system hierarchy is typically established to define the system structure in an orderly and comprehensible manner.

There are many different types of systems such as technical systems, political systems, eco systems, solar systems, economic systems, and cultural systems, just to name a few. There are natural and man-made (designed) technical systems. Natural systems may not have an apparent objective, but their purposes can be interpreted. Man-made systems are made with purposes that are achieved by the delivery of outputs. From a theoretical framework perspective, there are open systems, closed systems, and isolated systems. An open system exchanges matter and energy with its surroundings. Most systems are open systems such as a car, coffeemaker, or computer. A closed system exchanges energy, but not matter, with its environment, such as the Earth. An isolated system exchanges neither matter nor energy with its environment, a theoretical example of which would be the universe.

Examples of man-made technical systems are the automobile and the road system. The automobile is composed of many individual parts, such as the electrical subsystem, the hydraulic subsystem, and the steering subsystem. All of these subsystems work together in a planned design to form a system. The automobile and the road system are both open systems because they need (and supply) fuel and maintenance from outside sources in order to function.

Two distinguishing characteristic of systems theory are that (a) the whole is more than the sum of the parts and (b) what is best for the subsystems is not necessarily the best for the overall system, and vice versa. The concept that the system is more than the totality of its components is referred to as synergy. As might be expected, these two system characteristics are important in safety analysis and in identification of hazards. The discipline of system safety must evaluate the subsystems as well as the system as a whole.

A system is a construct or collection of different elements that together produce results not obtainable by the elements alone. The elements, or parts, can include people, hardware, software, facilities, policies, and documents; all things required to produce systems-level results. A system has unique qualities, properties, characteristics, functions, behavior, and performance. System safety works at the system level because it has been recognized that many hazards involve unique system interrelationships between the parts, rather than the parts in isolation.

See *System Hierarchy* for additional related information.

SYSTEM ACCEPTANCE REVIEW (SAR)

The SAR verifies the completeness of the specific end products in relation to their expected maturity level and assesses compliance to stakeholder expectations. The SAR examines the system, its end products and documentation, and test data and analyses that support verification. It also ensures that the system has sufficient technical maturity to authorize its shipment to the designated operational facility or launch site.

See *Critical Design Review (CDR)*, *Preliminary Design Review (PDR)*, and *Systems Engineering Technical Reviews (SETR)* for additional information.

SYSTEM BOUNDARY

A boundary is a line (visible or invisible) indicating the limit or extent of something. A system boundary is the set of limits designed into the system (for various reasons); the limits that define the discrete system and its operational capability. For example, a certain aircraft may be designed with an altitude boundary limit of 50,000 feet. A system typically has one or more boundaries that limit (and define) the scope, function, and capability of the system. Boundaries can be physical (e.g., hydraulic subsystem) or ethereal (e.g., software modules or functions).

See *System* for additional related information.

SYSTEM DEFINITION REVIEW (SDR)

The SDR examines the proposed system architecture and design and the flow down to all functional elements of the system. It is more detailed than the System Requirements Review (SRR).

See *Critical Design Review (CDR)*, *Preliminary Design Review (PDR)*, and *Systems Engineering Technical Reviews (SETR)* for additional information.

SYSTEM DEVELOPMENT MODEL

System development is the process of designing, developing, and testing a system design until the final product meets all requirements and fulfills all objectives. There are several different system development models by which a system can be developed. Each of these models has advantages and disadvantages, but they all achieve the same end—development of a system using a formal process. See *Engineering Development Model* for definition of these models.

SYSTEM ENVIRONMENT

System environment includes everything outside of the system boundary that may influence or directly impact the system. For technical systems, the system environment can include such things as weather, sand, ice, EMR, humidity, hurricanes, tornadoes, temperature, vibration, etc. Other types of systems will have different environments; for example, a political system's environment may include voter awareness, voter activism, and voter attitudes.

See *System* for additional related information.

SYSTEM FUNCTION

A system has one or more major functions that define its purpose or objective. For example, the primary function of a rail system is to transport passengers and cargo. In addition, systems are composed of many sub-functions that must be performed to support the primary function. For example, the trains in a rail system must have power functions, braking functions, control functions, and communication functions. System functions can be subdivided and classified into lower and lower levels until all the necessary functions are established. The hierarchical list of functions defines how the system operates.

See *System* for additional related information.

SYSTEM FUNCTIONAL REVIEW (SFR)

The process of defining the items or elements below system level involves substantial engineering effort. This design activity is accompanied by analysis, trade studies, modeling and simulation, as well as continuous developmental testing to achieve an optimum definition of the major elements that make up the system, with associated functionality and performance requirements. This activity results in two major systems engineering products: the final version of the system performance specification and draft versions of the performance specifications, which describe the items below system level (item performance specifications). These documents, in turn, define the system functional baseline and the draft allocated baseline. As this activity is completed, the system has passed from the level of a concept to a well-defined system design and, as such, it is appropriate to conduct another in the series of technical reviews.

SYSTEM HAZARD ANALYSIS (SHA)

The SHA is an analysis methodology for identifying hazards, evaluating risk and safety compliance at the system level, with a focus on interfaces and SCFs. The SHA ensures that identified hazards are understood at the system level, that all causal factors are identified and mitigated, and that the overall system risk is known and accepted. SHA also provides a mechanism for identifying previously unforeseen interface hazards and evaluating causal factors in greater depth.

The SHA is a detailed study of hazards resulting from system integration. This means evaluating all identified hazards and HCFs across subsystem interfaces. The SHA expands upon the SSHA and may use techniques, such as FTA, to assess the impact of certain hazards at the system level. The system level evaluation should include analysis of all possible causal factors from sources such as design errors, hardware failures, human errors, and software errors.

Overall, the SHA:

- Verifies system compliance with safety requirements contained in the system specifications and other applicable documents;
- Identifies hazards associated with the subsystem interfaces and system functional faults;
- Assesses the risk associated with the total system design, including software, and specifically of the subsystem interfaces; and
- Recommends actions necessary to eliminate identified hazards and/or control their associated risk to acceptable levels.

The SHA assesses the safety of the total system design by evaluating the integrated system. The primary emphasis of the SHA, inclusive of hardware, software, and HSI, is to verify that the product is in compliance with the specified and derived SSRs at the system level. This includes compliance with acceptable mishap risk levels. The SHA examines the entire system as a whole by integrating the essential outputs from the SSHAs. Emphasis is placed on the interactions and the interfaces of all the subsystems as they operate together.

The SHA evaluates subsystem interrelationships for the following:

1. Compliance with specified safety design criteria
2. Possible independent, dependent, and simultaneous hazardous events, including system failures, failures of safety devices, and system interactions that could create a hazard or result in an increase in mishap risk
3. Degradation in the safety of a subsystem or the total system from normal operation of another subsystem
4. Design changes that affect subsystems
5. Effects of human errors

System:			System Hazard Analysis		Analyst: Date:			
No.	TLM/SCF	Hazard	Causes	Effects	Initial MRI	Recommended Action	Final MRI	Status

Figure 2.84 Example SHA worksheet.

6. Degradation in the safety of the total system from COTS hardware or software
7. Assurance that SCFs are adequately safe from a total system viewpoint, and that all interface and CCF considerations have been evaluated

The SHA can be applied to any system; it is applied during and after detailed design to identify and resolve subsystem interface problems. The SHA technique, when applied to a given system by experienced safety personnel, is thorough in evaluating system-level hazards and causal factors, and ensuring safe system integration. Success of the SHA is highly dependent on completion of other system safety analyses, such as the PHA, SSHA, SRCA, and O&SHA.

As part of the SHA, it is beneficial if all identified hazards are combined under TLMs. The SHA then evaluates each TLM to determine if all causal factors are identified and adequately mitigated to an acceptable level of system risk. A review of the TLMs in the SHA will indicate if additional in-depth analysis of any sort is necessary, such as for a safety-critical hazard or an interface concern.

It is desirable to perform the SHA analysis using a worksheet. The worksheet will help to add rigor to the analysis, record the process and data, and help support justification for the identified hazards and safety recommendations. The format of the analysis worksheet is not critical, and typically columnar-type worksheets are utilized. An example recommended SHA columnar-type worksheet is shown in Figure 2.84.

SHA is an HA technique that is important and essential to system safety. For more detailed information on the SHA technique, see Clifton A. Ericson II, *Hazard Analysis Techniques for System Safety* (2005), chapter 7.

SYSTEM HIERARCHY

A system is a combination of subsystems interconnected together to accomplish a system objective. A subsystem is a subset of the system that could include equipment, components, personnel, facilities, processes, documentation, procedures, and software interconnected in the system to perform a specific function that contributes to accomplishing the overall system objective. Systems vary in size, shape, function, criticality, and complexity. A system can be small, such as a toaster that consists of less than 50 parts. A system can be very large, composed of hundreds of subsystems, thousands of assemblies, and millions of components, such as a commercial aircraft or a ship. Large complex systems can easily become overwhelming for human comprehension. In order to more easily understand a large system, the system is typically broken down or subdivided in a hierarchical manner into manageable pieces that can be easily understood.

A system hierarchy establishes nomenclature and terminology that support clear, unambiguous communication, and definition of the system, its functions, components, operations, and associated processes. A system hierarchy refers to the organizational structure defining dominant and subordinate relationships between subsystems, down to the lowest component/piece part level.

Several closely related approaches have been established that formulate a system hierarchy. The MEL is the systems engineering tool for exhibiting both system components and system hierarchy. The MEL is sometimes referred to as the IEL because it shows hierarchy via indenture level of the components and functions as defined in MIL-STD-1629, Procedures for Performing a Failure Modes, Effects and Criticality Analysis. The MEL is also sometimes referred to as the WBS because it identifies both components and tasks in an indenture list structure as defined in MIL-HDBK-881, Handbook on Work Breakdown Structure. Basically, the MEL is a list of all the systems, subsystems, units, assemblies, and components in the major system, with each item in the list indented to reflect its hierarchy and ownership level. The indenture level also identifies or describes the relative complexity of assembly or function. The levels progress from the more complex (system) to the simpler (part) divisions. System design data and drawings will usually describe the system's internal and interface functions beginning at system level and progressing to the lowest indenture level of the system.

The following is a typical breakdown of successive indenture level groupings in the system hierarchy:

- System—an integrated set of subsystems that accomplish a defined objective
- Subsystem—an integrated set of assemblies, components, and parts which performs a cleanly and clearly separated function
- Assembly—an integrated set of components and/or subassemblies that comprise a defined part of a subsystem, for example, the pilot's radar display console or the fuel injection assembly of an aircraft propulsion subsystem
- Subassembly—an integrated set of components and/or parts that comprise a well-defined portion of an assembly, for example, a video display with its related integrated circuitry

- Component—a cleanly identified item composed of multiple parts, for example, a cathode ray tube or the earpiece of the pilot's radio headset
- Part—the lowest level of separately identifiable items, for example, a bolt

When performing HA, all of the system components must be considered to ensure a complete analysis. A MEL aids the safety analyst in ensuring that all of the system hardware and functions have been adequately covered by the HAs. System hierarchy level can be used to set the level of detail for a particular HA. A system hierarchy table can also be a valuable tool for establishing the correct TLMs and level of risk for system hazards.

See *System*, *Master Equipment List (MEL)*, and *Work Breakdown Structure (WBS)* for additional related information.

SYSTEM INTERFACE

A system interface is a point where two or more systems connect across their physical and/or functional boundaries such that the output of one system is the input to the others or vice versa. System interfaces allow two systems (or subsystem) to interact. Interfaces provide the interconnectedness of systems and subsystems. An interface provides a common boundary between two things, such as between two systems, two subsystems, or a user and the system. A system interface is the method and place whereby two systems can communicate with one another across a common boundary. For example, an operating system interface is a set of commands or actions that an operating system can perform and a way for a computer program or person to activate them. The same concept applies to subsystem interfaces.

See *System* for additional related information.

SYSTEM LIFE CYCLE

The system life cycle refers to the actual phases of life a system goes through, from concept through disposal. This system life cycle is analogous to the human life cycle of conception, birth, childhood, adulthood, death, and burial. The life cycle of a system is very generic and is generally a universal standard. The system life-cycle stages are generally condensed and are summarized into the major phases of concept definition, preliminary design, final design, test, manufacture, operation, and disposal. All aspects of the system life cycle typically fit into one of these major phase categories.

See *System Life-Cycle Model* for additional related information.

SYSTEM LIFE-CYCLE MODEL

The system life cycle involves the actual phases a system goes through from concept through disposal. This system life cycle is analogous to the human life

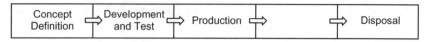

Figure 2.85 Major system life-cycle phases.

cycle of conception, birth, childhood, adulthood, death, and burial. The life cycle of a system is very generic and is generally a universal standard. The system life-cycle stages are typically condensed and are summarized into the five major phases as shown in Figure 2.85. All aspects of the system life cycle can be characterized by one of these major categories or phases in this life-cycle model.

Phase 1—Concept Definition

This phase involves defining and evaluating a potential system concept in terms of feasibility, cost and, risk. The overall project goals and objectives are identified during this basic concept evaluation phase. Design requirements, functions, and end results are formulated. The basic system is roughly designed, along with a thumbnail sketch of the subsystems required and how they will interact. During this phase, safety is concerned with hazardous components and functions that must be used in the system. The SSPP is generally started during this phase to outline the overall system risk and safety tasks, including HAs that must be performed.

Phase 2—Development and Test

This phase involves designing, developing, and testing the actual system. Development proceeds from preliminary through detailed tasks. The development phase is generally subdivided into the following three subphases:

- Preliminary design—initial basic design
- Detailed design—final detailed design
- Test—system testing to ensure all requirements are met

During preliminary design, the initial concept is translated into a workable design. During this phase subsystems, components and functions are identified and established. Design requirements are then written to define the systems, subsystems, and software. Some testing of design alternatives may be performed. During this phase, safety is concerned with hazardous system designs, hazardous components/materials, and hazardous functions that can ultimately lead to mishaps and actions to eliminate/mitigate the hazards.

During detailed design, the preliminary design evolves into the final detailed design. This phase involves completing development of the design specifications, sketches, drawings and system processes, and all subsystem designs. During the final design phase, safety is concerned with hazardous designs, failure modes, and human errors that can ultimately lead to mishaps during the life cycle of the system.

During test, the system is put through verification and validation testing of the design to ensure that all design requirements are met and are effective, and that the system performs as expected. During this phase, system safety is concerned that all safety requirements and design safety mitigation methods are verified and validated. In addition, safety is concerned with potential hazards associated with the conduct of the test and additional system hazards identified during testing. Test support equipment also requires system safety attention.

Phase 3—Production

This phase involves manufacturing production of the end product established from final design and testing. Production involves manufacturing numerous copies of the system or product that are all identical in quality and performance. Many different elements are involved in production, such as tooling, manuals, training, quality control, and parts logistics. During this phase, system safety is concerned with safe production procedures, human error, safe tools, tool calibration, safe support equipment, and HAZMATs.

Phase 4—Operation

The end product is put into actual operation by the user(s) during the operation phase. This phase includes use and support functions such as transportation/handling, storage/stowage, modification, and maintenance. The operational phase can last for many years, and during this phase performance and technology upgrades are likely. Safe system operation and support are the prime safety concerns during this phase. Safety concerns during this phase include operator actions, hardware failures, hazardous system designs, and safe design changes and system upgrades.

Phase 5—Disposal

This phase completes the useful life of the product or system; it entails disposing of the system in its entirety, or individual elements, following completion of its useful life. This stage involves phaseout, de-configuration, or decommissioning where the product is torn down, dismantled, or disassembled. Safe disassembly procedures and safe disposal of HAZMATs are safety concerns during this phase.

Normally, each of these life-cycle phases occurs sequentially, but occasionally, development tasks are performed concurrently, spirally, or incrementally to shorten and/or simplify the development process. Regardless of the development process used, sequential, concurrent, spiral, or incremental, the system life-cycle phases shown in Figure 2.85 basically remain the same. The life-cycle stages of a system are important divisions in the evolution of a product, and are therefore very relevant to the system safety process. System safety tasks are planned and referenced around these five phases. In order to proactively design safety into a product, it is essential that the system safety process start at the concept definition phase and continue throughout the entire life cycle of the system.

See *Concurrent Development Model*, *Engineering Development Model*, *Incremental Development Model*, and *Spiral Development Model* for additional information.

SYSTEM OBJECTIVE

The system objective is the purpose for a system's existence; it is the desired result to be accomplished by the system. The system objective, or objectives, defines the purpose for the system. In order to effectively and successfully develop a system, the system objective must be stated and well understood.

See *System* for additional related information.

SYSTEM OF SYSTEMS (SoS)

The term SoS is used to denote a system that is composed of a conglomeration of independent systems. An SoS is a super system that is built by interconnecting already existing and independently developed systems to work together for an SoS goal. An example of an SoS would be an antisubmarine warfare SoS consisting of submarines, surface ships, aircraft, static and mobile sensor systems, and additional systems. Although these systems can independently provide militarily useful capabilities, in collaboration they can more fully satisfy a more complex and challenging capability: to detect, localize, track, and engage submarines.

SoS is a collection of task-oriented or dedicated systems that pool their resources and capabilities together to obtain a new, more complex, "metasystem" which offers more functionality and performance than simply the sum of the constituent systems. Currently, SoS is a critical research discipline for which frames of reference, thought processes, quantitative analysis, tools, and design methods are incomplete. The methodology for defining, abstracting, modeling, and analyzing SoS problems is typically referred to as SoS engineering. While the individual systems constituting an SoS can be very different and operate independently, their interactions typically expose and deliver important emergent properties. These emergent patterns have an evolving nature that stakeholders for these problems must recognize, analyze, and understand. The SoS approach does not advocate particular tools, methods, or practices; instead, it promotes a new way of thinking for solving grand challenges where the interactions of technology, policy, and economics are the primary drivers. SoS study is related to the general study of designing, complexity, and systems engineering, but also brings to the forefront the additional challenge of design. SoS typically exhibits the behaviors of complex systems. But not all complex problems fall in the realm of SoS. Inherent to SoS problems are several combinations of traits, not all of which are exhibited by every such problem:

- Operational independence of elements
- Managerial independence of elements
- Evolutionary development
- Geographical distribution of elements
- Emergent behavior
- Interdisciplinary study
- Heterogeneity of systems
- Networks of systems

Applying system safety to SoS is a larger and more complex process than for a single system. Although the system safety process remains essentially the same, the necessary safety analyses and safety evidence on the various systems may not be done to the same level of detail or consider the SoS system interfaces. Integrating this safety knowledge into one super-system model may require a special SoS safety effort.

SYSTEM REQUIREMENTS REVIEW (SRR)

The SRR examines the functional and performance requirements established for the system and ensures that the requirements and the selected concept will satisfy the mission. The SRR examines the proposed requirements, the mission architecture, and the flow down to all functional elements of the mission to ensure that the overall concept is complete, feasible, and consistent with available resources. As the system passes into the acquisition process, that is, passes a Milestone B and enters System Development and Demonstration, it is appropriate to conduct an SRR. The SRR is intended to confirm that the user's requirements have been translated into system-specific technical requirements, that critical technologies are identified and required technology demonstrations are planned, and that risks are well understood and mitigation plans are in place. The SRR confirms that the system-level requirements are sufficiently well understood to permit the developer (contractor) to establish an initial system-level functional baseline. Once that baseline is established, the effort begins to define the functional, performance, and physical attributes of the items below system level and to allocate them to the physical elements that will perform the functions.

At an SRR, system safety typically makes a presentation summarizing the safety effort to date, the DSFs in the system design, and the current level of mishap risk which the design presents.

See *Critical Design Review (CDR)*, *Preliminary Design Review (PDR)*, and *Systems Engineering Technical Reviews (SETR)* for additional information.

SYSTEM SAFETY

System safety is an engineering discipline for developing safe systems and products, where safety is intentionally designed into the system or product. It involves the planned application of management and engineering principles, criteria, and techniques for the purpose of developing a system that presents acceptable mishap risk. System safety applies to all phases of the system life cycle and covers all system aspects, such as hardware, firmware, software, human operators, and procedures. System safety is the process for eliminating or reducing potential mishaps through a process of hazard identification, risk assessment, and risk control.

Hazards and mishap risk will always be with us because of the following natural laws:

- Eventually everything physical fails or wears out, thus causing a hazard potential.
- Human error will always occur, thus causing a hazard potential.
- Design errors occur, thus causing a hazard potential.
- Hazard sources are used within systems for needed system functions, and these hazard sources precipitate hazards.

Murphy's Law states "if anything can go wrong, it will." This basic truism illustrates that the unexpected and undesired must be anticipated and controlled in order to prevent mishaps, and this can only be achieved through the system safety process. Hazards and risk often cannot be eliminated; however, hazards and risk can be anticipated and mitigated, thereby preventing or reducing the likelihood of mishaps. If system safety is not applied, accidents and loss of life will not be prevented. If system safety is not applied, many system users will not be aware of the actual risk they are exposed to.

The overarching goals of system safety consist of the following:

- To save lives and preclude monetary losses by preventing product/system accidents and mishaps
- To protect the system and its users, the public, and the environment from mishap damage
- To identify and then eliminate or mitigate hazards
- To design and develop systems presenting minimal mishap risk
- To intentionally design safety into the overall system design
- To save costs by building-in safety from the start, rather than adding it later

System safety is a specialized engineering discipline for developing safe systems, products, processes, procedures, and operations. Safety itself is a somewhat invisible system quality that is inherent to a system or product. Safety is characterized by a metric called "mishap risk," which indicates the danger level (susceptibility to harm) presented by a system hazard. A system is considered "safe" when the inherent mishap risk it presents is known and considered acceptable. System safety involves applying proven engineering and management safety principles to intentionally design-in and build-in safety from the start of system development, as opposed to trying to add it in at a later time (after encountering incidents and mishaps). System safety anticipates the undesired effects that can result from failures, errors, and design flaws, and establishes design safety measures to counter these potentially hazardous situations.

The genesis of system safety, MIL-STD-882, defines system safety as "the application of engineering and management principles, criteria, and techniques to achieve acceptable mishap risk, within the constraints of operational effectiveness and suitability, time, and cost, throughout all phases of the system life cycle" (from version D dated 2000; however, all previous versions use the same definition). Through years of trial and error, a structured systematic system safety approach has been established, which is referred to as a best practice methodology. This methodology is built upon the following core elements:

1. Plan SSP
2. Hazard identification
3. Risk assessment
4. Risk mitigation
5. Mitigation verification
6. Risk acceptance
7. Hazard tracking

Figure 2.86 shows the core six steps as they form a closed-loop process around the overall SSP.

Figure 2.86 Core process, closed-loop view.

This core system safety process is a dynamic design safety methodology for two primary reasons: (1) it evolves or changes as the system design and development evolves, and (2) it is a closed-loop process where each step may be revisited as design data is updated until acceptable mishap risk is achieved. The overall core system safety process is a mishap risk management process, whereby safety is achieved through the identification of hazards, the assessment of hazard mishap risk, and the control or mitigation of hazards presenting unacceptable risk. Hazards are identified and continuously tracked until acceptable closure action is implemented and verified. This process should be performed in conjunction with actual system development, in order that the design can be influenced during the design process, rather than trying to implement costly design changes after the system is developed.

System safety is a "systems" approach to improving the safety property of a product, which accounts for the identifying name. System safety is the art and science of looking at all aspects and characteristics of a system, as an integrated whole, rather than looking at individual components in isolation from the system. System safety is a holistic approach that considers the subject as an integrated sum-of-the-parts combination, rather than looking at separate individual and solitary pieces of the system. This is necessary in order to fully comprehend and evaluate the many interactions and dependencies throughout the entire system.

System safety is an intentionally proactive process to design-in safety. When safety is intentionally designed into a system, mishap risk is significantly reduced. System safety is the discipline of identifying hazards, assessing potential mishap risk, and mitigating the risk presented by hazards to an acceptable level of risk. Risk mitigation is achieved through the implementation of a combination of design mechanisms safety features, warning devices, safety procedures, and safety training to counter the effect of HCFs.

System safety is a well-thought-out process that is planned, proactive, prudent, and preventive in nature. The primary objective of system safety is to avert mishaps by ensuring that safety is intentionally designed into a product or system. Designed-in safety leads to the inherent operational safety of a product or system. A system is considered safe when it presents acceptable mishap risk. Therefore, system safety is effectively a risk management process that deals in hazards and their associated mishap risk.

System safety is not prescriptive safety, which is rote compliance with standards and regulations. Designed-in safety is more than compliance safety; it involves a dynamic design safety process. This dynamic process follows a core system safety methodology that is carried out by an active SSP. The core system safety process involves six elements that are always performed by a credible, effective, and successful SSP. This core process is instituted on the safety metric of mishap risk and the system safety risk management process for reducing risk.

System safety is not occupational health safety. Occupational safety is an important safety discipline that deals with safety during the performance of a job or work activity. Occupational safety can apply system safety as part of its methodology, but system safety is much bigger and broader in scope. System safety attempts to design safety into a product or system before it is put into operational usage; operations safety is actually impacted during system design, long before the system becomes operational. System safety is not the same as reliability and cannot be supplanted or achieved strictly by reliability. Making a system reliable does not necessarily make it safe, and making a system safe does not necessarily make it reliable. Typically, system safety and reliability work well together; however, there are many situations where enhanced reliability actually degrades safety, and vice versa.

The fundamental goal of system safety is to develop a system with acceptable mishap risk, for all life-cycle phases, through a formal engineering and management process. This process is applied during the design development phase in order to impact all following phases, the operational phase in particular.

System safety is effectively a risk management process that deals on hazards, potential mishaps, and risk. System safety is involved in many different aspects of system/product development; however, it is structured around the six core elements that form the system safety process. These six elements are the building blocks that shape a foundation for the SSP. Figure 2.87 shows the core system safety elements and their interrelatedness. This viewpoint shows the core process as a sequence of tasks; however, in reality, they are quasi-sequential steps as the process has many iterations and interrelationships.

Note that the system safety process begins with step 1, which plans and documents the entire process, including establishing an HTS to accomplish step 6. As steps 2 through 4 are performed, their output is fed into the HTS, step 6, for data retention and report generation. The six core steps are firmly established; however, some of the tasks performed within each step may vary slightly, depending on the type of program and contracting involved. For example, on a military SSP, the risk acceptance authority persons are mandated by DoD policy, whereas in a corporate SSP, they may be delineated in corporate policy and look completely different.

Figure 2.87 Core elements, task view.

The following characteristics or qualities help characterize the system safety concept:

- Safety oriented—The primary goal is to save lives.
- Proactive—Identify and mitigate safety issues from the start of product design rather than trying to eliminate them after a mishap.
- Preventive—Intentionally design the safety quality into the product to prevent potential mishaps.
- Risk oriented—Apply risk management to control potential mishap risk.
- Hazard oriented—Concentrates on hazard identification and control because hazards are the key to potential mishaps and risk.
- System oriented—Focuses on the system as a whole, rather than just individual parts of the system, because of the many complex interactions involved between hardware, software, and humans.
- Life-cycle oriented—Focuses on the entire product/system life cycle for optimum product risk.
- Process oriented—Follows a defined and structured best practice methodology.

The objective of system safety is to develop a system that provides acceptable minimum mishap risk. The basic system safety philosophy for achieving this goal is to confront hazards, mishaps, and risk at three different levels of safety defense. These levels of defense are engineered into the system design. This layer of protection philosophy is summarized by the following three safety precepts:

1. Design the system to operate safely under normal operating conditions
2. Design the system to safely tolerate abnormal operations caused by faults and errors
3. Design the system to provide survival protection from credible mishaps

System safety is more than eliminating hardware failure modes; it involves designing the safe interactions of hardware, software, human,s and the environment under all failure and adverse conditions, as well as normal operating, testing, handling, or maintenance conditions. System safety involves anticipating potential failures and human errors, and designing to safety counter the threats they present.

SYSTEM SAFETY LEAD

A system safety lead is the single POC for SR matters on the government side of a program. The system safety lead is designated in writing by the PM and has the authority to speak for him on SR matters. A system safety lead is the technical authority regarding matters of system safety. The system safety lead may be referred to as the PFS on some programs.

SYSTEM SAFETY MANAGEMENT PLAN (SSMP)

On government programs, there are typically two program plans in effect: the SSPP prepared by the contractor and the SSMP prepared by the government procuring agency. The SSMP describes how the overarching SSP will be executed to meet government requirements and policies. In some cases, such as for larger projects, there may be a single SSMP for several different contractor SSPs and their associated SSPPs. Generally, the SSMP is developed prior to the individual contractor SSPPs that support it; however, this is not always the case, since systems are sometimes integrated into a larger system after they have been developed and proven. The SSMP is overarching and shorter than the SSPP; it provides general guidance and direction for the SSP. The SSPP is detailed and covers all aspects of the SSP that will be performed by the contractor.

As part of the contracting and system development process, the government develops the SSMP to guide all system safety aspects of the project. The contractor develops an SSPP that defines how the contractor plans to implement an SSP that meets the objectives and requirements set forth in the contract and the government SSMP. The SSPP should be a direct reflection of the SSMP, with more detail. In essence, the SSMP establishes the safety management process, and the SSPP establishes the safety program execution process by the contractor. Each plan will contain both management and technical aspects of the SSP.

The SSMP is the top-level plan for integrating the safety efforts of one or more contractor programs together into one overarching SSP. The purpose of an SSMP is to develop a blueprint for the overall SSP that defines and describes the management and engineering tasks and activities required to identify, evaluate, and eliminate/control hazards, and to reduce the residual risk to an acceptable level, throughout the system life cycle. The SSMP provides a formal, documented basis of understanding of the safety effort shared by the government and contractor. The SSMP provides a common framework in which individual activities or companies can work together for optimum consistency, effectiveness, and interoperability. The contractor SSPP must support and be consistent with all the requirements provided in the SSMP.

The SSMP delineates the scope, tasks, schedules, milestones, guidelines, responsibilities, and deliverables for the entire integrated system. The goal is to provide a common framework in which individual activities work together on a collective SSP, while avoiding any duplication of effort. The SSPP is a living document that is modified and updated as necessary during the life of the program; the SSMP is typically not updated. This is because it is a higher-level document that provides overall guidance and policy, which usually does not change during the life of the program. Although the SSMP format is not critical, it is essential that the SSMP contains the necessary and correct content, as well as be consistent with the SSPP. The basic program elements for a SSMP include the following:

- Introduction
- System overview
- Government safety policies and requirements
- Government SSP organization and interfaces
- Roles and responsibilities
- Schedules—SSP and program
- Hazard identification expectations
- Risk assessment expectations
- Government risk tables
- Risk mitigation and verification expectations
- Hazard tracking and closure expectations
- RAP
- SwS expectations
- Government safety data requirements
- SSP support activities
- COTS/NDI safety
- Program unique safety elements
- Review boards
- SSP audits
- SSWG and charter

See *System Safety Program (SSP)* and *System Safety Program Plan (SSPP)* for additional related information.

SYSTEM SAFETY ORGANIZATION

A system safety organization is the organization responsible for performing the system safety tasks and activities that formulate an SSP. The system safety organization applies the system safety process to develop a safe system. The system safety organization can be an individual, or group of individuals, depending on the size, complexity, and criticality of the particular system. The SSP is the process and the organization.

The system safety organization is part of the larger organization developing a product or system. The SSP organization should be a distinct organizational entity appearing on the program organizational chart with the responsibility and authority for the system safety function. The SSP organization should be part of the program decision-making process, with an organizational voice that can be heard by the PM so that system safety is not unduly constrained by competing organizational goals. The system safety organization is responsible for performing the tasks necessary to develop a system that presents minimal acceptable mishap risk, while considering competing factors and constraints, such as cost, schedule, technical complexity, and mission expediency. It is the responsibility of the system safety manager to establish and manage the SSP organization. This requires a person knowledgeable and skilled in the system safety discipline and process.

See *System Safety Program (SSP)* for additional related information.

SYSTEM SAFETY PROCESS

System safety process refers to the course of action required to effectively implement system safety. Figure 2.88 is a diagram depicting the basic steps in the core system safety process. It should be noted that the six core steps involve a firmly established best practice approach; however, some of the tasks performed within each step may vary slightly depending on the type of safety contracting involved. For example, in a military contract, the risk acceptance authority person is mandated by DoD policy and involves government personnel, whereas in a corporation, it may be delineated in corporate policy and involve a set of corporate personnel.

The objective of the system safety process is to achieve acceptable mishap risk through a systematic approach of hazard risk management, involving hazard identification, hazard risk assessment, and hazard risk mitigation. System safety is a well-thought-out process that is planned, proactive, prudent, and preventive in nature. The primary objective of system safety is to avert mishaps by ensuring that safety is intentionally designed into a product or system. Designed-in safety leads to the inherent operational safety of a product or system.

See *System Safety* for additional related information.

SYSTEM SAFETY PROGRAM (SSP)

An SSP is the combined set of people and tasks that implement and execute the system safety process on a development project or program. The SSP consists of an organization that performs the necessary system safety tasks and activities to realize the system safety objectives and obtain the necessary

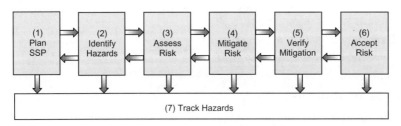

Figure 2.88 Core system safety process.

evidence of safety achievement. An SSP is the active, deliberate, and intentional application of the system safety process by an individual, or group of individuals, skilled in that process. The system safety organization is a part of the larger organization developing a product or system. The SSP should be a distinct organizational entity appearing on the program organization chart and it must be a part of the program decision-making process. The SSP should have an organizational voice that can be heard by the PM so that it is not constrained by competing organizational goals.

In general, the overarching objectives of the SSP are to:

- Manage and execute the system safety process
- Perform the core system safety elements
- Ensure that system design meets applicable safety requirements
- Ensure that all system hazards are identified and controlled
- Develop a system presenting minimal mishap risk
- Protect the system and its users, the public, and the environment from mishaps
- To intentionally design-in safety into the overall system design

The scope of an SSP includes hardware, software, firmware, and HSI for all system life-cycle phases. It should be noted that in addition to the core elements, an SSP also includes many program support tasks that are necessary, such as design reviews, technical review boards, SSWGs, etc. The support tasks may vary depending on the type, size, and safety-criticality of system. The SSP and organization can be tailored to a program's size, complexity, and safety-criticality. The SSP is planned and documented in the SSPP.

The intent of the SSP is to ensure that the system design meets applicable safety requirements, and that all hazards associated with the system are identified and eliminated, or controlled in a manner consistent with program objectives, constraints, and risks. The SSP also provides management visibility of safety risks inherent in the design and planned operations, and defines the process required for management to formally reduce and accept the system safety risks. Regardless of the type, size, or complexity of a system, there are specific items required to formulate an SSP. In order to exist as an entity, an SSP requires the following items, as a minimum:

- System safety organization
- Experienced safety manager or PFS
- Experienced system safety staff
- Budget
- Program authority
- Program safety policy
- Contractual safety requirements
- SSPP

- SSWG
- SSWG charter
- Program management recognition and support
- A safety culture

The SSP can be tailored to fit the needs of the program based on system size, complexity, safety-criticality, funding, development schedule, and other significant factors. Tailoring is recommended as there is no one-size-fits-all SSP; each SSP is unique and individual.

SYSTEM SAFETY PROGRAM PLAN (SSPP)

The SSPP is a document that defines and describes the SSP and how it will be implanted; it is a plan of purpose, organization, action, methodology, and schedule. It documents the management and engineering approach for applying the system safety methodology on a particular system or product development project. It has been shown through years of experience that it is more effective to design-in safety when there is a roadmap for the entire process.

Whereas the SSP consists of the actual people, activities, and products involved in implementing the system safety process, the SSPP is the document that formally defines the SSP tasks, products, interfaces, and milestones that will be required of the safety organization. The SSPP also defines the scope of the safety program and decision criteria to be used by the program. The SSPP should cover the essential core components of an SSP, plus all of the relevant SSP support activities that will be required. The SSP and SSPP can be tailored to fit the needs of the program based on system size, complexity, safety-criticality, funding, development schedule, and other significant constraints.

The SSPP is a formal documented plan that serves as a management tool for implementing an effective SSP. A well-prepared and documented SSPP is the key to a successful SSP. The SSPP should be written to cover all aspects of the SSP. It should also be written to cover all phases where system safety work is to be performed, that is, concept definition, design, test, deployment, operation, upgrade, and disposal. The SSPP describes and formalizes the system safety management and engineering tasks and activities; it is the how-to document that provides the "what, when, why, and who" for the SSP.

The depth, breadth, and quality of the SSPP reveals the value and importance placed on safety, and is an overall indication of management's commitment to system safety. It demonstrates how well system safety is understood by the safety manager, the PM, and the acquisition agency. The quality of the plan will be an indication of the quality of the SSP, which could also be an indication of relatively how safe the system will likely be when implemented. Although the SSPP format is not critical, it is essential that the SSPP contains the necessary and correct content. The basic program elements for an SSPP include the following, as a minimum:

- Introduction
- System description (short overview)
- System MEL
- Safety requirements, guidelines, and criteria
- Organization—SSP and program
- Roles and responsibilities
- Schedules—SSP and program
- Hazard identification approach (tied to the MEL)
- Risk assessment approach (including risk tables)
- Risk mitigation and verification approach
- Hazard tracking and closure approach
- RAP
- SwS approach
- Analysis methodologies
- Safety program compliance
- SSP products
- Safety data library
- SSP interfaces
- SSP support activities
- Resources
- COTS/NDI safety
- SAR
- Program unique safety elements
- Review boards
- SSP audits
- SSWG support

Although the content of each SSPP should contain the same basic elements, not all SSPPs are identical or necessarily look alike (nor should they). Each plan is unique to the particular type, size, and safety-criticality of system involved; it expresses the personality and management style for the organization developing the product. A tailored SSPP for a small project would normally look much different than a tailored plan for a large project. The SSPP for a non-safety-critical system would look different than one for an extremely safety-critical project.

The SSPP is a living document that is modified and updated as necessary during the life of the program. Updates may be required as each new development phase is entered to reflect new information and project changes. Also, during a particularly long development phase, the SSPP may require updating for various program-related reasons, such as design modifications and technology refresh.

On government programs, there are typically two program plans in effect, the SSPP for the contractor and the SSMP for the government procuring agency. The SSMP describes how the overarching SSP will be executed to meet government requirements and policies. Generally, the SSMP is developed prior to the individual contractor SSPPs that support it; however, this is not always the case, since systems are sometimes integrated into a larger system after they have been developed and proven. The SSMP is overarching and shorter than the SSPP; it provides general guidance and direction for the SSP. The SSPP is detailed and covers all aspects of the SSP that will be performed by the contractor.

As part of the contracting and system development process, the contractor develops an SSPP that defines how the contractor plans to implement an SSP that meets the objectives and requirements set forth in the contract and the government SSMP. The SSPP should be a direct reflection of the SSMP, with more detail. In essence, the SSMP establishes the safety management process and the SSPP establishes the safety program execution process by the contractor. Each plan will contain both management and technical aspects of the SSP. The contractor SSPP must support and be consistent with all the requirements provided in the SSMP.

See *System Safety Management Plan (SSMP)* and *System Safety Program (SSP)* for additional related information.

SYSTEM SAFETY REQUIREMENT (SSR)

An SSR is a design requirement specifically for purposes of safety (as opposed to a performance or reliability requirement). The objective of an SSR is to enhance the safety quality of a system by providing design guidance for intentionally designing safety into a system or product. SSRs generally focus on preventing, eliminating, or mitigating a hazard. An SSR may be a detailed design requirement developed specifically to mitigate a particular hazard, or a general requirement intended to provide guidance for a certain class of hazards. An SSR provides specific design direction for developing a safe system design. For example, an SSR might be "Weapon arming requires the sensing of two independent environments (set back and spin)."

Typically, the SSR specifies that a certain safety feature shall be implemented in the system design. The SSR requirement should be tagged or identified as safety-related or safety-critical in all specifications. Tagging notifies the designer (and maintainer) that the particular item or function is safety-related, and that it should be treated carefully and with the cognizance of the SSP.

Every identified hazard should have at least one corresponding SSR, as a minimum, which either eliminates or mitigates the hazard. Some hazards may require multiple SSRs for elimination or mitigation of the hazard. An SSR is the methodology by which system safety influences the design, and is thus the basis for the term "DFS." SSRs relate very closely with the hazard triangle for a particular hazard. When writing an SSR, the requirement should directly affect one or more of the hazard components.

There are many different sources for SSRs, some are prescribed and some are not. Hardware items known to be hazardous generally have a large amount of data, requirements, and guidelines available from lessons learned (e.g., high voltage, pressure, fuel, and lasers are all recognized as having inherent safety hazards). As a result, there exists considerable data for these items, as well as safety guidelines and requirements.

The process of developing SSRs is multifaceted and can be viewed in the form of a pyramid as shown in Figure 2.89. The concept behind this "requirement pyramid" is that the final list of design SSRs is gradually developed through a building process based on requirement analysis and design analysis. As the requirement analysis and design analysis stages progress, various types of requirements and guidelines are established in hierarchy levels. The final detailed design SSRs are the result of the lower levels of the process.

The initial foundational starting point for SSRs stems from analysis of customer safety requirements, guidelines, goals, and objectives. The customer states these requirements and objectives in the contract or SOW. Continued requirements analysis establishes safety precepts/principles for the development program. These are broad self-imposed rules or requirements that must be followed by the program. The next level in the pyramid consists of the regulatory safety requirements, standards, codes, laws, and criteria. These are the federal, state, and local codes that may be imposed upon a particular type of system or subsystem. For example, the National Electrical Code provides many design requirements for electrical subsystems within the United States. The next level in the pyramid consists of the equipment safety requirements that have been established through years of experience for different types of equipment (e.g., hydraulic, electrical). For example, MIL-STD-1316E specifies design safety requirements for fuze systems and MIL-STD-1472C specifies equipment and facilities safety requirements for human engineering safety concerns. The next level are generic SSRs that have already been established,

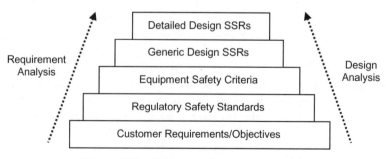

Figure 2.89 Safety requirements pyramid.

such as STANAG 4404 which provides design safety guidelines for software. Through the combined results of the requirement analysis and the design analysis, the final detailed design SSRs are established.

See *Hazard* and *Hazard Triangle* for additional related information.

SYSTEM SAFETY WORKING GROUP (SSWG)

The SSWG is a formally chartered group of persons that represent different organizations of an acquisition program. The SSWG is an integrated team of safety and non-safety personnel whose continuous activities monitor and guide the SSP throughout the program life cycle. The SSWG is responsible for ensuring that safety efforts are closely coordinated and safety tasks are completed on an effective and timely basis.

The SSWG is a government-conducted safety activity. The government principle for safety (PFS) typically serves as the chairman, or vice chairman, of the SSWG, and SSWG membership is composed of experienced government and contractor personnel involved in the development, installation, and life-cycle support of the system. SSWG meetings should be held on a regular basis (not less than once a year) to assess system safety issues. The SSWG reviews and reports on safety program status, resolves safety issues, and forwards recommendations to the PM. At a minimum, all open and monitor hazards will be reviewed at each SSWG. Newly proposed hazards will also be reviewed at SSWG meetings. SMEs will brief the SSWG on the technical aspects of safety issues. An SSWG is a prime component of the SSP. The primary function of the SSWG is to provide a complete overview of the safety program to ensure consistency and that nothing is overlooked. SSWGs are established in accordance with safety policy and the SSPP, and directly represent the PM for safety issues. The SSWG is also known as a system safety group (SSG).

The PFS is responsible for establishing the SSWG, the SSWG charter, and conducting regularly scheduled SSWG meetings. The PFS is also responsible for ensuring that SSWG meeting minutes are documented and all SSWG action items are resolved on a timely basis.

When a weapon system is being developed by multiple contractors, the opportunity for one subsystem to initiate hazards and UEs in other subsystems is enhanced. For this reason, it is important that SSWGs be established to determine if such possibilities exist. Such an examination by an SSWG constitutes a system safety review. An SSWG includes both safety engineers and engineers of other disciplines, representing each of the subsystems involved, plus similarly oriented representatives of the government PM and of any integrating contractor.

Some of the specific SSWG activities include, but are not limited to, the following:

- Presenting the contractor safety program status, including results of design or operations risk assessments
- Summarizing HAs, including identification of problems, status of resolution, and residual risk
- Presenting incident assessment results (especially mishaps and malfunctions of the system being acquired), including recommendations and action taken to prevent recurrences
- Responding to action items assigned by the chairman of the SSWG
- Developing and validating SSRs and criteria applicable to the program
- Identifying safety deficiencies of the program and providing recommendations for corrective actions or prevention of recurrence
- Planning and coordinating support for a required certification process
- Documenting and distributing meeting agendas and minutes

See *System Safety Working Group (SSWG) Charter* for additional information.

SYSTEM SAFETY WORKING GROUP (SSWG) CHARTER

An SSWG charter is the official document which provides guidelines and ground rules for the SSWG. The SSWG charter is an artifact of the SSP. The government PFS is responsible for establishing and documenting the SSWG charter.

The primary tasks or functions of the SSWG include, but are not limited to, the following:

- Review of proposed hazards for credibility and induction into the HTS
- Review and concurrence on the HRI of identified hazards
- Review and concurrence on the open, monitor, or close status of hazards
- Improve communication between all program safety participants
- Provide for the exchange of safety information
- Assemble pertinent safety data to aid management decisions
- Coordinate safety issues between interfacing contractors
- Discuss the resolution of identified hazards
- Collect and review data as necessary, such as accident/incident reports
- Ensure safety action items and issues are resolved and closed

As a minimum, the SSWG charter should include a discussion of the following items:

- Purpose
- Membership composition

- Chairperson
- Meeting rules
- Scope of SSWG
- Tasks, roles, and responsibilities
- Meeting minutes (responsibility, maintaining a minutes file)
- Action item list (responsibility, tracking, closure)
- Documenting and distributing meeting agendas and minutes
- Meeting and location schedule
- The process for reviewing, tracking, and closing hazards
- New business (new safety issues and/or concerns)

See *System Safety Working Group (SSWG)* for additional related information.

SYSTEMS ENGINEERING TECHNICAL REVIEW (SETR)

The SETR is a technical assessment process that evaluates the maturing design over the life of the program. The SETR provides a framework for structured systems engineering management, including assessment of predicted system performance. The SETR provides the PM with a better understanding of the program's technical health. SETRs are an iterative program process that maps the program technical reviews to the acquisition process timeline. A program risk assessment checklist is used for each SETR review.

SETRs are an integral part of the systems engineering process and life-cycle management, and are consistent with existing and emerging commercial/industrial standards. These reviews are not the place for problem solving, but to verify that problem solving has been accomplished. As a part of the overall systems engineering process, SETRs enable an independent assessment of emerging designs against plans, processes, and key knowledge points in the development process. SETRs also apply to post production, in-service improvements, and maintenance. An integrated team consisting of integrated program/product team (IPT) members and independent competency SMEs conducts these reviews. Engineering rigor, interdisciplinary communications, and competency insight are applied to the maturing design in the assessment of requirements traceability, product metrics, and decision rationale. These SETRs bring to bear additional knowledge to the program design/development process in an effort to ensure program success. Overarching objectives of these reviews are a well-managed engineering effort leading to a satisfactory technical evaluation (TECHEVAL), which will meet all of the required technical and programmatic specifications. This in turn will ensure a satisfactory operational evaluation (OPEVAL) and the fielding of a suitable and effective system.

U.S. Navy document NAVAIRINST 4355.19D establishes policy, outlines the process, and assigns responsibilities for the planning and conduct of SETRs

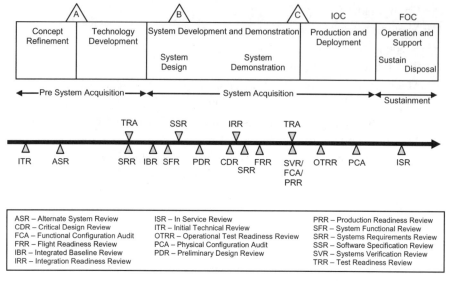

Figure 2.90 SETR process diagram.

for Naval Air Systems Command (NAVAIR) programs. Figure 2.90 shows the overall SETR process.

SYSTEM TYPE

Understanding system type and scope is very important in system safety and HA. The system type can be an indication of the safety-criticality involved. The scope of the system boundaries establishes the size and depth of the system. The system limitations describe basically what the system can and cannot safely do. Certain limitations may involve the ability of the system to include DSFs. Every system operates within one or more different environments. The specific environment establishes what the potential hazardous impact will be on the system. System criticality establishes the overall safety rating for the system. A nuclear power plant system has a high consequence safety rating, whereas a TV set as a system has a much lower safety-criticality rating.

There are four basic "system-type" models that describe almost all types of systems and their relative complexity. Each model varies depending upon factors such as composition, relationships, intent, and environment. As shown in Figure 2.91, these system types are (a) static, (b) dynamic, (c) homeostatic, and (d) cybernetic (Systematic Systems Approach, Thomas H. Athey, Prentice-Hall Inc., 1974).

System complexity has a very direct relationship upon system safety. Generally, the more complex the system, the more likely it is to have safety concerns or problems. Also, more complex systems are more difficult to analyze

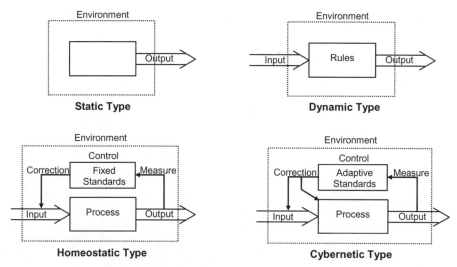

Figure 2.91 System types.

and verify for safety. Understanding these system types and their overall complexity is part of the system safety process.

SYSTEMS THEORY

Systems theory is a cross-disciplinary approach, which abstracts and considers a system as a set of independent and interacting parts. Systems theory, or systems thinking, is the process of understanding how things influence one another within a whole. In nature, systems thinking examples include ecosystems in which various elements such as air, water, movement, plant, and animals work together to survive or perish. In organizations, systems consist of people, structures, and processes that work together to make an organization healthy or unhealthy. Systems thinking is an approach to problem solving, by viewing problems as parts of an overall system, rather than reacting to a specific part, outcome, or event, and potentially contributing to further development of unintended consequences. Systems thinking is based on the belief that the component parts of a system can best be understood in the context of relationships with each other and with other systems, rather than in isolation. Systems thinking is the genesis for systems engineering and system safety.

TACTICAL DIGITAL INFORMATION LINK (TADIL)

TADIL is a standardized radio communication data links used by the U.S. armed forces. Army INFOSYS uses TADILs to transmit and receive data. TADILs are characterized by their standard message and transmission formats.

See *Datalink* for additional related information.

TAILORING

A formal SSP involves the conduct of many tasks and activities necessary to identify hazards and mitigate potential mishap risk. The success of the system safety effort depends on definitive statements of safety objectives and requirements. Selective tailoring of an SSP is often necessary to effectively achieve all of the safety objectives within the constraints of time, cost, schedule, and prevention of potential mishap loss. As such, tailoring becomes an important aspect of conducting an effective and successful SSP.

There is not a one-size-fits-all SSP, and tailoring may be needed to account for differences in program size, cost, complexity, and safety-criticality. SSP tailoring should be done judiciously with the appropriate analysis, justification, and rationale to support the tailoring decisions.

Tailoring is the process of establishing the specific tasks that must be performed by an SSP. It also involves establishing the specific risk tables and risk criterion for the program. Some of the important aspects to consider for SSP tailoring include:

- Ensure tailoring is consistent
- Establish a tailoring process to follow prior to tailoring an SSP
- Ensure tailoring is in compliance with DoD instructions
- Ensure that tailoring does not degrade the system safety effort
- Ensure that tailoring makes an SSP as stringent (or more) as one defined by MIL-STD-882
- Ensure that all stakeholders concur with the tailored SSP aspects

TECHNICAL DATA

Data required for the accomplishment of logistics and engineering processes in support of the contract end item. It includes drawings, operating and maintenance instructions, provisioning information, specifications, inspection and test procedures, instruction cards and equipment placards, engineering and support analysis data, special-purpose computer programs, and other forms of audiovisual presentation required to guide personnel in the performance of operating and support tasks.

TECHNICAL DATA PACKAGE (TDP)

A TDP is packet of technical (and some management) material that is provided to a review board that is evaluating the safety of a product or system. The review board uses the information and data contained in the data package to draw conclusions regarding the safety of a product or system. The TDP must contain correct, current, and appropriate information in order for the review board to make a favorable determination.

As a minimum, the following information and data should be provided in the data package:

- System description (including hardware, software, and human interface design)
- Description of the SSP
- Program schedule with safety milestones
- Copy of the SSPP
- List of the SCFs
- Discussion and summary of the safety analyses performed
- Description and summary of the tests performed
- Discussion of the SSWG
- List of safety features
- SAR
- Summary of residual risk
- Copies of the detailed HAs performed (as appendices)

A TDP may contain or reference all applicable technical data such as drawings or automated models and associated lists, specifications, standards, performance standards, quality assurance requirements, software, and packaging details.

TECHNIQUE FOR HUMAN ERROR RATE PREDICTION (THERP)

THERP is a methodology for generating human error probabilities for use in safety and reliability analyses.

TECHNOLOGY REFRESH

Replacing obsolete or no longer available hardware/software components with components having identical or similar functions with newer technology is known as technology refresh. It is very likely that since the original development of the component, the technology involved has changed or improved, and new replacement parts have been developed using new processes or materials. Therefore, when performing technology refresh, an essentially new component that has not been not developed or qualified with the original system design is placed into the system design (which may have unknown collateral effects).

System safety should always be concerned about the impact of technology refresh and should be involved in the process. HA should be performed on technology refresh items.

See *Technology Insertion* for additional information.

TECHNOLOGY INSERTION

Replacing existing, but not necessarily obsolete, hardware/software components, with newer technology that enhances system capabilities is known as technology insertion. This is a case of intentionally replacing a component because the replacement component utilizes newer and better technology that should provide a benefit to system operation. In this case also, the new component has not been developed or qualified with the original system design.

System safety should always be concerned about the impact of technology refresh and should be involved in the process. HA should be performed on technology refresh items.

See *Technology Insertion* for additional information.

TELE-OPERATION

Tele-operation is a mode of operation of a UMS wherein the (remote) human operator, using video feedback and/or other sensory feedback, either directly controls the actuators or assigns incremental goals, waypoints in mobility situations, on a continuous basis, from off the vehicle and via a tethered or radio-linked control device. In this mode, the UMS may take limited initiative in reaching the assigned incremental goals.

TELEPRESENCE

Telepresence is the capability of a UMS to provide the remotely located human operator with some amount of sensory feedback similar to that which the operator would receive if he were in the vehicle.

TEMPERATURE CYCLE

A temperature cycle is the transition from some initial temperature condition to temperature stabilization at one extreme and then to temperature stabilization at the opposite extreme and returning to the initial temperature condition.

TEST READINESS REVIEW (TRR)

A TRR ensures that the test article (hardware/software), test facility, support personnel, and test procedures are ready for testing and data acquisition, reduction, and control.

See *Critical Design Review (CDR)*, *Preliminary Design Review (PDR)*, and *Systems Engineering Technical Reviews (SETR)* for additional information.

TEST WITNESS

A test witness is a person acting as an on-the-scene observer of the performance of a test with the purpose of verifying compliance with project requirements. On some programs, a system safety engineer is required to witness testing hardware and/or software that is safety-related.

THERMAL BALANCE TEST

A test conducted to verify the adequacy of the thermal model, the adequacy of the thermal design, and the capability of the thermal control system to maintain thermal conditions within established mission limits.

THERMAL CONTACT HAZARDS

The thermal surface temperatures of surfaces which a human will contact are a system safety concern when designing systems that are safe for human use. Thermal temperatures in a system are primary hazard sources for many different types of hazards, and high surface contact temperature can be hazardous for surfaces or controls that a human operator will touch.

MIL-STD-1472, Human Engineering, provides the following guidance for safe surface temperatures:

> 5.13.4.6 Thermal Contact Hazards
> Equipment which, in normal operation, exposes personnel to surface temperatures greater or less than those shown below (Table 2.22 here) shall be appropriately guarded. Surface temperatures induced by climatic environment are exempt from this requirement. Cryogenic systems shall also be appropriately guarded.

THERMAL-VACUUM TEST

Thermal-vacuum test is a test conducted to demonstrate the capability of the test item to operate satisfactorily in vacuum at temperatures based on those

TABLE 2.22 Thermal Contact Limits from MIL-STD-1472

Exposure	Temperature Limits		
	Metal	Glass	Plastic or Wood
Momentary contact	60°C (140°F)	68°C (154°F)	85°C (185°F)
Prolonged contact or handling	49°C (120°F)	59°C (138°F)	69°C (156°F)
Momentary or prolonged contact or handling	0°C (32°F)	0°C (32°F)	0°C (32°F)

expected for the mission. The test, including the gradient shifts induced by cycling between temperature extremes, can also uncover latent defects in design, parts, and workmanship.

THINGS FALLING OFF AIRCRAFT (TFOA)

TFOA refers to items that fall off an aircraft during taxi, takeoff, flight, or landing. These items can cause damage or injury, depending upon their size. When they fall on an aircraft runway, they can become foreign object debris (FOD) for another aircraft. Recall the Concorde mishap where a FOD item on the runway was the basic cause of the crash, and the FOD item was a TFOA item from a previous aircraft that had used the same runway.

THREAT HAZARD ASSESSMENT (THA)

The threat hazard assessment is an evaluation of a munition and its life-cycle environmental profile to determine the threats and hazards to which the munition may be exposed. The assessment includes threats posed by friendly munitions, enemy munitions, accidents, handling, transportation, storage, and so on. The assessment is based on analytical or empirical data to the best extent possible. A THA is a mandatory requirement specified in MIL-STD-2105 (Series), Hazard Assessment Tests for Non-Nuclear Munitions. A THA covers the life cycle of the munition item, including friendly and hostile environments starting with production delivery and extending until the item is expended, or properly disposed. The THA identifies threats and hazards, both qualitatively and quantitatively, along with their causes and effects.

The THA evaluates potential threats and hazards throughout all the of the weapon system's life-cycle scenarios, including combat threats and normal operational threats. Scenarios include transportation, handling, storage, and operational use. Potential threats are evaluated as hazards, and design action is taken to eliminate or mitigate these hazards. Identified hazardous scenarios are matched with insensitive munitions design and testing. Testing required by

MIL-STD-2105 (Series) should be modified to address the hazards identified in the THA.

The intent of the THA is to identify and evaluate hazards and threats to a weapon system containing non-nuclear munitions. Potential threats can result from enemy action or self-induced hazards. The identified hazards and threats are used as input into the munitions test program. The THA provides a framework for the development of a consolidated safety and IM assessment test program for non-nuclear munitions. It should be noted that the THA could also be applied to nuclear weapons and even other types of systems. THA is an aid in designing a suitable hazard assessment test program for system-containing munitions that will influence design early and give management the information necessary to determine the risk associated with the weapon system. The final product should be a list of hazards that prescribe safety tests and IM tests. The THA should provide inputs to other program safety analyses, such as the PHA, SSHA, and SHA. Some THA identified hazards may require more detailed analysis by other techniques (e.g., FTA) to ensure that all causal factors are identified and mitigated.

THA helps in establishing guidelines for the sensitivity assessment of munitions. The purpose of an IM program is to increase the survivability of ships and aircraft by making munitions less sensitive to unplanned stimuli. MIL-STD-2105 provides a series of tests to assess the reaction of energetic materials to external stimuli representative of credible exposures in the life cycle of a weapon and requires the use of a THA in developing test plans.

TOP-LEVEL HAZARD (TLH)

A TLH is a generic hazard category where hazards with similar outcomes are collected together. A TLH is very similar to a TLM and is also sometimes used interchangeably with the TLM.

The purpose is to group hazards together that share a common outcome in order to reduce hazard clutter and provide better visibility when many different hazards exist. Typically, several TLHs will fall under a TLM category, and each TLH would have multiple hazards beneath them, thus creating a hazard hierarchy for mapping hazard concerns.

See *Top-Level Mishap (TLM)* for additional related information.

TOP-LEVEL MISHAP (TLM)

A TLM is a generic mishap category for collecting together various hazards that share the same general outcome or type of mishap. A TLM is a common mishap outcome that can be caused by one or more hazards; its purpose is to serve as a collection point for of all the potential hazards that can result in the same outcome, but have different causal factors. TLMs provide a design safety focal point for a particular safety concern (i.e., the TLM outcome). Each contributing hazard has different IMs or causal factors, but a common TLM outcome event. This common outcome is extracted from the hazards and is used as a common TLM to unite the hazards.

During HA of a large system, the number of potential hazards can become so large and diverse that the problem becomes one of how to easily and accurately represent the safety risks of the system design. When several different hazards can result in the same mishap, that mishap is categorized as a TLM. The TLM becomes a generic mishap category for collecting various hazards contributing to it. It is referred to as a TLM rather than a TLH because it is a collection of several different hazards, each with the same overall mishap.

Figure 2.92 illustrates the TLM concept. In this example, there are five different hazards resulting in an uncontrolled aircraft fire. Each hazard has different causal factors, but a common outcome, which is an uncontrolled fire in the aircraft. This common outcome is extracted from the hazards and used as a common TLM to unite the hazards. A different hazard such as "Landing gear fails to lower" could not be placed directly under this TLM (it would fall under a different TLM).

Systems typically have several different TLMs, depending on the size and safety-criticality of the system and the desired safety focal points. Also, different types of systems have different types of TLMs, although there may be a few similar TLMs between some system types. Some example TLMs for three different system types are shown in Table 2.23.

"Top level" in TLM does not necessarily imply a particular level of safety importance, but rather the common category visible at the system level (i.e., a hazard should fall within a particular TLM category). It should be noted, however, that by their very nature, TLMs have an implied level of safety-criticality. For example, the TLM "Inadvertent missile launch" has a greater safety-criticality than the TLM "Personnel injury due to electrical contact."

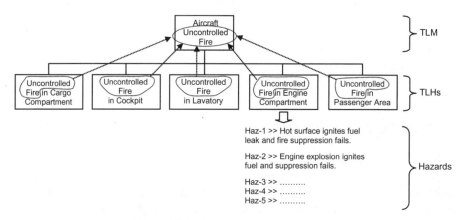

Figure 2.92 Top-level mishap concept.

TABLE 2.23 Example TLMs for Different System Types

Missile System	Aircraft System	Spacecraft System
Inadvertent missile launch	Controlled flight into terrain	Loss of astronaut oxygen
Inadvertent warhead initiation	Loss of all engines	System fire
Incorrect missile target	Loss of all flight controls	Loss of reentry capability
Inability to destroy errant test missile	Loss of landing gear	Loss of astronaut temperature control
Personnel electrical injuries	Inadvertent thrust reverser operation	Loss of earth communications
Personnel mechanical injuries	Personnel electrical injuries	Personnel electrical injuries
Personnel RF radiation injuries	Personnel mechanical injuries	Personnel mechanical injuries
Weapon-ship fratricide		

The real value and need for TLMs is based on the need for system hazard clarity and focus. The use of TLMs helps to resolve some safety programmatic issues, such as (a) hazard abundance, (b) hazard confusion, (c) subsystem confusion, and (d) total mishap risk. As HAs are performed on a system design, many hazards are identified, sometimes in the thousands. With a large number of hazards, it often becomes difficult to maintain hazard visibility. Sometimes hazards are inadvertently repeated; sometimes hazards are stated as causal factors rather than hazards. Hazard risk can be assessed against TLMs to determine if a thread exists that creates increased risk for that TLM category.

TLMs should be established early in the SSP, generally during the PHL analysis or the PHA phases. Each unique system will have its own unique set of TLMs. As TLMs are established, DSPs and principles can be derived for TLMs and the hazards within a TLM. The safety precepts serve as a mechanism for facilitating the derivation of specific detailed safety requirements for hazard mitigation.

TOXICITY

Toxicity is the degree to which something is poisonous. It is the harmful effect of a chemical or physical agent on the physiological functions of a biological system.

TRACK CORRELATION

Track correlation, or correlation, is the process of combining one track with another track. Only one track retains its track number, and the other is dropped. This is in reference to radar tracking from one or more tracking systems watching an incoming enemy system of some type.

TRANSPONDER

A transponder (sometimes abbreviated as XPDR) is a telecommunication device that receives, amplifies, and retransmits a signal on a different frequency. It is a device that transmits a predetermined message in response to a predefined received signal. It is also a receiver-transmitter that will generate a reply signal upon proper electronic interrogation.

In general aviation, depending on the type of interrogation, the transponder sends back a transponder code (mode A) or altitude information (mode C) to help ATCs to identify the aircraft and to maintain separation. Another mode called mode S (mode select) is designed to help air traffic control in busy areas and allow automatic collision avoidance. Mode S transponders are "backwards compatible" with modes A and C. Mode S is mandatory in controlled airspace in many countries.

In satellite communications, a communications satellite's channels are called transponders because each is a separate transceiver or repeater. In road transportation systems, E-ZPass tags are RFID transponders, which when queried provide the vehicle's data for automatic billing for road and bridge tolls.

ULTRAVIOLET (UV) RADIATION

UV radiation is EMR beyond the visible spectrum and having a wavelength shorter than visible light and longer than that of an X-ray.

UNATTENDED SYSTEM

Unattended system is any manned, unmanned, mobile, stationary, active, and/or passive system, with or without power that is designed to not be watched, or lacks accompaniment by a guard, escort, or caretaker.

UNDETECTABLE FAILURE

A failure mode for which there is no failure detection method by which the operator is made aware of the failure. Typically, undetected failures are identified in the FMEA.

UN-EXECUTABLE CODE

Un-executable code is software code that cannot be executed for various reasons. Un-executable code is often also referred to as "dead code."
See *Dead Code* for additional related information.

UNEXPLODED ORDNANCE (UXO)

The term refers to ordnance that has been used but has not exploded as intended, or ordnance that has not been used but is intended for decommission.

UNIFIED MODELING LANGUAGE (UML)

UML is a notation used to document system specifications. It is not a methodology; it is a special notation used to guide an analyst in structuring system documentation and design. It is primarily used as a software development tool.

UNINTENDED FUNCTION

Systems are designed to contain an intended functionality that reflects the design requirements. As a system is designed with intended functions, it is also sometimes inadvertently designed with built-in unintended functions. An unintended function is a function that is not specified by the design requirements and that is not desired (or intended) by the user. Unintended functions are typically not readily apparent and are usually discovered only by system test or operation. An unintended function is similar to a sneak circuit in electronics. It is a path of operation that might possibly occur under certain circumstances but was never intentionally designed-in. Unintended functions can be the source of latent hazards. These type of hazards are generally not so obvious and are very difficult to identify or postulate. One objective of testing is to ensure that the design performs all of the intended functions and that it does not contain any unintended functions.

UNIT

A unit is an assemblage of parts that is regarded s a single entity within a system. A unit is typically a subsystem or major assembly within the system hierarchy. A unit is a system item that can be removed and replaced as a single individual functional item.

UNIT TESTING

Unit testing is a software verification and validation method in which a programmer tests if individual units of source code are fit for use. A unit is the smallest testable part of an application. In procedural programming, a unit may be an individual function or procedure. Unit tests are typically written and run by software developers to ensure that code meets its design and behaves as intended (or specified). Its implementation can vary from being very manual (pencil and paper) to being formalized as part of build automation.

The goal of unit testing is to isolate each part of the program and show that the individual parts are correct. Unit tests find problems early in the development cycle. Unit testing by definition only tests the functionality of the units themselves. Therefore, it will not catch integration errors or broader system-level errors, such as functions performed across multiple units, or nonfunctional test areas such as performance. Unit testing must be done in conjunction with other software testing activities. Like all forms of software testing, unit tests can only show the presence of errors; they cannot show the absence of errors.

UNMANNED AIRCRAFT (UA)

A UA is an unmanned aircraft; it is one component of an unmanned aircraft system (UAS). A UA can fly autonomously or be piloted remotely. It is also sometimes referred to as an unmanned aircraft vehicle (UAV) or unmanned aerial vehicle (UAV).

See *Unmanned Aircraft System (UAS)* for additional related information.

UNMANNED AIRCRAFT SYSTEM (UAS)

A UAS is an aircraft that flies without a human crew on board the aircraft. A typical UAS consists of the unmanned aircraft (UA), the control system, the datalink, and other related support equipment. A UAS can fly autonomously or be piloted remotely, can be expendable or recoverable, and can carry a lethal or nonlethal payload. Their largest uses are in military applications. A UAS is also known as a remotely piloted vehicle (RPV) and unmanned aerial system; however, these are essentially obsolete terms.

Ballistic or semi-ballistic vehicles, cruise missiles, and artillery projectiles are not considered UASs. To distinguish UASs from missiles, a UAS is defined as a reusable, uncrewed vehicle capable of controlled, sustained, level flight, and powered by a jet or reciprocating engine. Therefore, cruise missiles are not considered UAVs because, like many other guided missiles, the vehicle itself is a weapon that is not reused, even though it is also unmanned and in some cases remotely guided.

UNMANNED AIRCRAFT VEHICLE (UAV)

A UAV is an unmanned aircraft. It is one component of a UAS. It can fly autonomously or be piloted remotely. It is also referred to as an unmanned aircraft (UA).
  See *Unmanned Aircraft System (UAS)* for additional related information.

UNMANNED GROUND VEHICLE (UGV)

A UGV is a powered, mobile, ground conveyance that does not have a human aboard; it can be operated in one or more modes of control (autonomous, semiautonomous, tele-operation, remote control); it can be expendable or recoverable; and it can have lethal or nonlethal mission modules. It is often referred to as a robotic platform that is used as extensions of human capability. A UGV is generally capable of operating outdoors and over a wide variety of terrain, functioning in place of humans. UGVs are used for both civilian and military use to perform dull, dirty, and dangerous activities. There are two general classes of UGVs: tele-operated and autonomous.

UNMANNED SYSTEM (UMS)

UMS is an electromechanical system that is able to exert its power to perform designed missions and includes the following four common characteristics:

- There is no human operator aboard
- The system can operate in a full or partial autonomous mode
- The system can be operated remotely by a human operator
- The system is designed to be recoverable

A UMS may be mobile or stationary, and includes the vehicle/device and its associated control station. UMSs include, but are not limited to UGVs, UASs, unmanned underwater vehicles, unmanned surface vessels, and unattended systems. Missiles, rockets, submunitions, and artillery are not considered UMSs.
  See *Autonomous System* for additional related information.

UNSAFE ACTION

Unsafe actions are human actions inappropriately taken, or not taken when needed, by system personnel that result in a degraded system safety condition.

USE CASE

Use case is a software development tool used for establishing software design and design requirements. A use case consists of (a) a diagram and (b) descriptive text. The use case diagram provides an overview of important interactions, and the use case text details the requirements.

USER INTERFACE (UI)

The term *UI* is used to describe the controls and displays that interface a human operator with a system. In today's environment, many UIs involve computer controlled electronic devices. A basic goal of HSI and system safety is to improve the interactions between users and computer-controlled systems by making the interface more usable and receptive to the user's needs, as well as making it less error prone. HSI applies the principles of human perception and information processing to create an effective UI. A few of the many potential benefits that can be achieved through utilization of HSI principles include a reduction in errors, a reduction in required training time, an increase in efficiency, and an increase in user satisfaction. Usability is the degree to which the design of a particular UI takes into account the human psychology and physiology of the users, and makes the process of using the system effective, efficient, satisfying, and safe.

On the surface it might seem that UI designs are relatively benign and would not pose any safety concerns. However, clearly the opposite is true, as demonstrated by past mishaps in a variety of different systems. A mechanical UI switch that fails *on* or *off* can affect an SCF. The primary safety concerns with UIs tend to deal with user confusion and user overload (too many controls and control options), which could cause a safety error to be committed by the user.

See *Graphical User Interface (GUI)* for additional related information.

VALIDATION

Validation is the determination that the requirements for a product or system are sufficiently correct and complete. Validation implies that a solution or process is correct or is suited (when it meets its requirements) for its intended use by the end user.

VALIDATION BY SIMILARITY

Validation by similarity is a procedure of comparing an item to a similar one that has already been verified.

VALID COMMAND

A valid command is a command that meets the following criteria:

1. The command originates from an authorized entity.
2. The received command is identical to the sent command.
3. The command is a valid executable command for the system.

VALID MESSAGE

A message that meets the following criteria:

1. The message originates from an authorized entity.
2. The received message is identical to the sent message.
3. The message meets valid message structure criteria for the system.

VERIFICATION

Verification is the determination that an implementation meets applicable requirements. This is typically achieved through test, analysis, inspection, or a combination of these.

VERY LARGE-SCALE INTEGRATION (VLSI) HARDWARE DESIGN LANGUAGE (VHDL)

A VDHL is HDL that is particularly suited as a language for describing the structure and behavior of digital electronic hardware designs, such as ASICs and FPGAs, as well as conventional digital circuits.

VIBROACOUSTICS

Vibroacoustics is an aircraft environment induced by high-intensity acoustic noise associated with various segments of the flight profile; it manifests itself throughout the structure in the form of directly transmitted acoustic excitation and as structure-borne random vibration.

VISIBLE RADIATION (LIGHT)

Visible radiation is that part of EMR that can be detected by the human eye. This term is commonly used to describe wavelengths which lie in the range 0.4–0.7 mm.

See *Electromagnetic Radiation (EMR)* for additional related information.

WASH-1400

WASH-1400, "The Reactor Safety Study," was a report produced in 1975 for the Nuclear Regulatory Commission by a committee of specialists under Professor Norman Rasmussen; it is often referred to as the Rasmussen Report. WASH-1400 considered the course of events which might arise during a serious accident at a large modern light water reactor. It estimated the radiological consequences of these events, and the probability of their occurrence, using an FTA and ETA approach. This technique is called PRA. The report concluded that the risks to the individual posed by nuclear power stations were acceptably small, compared with other tolerable risks. Specifically, the report concluded, using the methods and resources and knowledge available at the time, that the probability of a complete core meltdown is about 1 in 20,000 per reactor per year. The PRA methodology became generally followed as part of the safety assessment of all modern nuclear power plants. ETs made their first appearance in risk assessment in WASH-1400, where they were used to generate, define, and classify scenarios specified at the PE level.

WARNING

A warning is an operating procedure, practice, or condition, which may result in injury or damage to equipment if not carefully observed or followed. The intent of a warning is to prevent a potential mishap. A warning is much more critical than a caution. The need for warnings are typically identified through HAs, and the warnings are incorporated into the appropriate manuals and procedures for operation, maintenance, repair, testing, and so on, of a system or product.

See *Caution* for additional related information.

WATT

The watt is a derived unit of power, named after the Scottish engineer James Watt (1736–1819). The symbol for a watt is W. The unit measures the rate of energy conversion, as shown by the following.

- One watt is equal to 1 joule (J) of energy per second.
- In terms of mechanical energy, one watt is the rate at which work is done when an object is moved at a speed of one meter per second against a force of one newton.

By the definitions of the units for measuring electric potential (volt V) and current (ampere A), work is done at a rate of one watt when one ampere flows through a potential difference of one volt, as shown by the following equation:

$$1 \text{ W} = 1 \text{ V} \times 1 \text{ A}.$$

WAVELENGTH

The distance between two points in a periodic wave that have the same phase is termed one wavelength. The velocity of light in centimeters per second divided by frequency (in hertz) equals the wavelength (in centimeter).

WAYPOINT

A waypoint is an intermediate location through which a UMS must pass, within a given tolerance, en route to a given goal location.

WAYPOINT NAVIGATION

Waypoint navigation is the process whereby a UMS makes its way along a route of planned waypoints that it planned itself or that were planned for it.

WEAPON REPLACEABLE ASSEMBLY (WRA)

A WRA indicates the operations or maintenance level at which a system element can be replaced or repaired. A WRA is typically a subsystem that can be replaced on the operational line from a spare that is in inventory. It is usually a functional subsystem that is at a higher level in the system hierarchy, such as an HR radio in an aircraft or a helicopter transmission unit.
   See *System Hierarchy* for additional related information.

WEAPON

A weapon is any instrument or instrumentality used in fighting or hunting.

WEAPON SYSTEM

A weapon system consists of a weapon and those components required for its operation and support. This includes all conventional weapons, ammunition, guns, missiles, rockets, bombs, flares, powered targets, depth charges, mines, torpedoes, and explosive-operated devices. It includes all explosive items, packaging, handling, stowage, test equipment, guidance systems, fire control systems, and launchers and their components. Software and firmware related to monitoring, arming, initiation, or deployment of a weapon is included. This definition also encompasses the manufacturing, processing, packaging, handling, transport, and storage of explosive items and related components ashore.

WEAPONS SYSTEMS EXPLOSIVES SAFETY REVIEW BOARD (WSESRB)

The WSESRB provides independent oversight of all ordnance items, weapon devices, systems, ordnance or weapons programs, or weapons support test systems used, handled, stored, or tested aboard U.S. Naval vessels, and ensures that appropriate explosives safety, system safety, and environmental requirements are adhered to. Every DON weapon system acquisition program and all other services weapons system programs representing a system destined to be deployed aboard a U.S. Naval vessel must be reviewed by the WSESRB.

WEAROUT

Wearout is the process where an item experiences an increasing rate of failure as a result of equipment deterioration due to age or use. For example, mechanical components such as transmission bearings will eventually wear out and fail, regardless of how well they are made. Early failures can be postponed and the useful life of equipment extended by good design and maintenance practices. The only way to prevent failure due to wearout is to replace or repair the deteriorating component before it fails.

WEB-ENABLED SAFETY SYSTEM (WESS)

WESS is a web-based safety mishap data collection and reporting system developed for the Navy and Marine Corps WESS, and the disconnected system WESS-DS in Microsoft Access format, provide a real-time data entry and retrieval system with 20 years of data in a consolidated database.

WHAT-IF ANALYSIS

What-if analysis is a systematic approach to HA that asks "What if this, that or some other event might occur?" in order to identify every conceivable hazard and HCFs. What-if analysis is a structured brainstorming method of determining what things can go wrong and judging the likelihood and consequences of those situations occurring. The answers to these questions form the basis for making judgments regarding the acceptability of those risks and determining a recommended course of action for risks judged to be unacceptable. An experienced review team can effectively and productively discern major issues concerning a system or process. What-if analysis is very similar to HAZOP analysis, except that is a little less formal in methodology. What-if analysis has typically been applied to process plant designs and operations.

The analysis is led by an experienced facilitator; each member of the team participates in assessing what can go wrong based on their past experiences and knowledge of similar situations. Team members usually include operating and maintenance personnel, design and/or operating engineers, specific skills as needed (chemist, structural engineer, radiation expert, etc.), and a safety representative. At each step in the design, procedure, or process, what-if questions are asked and answers are generated. To minimize the chances that potential problems are not overlooked, moving to recommendations is held until all of the potential hazards are identified. The review team then makes judgments regarding the likelihood and severity of the "what-if" answers. If the risk indicated by those judgments is unacceptable, then a recommendation is made by the team for further action. The completed analysis is then summarized and prioritized, and responsibilities are assigned.

The major steps involved in performing an effective analysis include establishing the boundaries of the review, involving the right individuals, and using the right information. The boundaries of the review may be a single piece of equipment, a collection of related equipment, or an entire facility. Assembling an experienced, knowledgeable team is probably the single most important element in conducting a successful analysis. Individuals experienced in the design, operation, and servicing of similar equipment or facilities is essential. Their knowledge of design standards, regulatory codes, past, and potential operational errors as well as maintenance difficulties brings a practical reality to the review. The analysis is based on the information and data of the product, system, or process that is being analyzed. For this reason, it is important that all of the appropriate design and operational information is available and up to date.

A typical what-if analysis worksheet is shown in Figure 2.93.

Description of system or process:				
What If?	Answer	Likelihood	Consequences	Recommendations

Figure 2.93 Example what-if analysis worksheet.

Using the documents available and knowledge of the review team, "what-if" questions can be formulated around human errors, process upsets, and equipment failures. These errors and failures can be considered during normal production operations, during construction, during maintenance activities, as well as during debugging situations. Some example questions include the following:

- Equipment failure occurs
- The procedures are not followed incorrectly
- The procedures are not correct or up to date
- The operator is not properly trained
- Instruments or support equipment is incorrectly calibrated
- Utility failures occur, such as power, steam, and gas
- External influences occur, such as weather, vandalism, and fire

What-if analysis is simple to use and has been effectively applied to a variety of processes. It can be useful with mechanical systems such as production machines, with simple task analysis such as assembly jobs, as well as with reviewing tasks in chemical processing. Individuals with little hazard analysis training can participate in a full and meaningful way. It can be applied at any time of interest, such as during construction, during debugging, during operations, or during maintenance. The results of the analysis are immediately available and usually can be applied quickly.

WHITE BOX

White box refers to having visibility of the internal architecture, structures, features, and implementation as well as the externally visible performance and interfaces of an item, unit, subsystem, and so on.

WHITE BOX TESTING

White box and black box testing are terms used to describe the point of view a test engineer takes when designing test cases; black box testing takes an external view of the test object, while white box testing takes an internal view. Both methods have advantages and disadvantages, but it is only when black box and white box testing methodologies are combined that comprehensive test coverage is achieved.

White box testing uses an internal perspective of the system to design test cases based on internal structure. It requires programming skills to identify all paths through the software. The tester chooses test case inputs to exercise paths through the code and determines the appropriate outputs. In electrical hardware testing, every node in a circuit may be probed and measured, for example, in-circuit testing (ICT). White box testing is also referred to as clear box testing, glass box testing, transparent box testing, or structural testing.

Since white box tests are based on the actual implementation, if the implementation changes, the tests probably will need to change also. This adds financial resistance to the change process, thus buggy products may stay buggy. Although white box testing is applicable at the unit, integration, and system levels of the software testing process, it is typically applied to the unit. While it normally tests paths within a unit, it can also test paths between units during integration, and between subsystems during a system-level test. Though this method of test design can cover an overwhelming number of test cases, it might not detect unimplemented parts of the specification or missing requirements; however, all paths through the test object are executed.

Typical white box test design techniques include:

- Control flow testing
- Data flow testing
- Branch testing
- Path testing

White box testing looks under the covers and into the subsystem of an application, whereas black box testing concerns itself exclusively with the inputs and outputs of an application. White box testing enables the tester to see what is happening inside the application. White box testing provides a degree of sophistication that is not available with black box testing as the tester is able to refer to and interact with the objects that comprise an application rather than only having access to the UI. In situations where it is essential to know that every path has been thoroughly tested, and that every possible internal interaction has been examined, white box testing is the only viable method. As such, white box testing offers testers the ability to be more thorough in terms of how much of an application they can test.

The main difference between black box and white box testing is the areas on which they choose to focus. In simplest terms, black box testing is focused on results. If an action is taken and it produces the desired result, then the process that was actually used to achieve that outcome is irrelevant. White box testing is concerned with the details; it focuses on the internal workings of a system and only when all avenues have been tested and the sum of an application's parts can be shown to be contributing to the whole is testing considered complete.

See *Black Box Testing* for additional related information.

WHY-BECAUSE ANALYSIS (WBA)

WBA is a method developed for the failure analysis of complex, open, heterogeneous systems. The adjective "open" means that the behavior of the system is highly affected by its environment. Aviation operations are significantly affected by the weather through which aircraft fly, for example, and landing risks are significantly affected by obstacles and their clearance on the approach and go-around paths. The adjective "heterogeneous" means that the system has components of different types that are all supposed to work together: digital, physical, human, and procedural components, and combinations of some or all of these. Modern aviation operations have all of these components. Aviation operations thus form a complex, open, heterogeneous system. Using WBA allows causal aspects of the faulty behavior of the complex system to be specified and analyzed in a uniform manner.

Why-because analysis (WBA) is a method for accident/mishap analysis. It is independent of application domain and has been used to analyze, among others, aviation-, railway-, marine-, and computer-related accidents and incidents. It is mainly used as an after the fact (or a posteriori) analysis method. WBA strives to ensure objectivity and reproducibility of results. The result of a WBA is a why-because graph (WBG). The WBG depicts causal relations between factors of an accident. It is a directed acyclic graph where the nodes of the graph are factors. Directed edges denote cause–effect relations between the factors. At each graph node (factor), each contributing cause (related factor) must have been necessary, and the totality of causes must be sufficient: it gives the causes, the whole causes (sufficient), and nothing but the causes (necessary).

WILL

With regard to design requirements, the term "will" indicates futurity and does not infer required action. It typically does not indicate any degree of necessity for the application of a requirement or procedure. Requirements defined using "will" statements express intent, are not binding, and therefore do not necessarily require formal verification.

See *Shall* and *Should* for additional related information.

WIND SHEAR

Wind shear is a sudden and violent change in wind speed and/or direction. Low-altitude wind shear is caused by the strong downdraft of a shower or thunderstorm which, as it hits the ground, spreads out in all directions, producing dangerous air currents. Low-altitude wind shear can have devastating effects on an aircraft during taking or landing because there is usually insufficient altitude for the pilot to prevent hitting the ground. Wind shear is also referred to as a microburst.

WORK BREAKDOWN STRUCTURE (WBS)

The WBS displays and defines the product to be developed or produced by hardware, software, support, and/or service element, and relates the work scope elements to each other and to the end product. The framework of the WBS defines all contractual authorized work; it defines the work scope and assignments to the functional organizations responsible for performing the work. The WBS includes the levels at which required reporting information is summarized for submittal to the customer. An initial WBS is developed during the system acquisition proposal; it is expanded and finalized after contract award.

One aspect of the WBS is a product-oriented family tree composed of hardware, software, services, data, and facilities. The family tree results from systems engineering effort required to develop the system or product. The family tree is very similar to a MEL which is a hierarchical breakdown of the system by subsystems, sub-subsystem, assemblies, and so on.

MIL-HDBK-881, DOD Handbook—Work Breakdown Structure, 1998, defines the WBS process and structure. NASA document "Work Breakdown Structure Reference Guide," 1994, describes how to develop a WBS.

See *System Hierarchy* for additional related information.

WORKMANSHIP TESTS

Workmanship tests are tests performed during the environmental validation program to verify adequate workmanship in the construction of a test item. It is often necessary to impose stresses beyond those predicted for the mission in order to uncover defects. Thus, random vibration tests are conducted specifically to detect bad solder joints, loose, or missing fasteners, improperly mounted parts, and so on. Cycling between temperature extremes during thermal-vacuum testing and the presence of EMI during EMC testing can also reveal the lack of proper construction and adequate workmanship.

WORST-CASE SCENARIO

A worst-case scenario is the most credible and reasonable overall effect of the hazard occurring on the system. A "credible" assessment is based on the engineering judgment of the safety engineer.

WORST-CREDIBLE HAZARD

A worst-credible hazard is a hazard that is credible and where the hazard risk parameters are taken at the maximum, or worst-case values. For example, the hazard probability is taken at worst-case possibility, and the hazard severity is taken at worst-case scenarios.

WRAPPER

A wrapper is a layer of software that encloses application software and isolates it from external software. In safety-critical applications, the wrapper ensures that external software cannot adversely affect the safety-critical functionality of the applications software.

X-TREE ANALYSIS

X-tree analysis is an analysis technique that combines FTA and ETA together to evaluate multiple possible outcomes from a UE. The various outcomes result from the operation or failure of barriers intended to prevent a mishap. The analysis begins with identification of the IE of concern in the center. An FTA is performed to identify the causal factors and probability of this event. Then an ETA is performed on all the barriers associated with the IE, and the possibilities of each barrier function failing. The various different failure combinations provide the various outcomes possible, along with the probability of each outcome. X-tree analysis is also referred to as bow-tie analysis.

See *Bow-Tie Analysis*, *Event Tree Analysis (ETA)*, and *Safety Barrier Diagram* for additional related information.

X-RAYS

X-rays are EMR of short wavelength that can penetrate opaque or solid substances. They are similar to gamma rays; however, the two types of radiation are usually distinguished by their origin: X-rays are emitted by electrons outside the atomic nucleus, while gamma rays are emitted by the atomic nucleus. X-rays are made artificially, and they can be detected photographically or by the ionization they produce in gases. X-rays have a wavelength in the range of 10–0.01 nm, corresponding to frequencies in the range from 30 PHz to 30 EHz (from 3×10^{16} Hz to 3×10^{19} Hz) and energies in the range 120 eV–120 keV.

X-rays are a hazard source; they are a form of ionizing radiation, and exposure to them can be a health hazard.

See *Electromagnetic Radiation (EMR)* for additional related information.

YOKE

A yoke is a connection such as a clamp or vise, between two things so they move together. The control stick in an aircraft is also referred to as a yoke.

ZONAL SAFETY ANALYSIS (ZSA)

ZSA is an HA technique that identifies hazards that are created by failures that cross system zones and violate design safety independence requirements. ZSA is part of a comprehensive CCF analysis. It was developed for analysis of aircraft systems and is described in SAE/ARP-4761, Guidelines and Methods for Conducting the Safety Assessment Process on Civil Airborne Systems and Equipment, 1996.

ZSA is part of a common cause analysis to discover where design independence is required but violated; it is primarily a qualitative analysis. ZSA should be carried out during the development process of a new aircraft or of any major modification to an existing aircraft. The objective of the analysis is to ensure that the equipment installation meets the applicable safety requirements, particularly those relating to CCF. Zonal safety concerns typically arise from design errors, installation errors, or failures that cross physical system zones.

The ZSA approach recognizes the implications of the physical installation of the system hardware which could significantly impair the independence between SR items. ZSA analysis considers installation aspects of individual systems/items and the mutual influence between several systems/items installed in close proximity on the aircraft. A ZSA should be carried out for each zone of the aircraft. The partitioning of an aircraft into zones is a task which is accomplished in order to perform the ZSA, and also to evaluate maintenance operations.

CHAPTER 3

System Safety Specialty Areas

THE BROADNESS OF SAFETY

There are many different aspects to system safety that must be considered when conducting a system safety program (SSP). The primary tasks of a SSP are to identify hazards and mitigate them to an acceptable level of risk. However, there are many secondary SSP tasks that must be performed, such as supporting the special safety requirements in the specialized areas of safety (e.g., laser safety, explosives safety, fire safety, battery safety). The system safety engineer is truly a systems engineer, and a jack-of-all-trades, who must be familiar with all aspects of the system involved, as well as all aspects of safety applicable to the system.

The system safety engineer/analyst must be familiar with many different safety specialty areas, their unique hazards, and their unique safety requirements and standards. Some of these specialized disciplines (or domains) fall under the system safety umbrella, while others are somewhat standalone but interface with system safety. The following list is not necessarily a complete list, but it does cover the major areas and demonstrates the extensive number of safety specialty areas that exist. Examples of established safety specialty areas include:

- Aircraft Safety
- Automobile Safety
- Aviation Safety
- Battery Safety
- Battle Short Safety
- Biomedical Safety
- Chemical/Biological Safety
- Child Safety
- Construction Safety

Concise Encyclopedia of System Safety: Definition of Terms and Concepts, First Edition. Clifton A. Ericson II.
© 2011 by John Wiley & Sons, Inc. Published 2011 by John Wiley & Sons, Inc.

- Driver Safety
- Drug Safety
- Electrical Safety
- Electromagnetic Radiation (EMR) Safety
- Environment Qualification Safety
- Environmental Safety
- Electrostatic Discharge (ESD) Safety
- Explosives Safety
- Facilities Safety
- Fire Safety
- Flight Test Safety
- Food Safety
- Fuel Safety
- Functional Safety
- Fuze Safety
- Hazardous Materials (HazMat) Safety
- Highway Safety
- Home Safety
- Hydraulics Safety
- Insensitive Munitions (IM) Safety
- Industrial Safety
- Intrinsic Safety (IS)
- Job Safety
- Laser Safety
- Lightning Safety
- Manufacturing Safety
- Marine Safety (Shipping)
- Materials Safety
- Medical Equipment Safety
- Mining Safety
- Nuclear Power Safety
- Nuclear Weapon Safety
- Occupational Safety
- Patient Safety
- Personnel Safety
- Packaging, Handling, Storage, and Transportation (PHS&T) Safety
- Pipeline Safety
- Pneumatics Safety
- Process Safety

- Product Safety
- Radiation Safety
- Radiological Safety
- Rail Transportation Safety
- Railroad Safety
- Range Safety
- Road Safety
- Robotic Safety
- Software Safety (SwS)
- Space Safety
- Tire Safety
- Unmanned Systems (UMS) Safety
- Weapons Safety

SPECIALTY AREAS

This section provides safety concerns and safety standards that are applicable to some key safety specialty areas.

System Safety (SyS)

System safety covers all safety aspects of a system or product. It applies a systems engineering viewpoint to identify hazards, assess hazard risk, and mitigate hazard risk. It also ensures the applicable safety laws, regulations, and standards are met. The objective of system safety is to intentionally design safety into the system or product from the very beginning of development. Safety coverage includes design, manufacture, test, support equipment, tools, materials, and personnel.

System safety concerns include:

- Hazards that can lead to mishaps resulting in
  - Death/injury of system personnel (or operators)
  - System loss of significant damage
  - Collateral damage (e.g., bystanders)
  - Environmental damage
- Hazard–Mishap risk
- Hazard causal factors (e.g., failures, human error, software errors, design flaws, environment)
- Correct operation of safety functions and safety-critical functions (SCFs)
- Safe processes, procedures, training, and certification
- Fault tolerance, fault recovery, and fail-safe design

- Single-point failures impacting SCFs
- Compliance with safety guidelines, standards, regulations, and laws

Although there are many tasks in an SSP, the following tasks are the most basic and essential:

- Plan the SSP tasks and schedule
- Identify hazards
- Assess the risk presented by the identified hazards
- Mitigate the risk as necessary and verify implementation
- Track hazards and risk
- Accept the risk presented by the identified hazards

Key system safety reference standards include:

1. Military Standard (MIL-STD)-882D, Standard Practice for System Safety, February 10, 2000.
2. American National Standards Institute (ANSI)/Government Electronic and Information Technology Association (GEIA)-STD-0010-2009, Standard Best Practices for System Safety Program Development and Execution, February 12, 2009.
3. Naval Sea Systems Command (NAVSEA) SWO20-AH-SAF-010, Weapon System Safety Guidelines Handbook, February 1, 2006.
4. Air Force System Safety Handbook, July 2000.
5. Military Handbook (MIL-HDBK)-764, System Safety Engineering Design Guide for Army Materiel, January 12, 1990.
6. National Aeronautics and Space Administration (NASA) Procedural Requirement (NPR) 8715.3, NASA General Safety Program Requirements, March 12, 2008.

Software Safety (SwS)

SwS covers the safety of software (SW) in a system or product. SwS has unique characteristics that require that different methods be applied over and above those methods used for typical hardware (HW) safety.

Some of the safety considerations for SwS include:

- SW fails to perform SCFs
- SW performs SCFs erroneously
- SW performs SCFs inadvertently
- SW performs (unexpected) unintended function
- HW failure changes a bit, which in turn changes SW instruction

- Sneak path in SW
- SW corrupts safety-related data

The following are the basic tasks in an SwS program:

- Plan the SwS program tasks and schedule
- Ensure that SW-related hazards are identified and mitigated
- Establish safety integrity levels (SILs) for all SW modules
- Apply more stringent development rigor to SW with more critical SILs
- Ensure all safety design requirements are verified through testing
- Ensure all safety-critical (SC) requirements are verified through testing
- Tag SC requirements in specifications and code
- Apply generic safety guidelines and coding standards to SW design

Key SwS reference standards include:

1. Department of Defense (DoD) Joint Software Systems Safety Handbook, December 1999.
2. DoD Joint Software Systems Safety Engineering Handbook, September 30, 2009 (draft).
3. Radio Technical Commission for Aeronautics (RTCA)/DO-178B, Software Considerations in Airborne Systems and Equipment Certification, 1992.
4. U.S. Army Aviation and Missile Command (AMCOM) Regulation 385-17, AMCOM Software System Safety Policy, March 15, 2008.
5. MIL-STD-882C, System Safety Program Requirements, January 1993 (includes SwS section).
6. ANSI/GEIA-STD-0010-2009, Standard Best Practices for System Safety Program Development and Execution, February 12, 2009 (includes SwS section).
7. Standardization Agreement (North Atlantic Treaty Organization [NATO]) (STANAG) 4404, Safety Design Requirements and Guidelines for Munition Related Safety Critical Computing Systems, 1997.
8. Institute of Electrical and Electronic Engineers (IEEE) Std-1228, IEEE Standard for Software Safety Plans, March 17, 1994.
9. NASA-Guidebook (GB)-1740.13-96, NASA Guidebook for Safety Critical Software—Analysis and Development, 1996.
10. NASA-STD-8719.13A, Software Safety, September 15, 1997.
11. Underwriters Laboratory (UL) 1998, Software in Programmable Components, May 29, 1998.
12. Electronic Industries Association (EIA) SEB-6A, System Safety Engineering in Software Development, April 1990.

Aircraft Safety

Aircraft safety covers all aspects of aircraft design and operation. It includes consideration of SW, human interface design, human error, latency, and redundancy.

Some of the safety considerations for aircraft safety include:

- Loss of propulsion
- Loss of flight controls
- Aircraft fire
- Aircraft collision
- Aircraft controlled flight into terrain (CFIT)
- Probability of loss of aircraft
- Design to prevent or counter human operator error
- Verifying (with evidence) compliance with applicable aircraft safety standards

Key aircraft safety reference standards include:

1. Title 14 Code of Federal Regulations (14 CFR) Part 25, Advisory Circulars (AC), AC 25.1309-1A (Joint Aviation Requirements [JAR] 25.1309), System Design and Analysis, June 1, 1988.
2. Title 14 Code of Federal Regulations (14 CFR) Part 23 Airplanes, Advisory Circulars (AC), AC 23.1309-1C, Equipment, Systems, and Installations, March 12, 1999.
3. Society of Automotive Engineers (SAE)/Aerospace Recommended Practice (ARP)-4754, Certification Considerations for Highly-Integrated or Complex Aircraft Systems, Aerospace Recommended Practice, 1996.
4. SAE ARP-4761, Guidelines and Methods for Conducting the Safety Assessment Process on Civil Airborne Systems and Equipment, 1996.
5. RTCA/DO-178B, Software Considerations in Airborne Systems and Equipment Certification, 1992.
6. RTCA/DO-254, Design Assurance Guidance for Airborne Electronic Hardware, 2000.
7. Federal Aviation Administration (FAA) AC 25-22, Certification of Transport Airplane Mechanical Systems, March 14, 2000.

Laser Safety

Laser safety covers all aspects of laser design and operation. Each system that uses a laser is unique and must be assessed for hazards unique to that system. In the United States, laser safety is a matter of laws and regulations.

Some of the safety considerations for laser safety include:

- Inadvertent laser operation
- Unknown laser operation
- Faults causing laser operation outside of safe or intended parameters
- Eye damage from exposure to laser beam
- Ignition source from high-energy laser beam
- Cutting source from high-energy laser beam
- Fire/explosion from laser system components
- Toxic chemicals used in laser systems
- Electrocution from high-voltage electronics in laser systems
- Skin burns from high-energy laser beam
- Hearing damage from high noise levels in some laser systems
- Injury from cryogenics used in some laser systems
- Verifying compliance with applicable laser safety standards

Key laser safety reference standards include:

1. 21 CFR 1040, Part 1040—Performance Standards For Light-Emitting Products, Sec. 1040.10 Laser products, April 1, 2009.
2. ANSI Z136.1, American National Standards Institute, American National Standard for Safe Use of Lasers, 2000.
    - ANSI Z136.1—Safe Use of Lasers
    - ANSI Z136.2—Optical Fiber Systems Utilizing Laser Diode and Light-Emitting Diode Sources
    - ANSI Z136.3—Safe Use of Lasers in Health Care Facilities
    - ANSI Z136.4—Recommended Practice for Laser Safety Measurements for Hazard Evaluation
    - ANSI Z136.5—Safe Use of Lasers in Educational Institutions
    - ANSI Z136.6—Safe Use of Lasers Outdoors
    - ANSI Z136.7—Testing and Labeling of Laser Protective Equipment
3. MIL-STD-1425 (Series), Safety Design Requirements for Military Lasers and Associated Support Equipment, August 30, 1991.
4. Office of the Chief of Naval Operations Instruction (OPNAVINST) 5100.27A/Marine Corps Order (MCO) 5104.1B, Navy Laser Hazards Control Program, September 24, 2002.
5. NAVSEA Technical Manual EO-410-BA-GYD-010, Laser Safety.
6. U.S. Army Technical Bulletin (TB) Med 524, Control of Hazards to Health from Laser Radiation.

7. U.S. Navy Space and Naval Warfare Systems Command (SPAWAR) Instruction 5100.128, Navy Laser Hazards Control Program.
8. U.S. Air Force Occupational Safety and Health (AFOSH) 48-139, Laser Radiation Protection Program.
9. European Standard, European Norm EN 60825-1, Safety of Laser Product—Part 1: Equipment Classification, Requirements, and User's Guide, October 1996. (International Electrotechnical Commission [IEC] 60825-1 is the comparable international standard).
10. Document 316-91, Laser Range Safety, October 1991, Range Commanders Council, U.S. Army White Sands Missile Range.
11. DOD Instruction 6055.11, "Protection of DoD Personnel from Exposure to Radio Frequency Radiation and Military Exempt Lasers," February 21, 1995.
12. MIL-HDBK-828 (Series), Laser Range Safety.
13. Secretary of the Navy Instruction (SECNAVINST) 5100.14C, Military Exempt Lasers.
14. Bureau of Medicine and Surgery Instruction (BUMEDINST) 6470.19, Laser Safety for Medical Facilities.
15. OPNAVINST 5100.23D, October 11, 1994, Chapter 22, Non-Ionizing Radiation.
16. IEC 60825-1, Safety of Laser Products—Part 1: Equipment classification and requirements, 2007.

Functional Safety

Functional safety is essentially the same as system safety and utilizes most of the same tools and techniques. Functional safety applies the SIL methodology to system design and development. Functional safety provides a major focus on safety instrumented systems (SISs).

Key functional safety reference standards include:

1. IEC 61508, Functional Safety of Electrical/Electronic/Programmable Electronic Safety-Related Systems, Parts 1–7, September 2005.
2. IEC 61511, Functional Safety of Safety Instrumented Systems for the Process Industry Sector, 2003.
3. Independent Safety Auditor (ISA) 84.01-2003, Functional Safety: Safety Instrumented Systems for the Process Industry Sector, 2003.

Human Engineering Safety

Human engineering safety deals with providing a safe system–operator interface. This includes protecting the operator from the system and protecting the system from the operator.

Typical hazards and safety concerns include, but are not limited to:

- Electrocution
- Environmental temperature
- Thermal contact temperature
- Lifting injuries
- Vibration injuries
- Noise damage to hearing
- Lighting
- Emergency egress (doors, exits, etc.)
- Operator stress, fatigue, skills, and cognition
- Operator experience and training
- Design complexity (allows or forces an operator to make errors)
- Compliance with safety guidelines, standards, regulations, and laws

Key human engineering safety reference standards include:

1. MIL-STD-1472E, Human Engineering, March 31, 1998.
2. MIL-HDBK-454, General Guidelines for Electronic Equipment, Guideline 1, Safety Design Criteria- Personnel Hazards, April 28, 1995.
3. MIL-HDBK-46855A, Human Engineering Program Process and Procedures, May 17, 1999.
4. DOD-HDBK -763, Human Engineering Procedures Guide, February 27, 1987.
5. MIL-HDBK-1908A, Definitions of Human Factors Terms, 1996.
6. MIL-HDBK-759C, Human Engineering Design Guidelines, July 31, 1995.

Robotic Safety

Robotic safety deals with all aspects of robotic systems, where robotic systems primarily involve manufacturing and industrial applications. Traditional industrial robots are programmed autonomous or semiautonomous systems, in that they perform a programmed task and an operator interfaces with them in some manner. The operator starts/stops the robotic process and oversees operation; in some applications the operator actually guides some of the operations. These types of robots are also stationary, or fixed to a permanent work site, and do not physically traverse. The service robot performs simple work tasks, such as floor sweeping, floor washing, and polishing. This is typically a mobile robot that is on a fixed path or an autonomous path that changes course when it bumps into something. Robotic systems also include search and rescue, fire fighting, and explosives handling and explosive ordnance disposal (EOD) applications. These systems perform the dull, dangerous, and tedious tasks; they tend to be stationary or don't move significant distances. Unmanned vehicles and aircraft tend to fall into the UMS category rather than the robotic category.

Typical hazards and safety concerns include, but are not limited to:

- SwS
- Humans safely entering the robots' work area (interlocks fail, power shutoff fails, human error)
- Injury to operators or other personnel within the robot work space
- Loss of robot control causing injury/damage within the robots' fixed radius work area (unanticipated movements, incorrect movements, inadvertent operation, etc.)
- Loss of robot control causing injury/damage outside robots' fixed radius work area
- Missile-like parts released from a disintegrating robot
- Human operator errors (controls and displays confuse or mislead operator, procedures, warnings, emergency safety controls, training, etc.)
- HW energy sources (hydraulics, pneumatics, electricity/voltage, fuel, engines, moving arms, etc.)

Key robotic safety reference standards include:

1. ANSI/Robotic Industries Association (RIA)/International Standards Organization (ISO) 10218-1:2007, Robots for Industrial Environment—Safety Requirements—Part 1—Robot
2. ANSI/RIA R15.06:1999, American National Standard for Industrial Robots and Robot Systems—Safety Requirements, 1999.
3. ANSI/RIA R15.06:2011, Robot Safety Standard (planned update in 2011).

Safety of UMSs

UMS safety deals with all aspects of autonomous and semiautonomous systems. These types of systems tend to be military in nature and have the capability to move great distances. A UMS is an electromechanical system that is able to exert its power to perform designed missions and includes the following: (1) there is no human operator aboard, (2) manned systems that can be fully or partially operated in an autonomous mode, and (3) the system is designed to return or be recoverable. The system may be mobile or stationary, and includes the vehicle/device and the control station. Missiles, rockets and their submunitions, and artillery are not considered UMSs. UMSs include, but are not limited to: unmanned ground vehicles (UGVs), unmanned aerial/aircraft systems, unmanned underwater vehicles (UUVs), unmanned surface vessels, unattended munitions, and unattended ground sensors. UMSs include unmanned air vehicles (UAVs), remotely piloted vehicles (RPVs), UUVs, and UGVs.

Typical hazards and safety concerns include, but are not limited to:

- Loss of control and/or communication with UMS
- Missile-like parts released from a disintegrating component
- False triggering of safety devices
- Inadvertent launch/release of weapons
- Inadvertent release of fuel
- SwS considerations
- Loss of control over the UMS
- Loss of communications with the UMS
- Loss of UMS ownership (lost out of range or to the enemy)
- Loss of UMS weapons
- Unsafe UMS returns to base (failures, biological exposure, etc.)
- UMS in an indeterminate or erroneous state
- Knowing when a UMS potentially is in an unsafe state
- Unexpected human interaction with the UMS
- Inadvertent/erroneous firing of UMS weapons
- Erroneous target discrimination
- UMS or equipment injures operators, own troops, and so on
- Enemy jamming or taking control of UMS
- Loss of, or inadequate, situational awareness
- Provision for emergency operator stop
- Battle damage to UMS
- UMS exposure to radiation, biological contamination, and so on

Key UMS safety reference standards include:

1. Unmanned Systems Safety Guide for DoD Acquisition, June 27, 2007.
2. FAA AFS-400 Unmanned Aircraft Systems (UAS) Policy 05-01, Unmanned Aircraft Systems Operations in the U.S. National Airspace System—Interim Operational Approval Guidance, September 16, 2005.
3. NASA/Technical Memorandum (TM)-2007-214539, Preliminary Considerations for Classifying Hazards of Unmanned Aircraft Systems, February 2007.
4. American Institute of Aeronautics and Astronautics, Inc. (AIAA) R-103-2004, Terminology for Unmanned Aerial Vehicles and Remotely Operated Aircraft.

5. American Society for Testing and Materials (ASTM) F2395, Standard Terminology for Unmanned Air Vehicle Systems.
6. ASTM F2411, Standard Specification for Design and Performance Requirements for an Unmanned Air Vehicle Sense-and-Avoid System.

PHS&T Safety

PHS&T safety covers those safety aspects involved with PHS&T issues for the system, including support equipment and personnel.

Some of the safety considerations for packaging include:

- Protective systems for cargo
- Protection and warnings for HazMat
- Safe procedures
- Proper handling of equipment
- Correct use of equipment (training, qualification, etc.)
- Storage facility electrical, lighting, ventilation, alarms, and fire protection
- Emergency preparedness (procedures, contacts, cargo info, etc.)
- Spills/leaks
- HazMat
- Protection from earthquakes, floods, tornadoes, terrorism, and so on
- Cargo needs, such as electrical, lighting, ventilation, alarms, fire protection, heating, cooling, and so on
- Effect on cargo if an accident occurs

Facilities Safety

Facilities safety covers those safety aspects involved with the design, construction, and operation of a facility, including support equipment and personnel. Facilities often house expensive systems and equipment, and systems of high safety concern, such as explosives.

Some of the considerations for facilities safety include:

- Electrical, lighting, ventilation, alarms, fire detection, and fire suppression
- Emergency preparedness (procedures, contacts, checklists, etc.)
- Emergency access/egress, communication systems
- Emergency shutdown
- Facility training and procedures
- Spills, leaks
- HazMat
- Structural safety

- Federal, state, and local safety codes and regulations
- Protection from earthquakes, floods, tornadoes, terrorism, and so on
- Protection for hazardous components such as fuel, chemicals, explosives, and so on
- Special facilities (vacuum chamber, clean rooms, laser rooms, computer rooms, etc.)
- Occupational Safety and Health Administration (OSHA) safety compliance
- Explosives safety quantity distance (ESQD) requirements
- Facility explosives and personnel operating limits

Construction/Disposal Safety

Construction safety covers those safety aspects involved with the construction of a site and/or facility for a system, including construction support equipment, tools, materials, and personnel. Since many of the tasks and safety concerns for deconstruction or disposal/demilitarization of a site or facility are the same as construction, it is included here.

Construction safety concerns and hazards include:

- Personnel safety
- Safe processes, procedures, training, and certification
- Safe tools and equipment
- Emergency preparedness (procedures, contacts, etc.)
- Emergency medical equipment
- Protective devices (barriers, lockouts, helmets, fall ropes, etc.)
- OSHA safety compliance
- ESQD requirements
- Facility explosives and personnel operating limits
- Environmental safety
- Pollution prevention
- OSHA safety compliance
- Protection systems and equipment
- HazMat safety

Manufacturing Safety

Manufacturing safety covers those safety aspects involved with the manufacture and production of a system, including manufacture support equipment, tools, materials, and personnel.

Manufacturing safety concerns and hazards include:

- Safe procedures, training, and certification
- Safe tools and equipment
- Emergency preparedness (procedures, contacts, etc.)
- Emergency medical equipment
- System damage that may affect operational system (e.g., dent in missile body)
- OSHA safety compliance
- ESQD requirements
- Facility explosives and personnel operating limits

Process Safety

Plant processing is similar to manufacturing in nature, but has some unique differences. A plant process primarily deals with a continuous flow of production of a product, as opposed to the manufacture of a discrete item. Process safety covers those safety aspects involved with the processing or production of chemicals, explosives, and biological agents. This includes production plants and processes for products such as gasoline, natural gas, and fertilizer.

Process safety-related hazards and safety concerns include:

- Personnel safety
- Personnel qualification and certification
- Safe processes and procedures
- Safe equipment
- Emergency preparedness (procedures, contacts, etc.)
- Emergency egress and escape
- Protection systems and equipment
- Readiness of SISs

Key process safety reference standards include:

1. IEC 61511, Functional Safety of Safety Instrumented Systems for the Process Industry Sector, 2003.
2. ISA 84.01-2003, Functional Safety: Safety Instrumented Systems for the Process Industry Sector, 2003.
3. IEC 61508, Functional Safety of Electrical/Electronic/Programmable Electronic Safety-Related Systems, Parts 1–7, September 2005.

Electrical Safety

Electrical safety involves analysis of systems for electrical hazards, and compliance with the National Electrical Safety Code (NESC). Electrical safety includes all electrical aspects of a system or a facility, which may include electrical power generation, distribution, and transmission.

Some examples of electrical-related safety concerns include the following:

- Personnel exposure to hazardous voltages and currents
- Electrical arc/sparks as a fire ignition source
- Design compliance with the National Electrical Code (NEC)
- Safe electrical design (circuit breakers, fuses, physical separation of wires, etc.)
- Equipment damage from overvoltages and power surges
- Undesirable EMR from electrical power sources
- Safe power transmission
- Safety of operational equipment
- Backup safety systems
- Safety procedures
- Emergency preparedness (procedures, contacts, etc.)
- The inadvertent application of electrical power in SCFs
- The failure of power in SCFs

Key electrical safety reference standards include:

1. National Fire Protection Association (NFPA) 70E-1995, Standard for Electrical Safety Requirements for Employee Workplaces.
2. OSHA 29 CFR 1910, Subpart S—Electrical, Electrical Industry Safe Occupational Working Standards.
3. NESC.
4. NEC.

Electromagnetic Radiation (EMR) Safety

EMR is the energy radiated from an electromagnetic field, consisting of alternating electric and magnetic fields that travel through space at the velocity of light. Radiation includes gamma radiation, X-rays, ultraviolet light, visible light, infrared radiation, and radar and radio waves. An EMR source can be natural, such as from sunlight or lightning. Most EMR sources of concern are man made and propagate from electronic equipment, such as radar, microwaves, and computers.

Some examples of EMR-related safety concerns include the following:

- Eye damage
- Skin burn
- Organ damage
- Fuel ignition source
- Material ignition source
- Explosives ignition source
- Electronics disruption
- SW disruption
- Electromagnetic interference (EMI)
- Electromagnetic compatibility (EMC)
- Hazards of electromagnetic radiation to ordnance (HERO)
- Electromagnetic pulse (EMP)
- Hazards of electromagnetic radiation to personnel (HERP)

Key EMR safety reference standards include:

1. United States Air Force (USAF) Technical Order 31Z-10-4, Electromagnetic Radiation Hazards, McClellan Air Force Base (AFB), CA, October 15, 1981.
2. MIL-STD-461(Series), Requirements for the Control of Electromagnetic Interference Characteristics of Subsystems and Equipment.
3. MIL-STD-464 (Series), Electromagnetic Environmental Effects, Requirements for Systems.

Lightning Safety

Lightning safety covers those safety aspects involved with the effects of lightning on personnel, systems, and equipment.

Some typical hazards and safety concerns related to lightning include:

- Injury to exposed personnel
- Proper bonding and grounding
- Fire ignition source
- Ignition source for explosives
- Equipment damage preventing proper operation of critical electronics

Fuel Safety

Fuel safety covers all aspects of fuel systems that can result in a mishap. Fire and explosion hazards and compliance with the appropriate regulatory safety criteria are major concerns; however, there are typically many other fuel-related hazards in a system that must be identified and evaluated.

Some examples of fuel-related safety concerns include the following:

- Fire
- Toxicity
- Explosive atmosphere (and ignition)
- The use of explosion-proof equipment
- Fuel and oxidizer interactions
- Fuel tank grounding
- Fuel transfer safety procedures (including proper grounding)
- Fuel transportation and storage
- Identifying fuel safety zones in the system and providing the appropriate safety measures in the applicable zones
- Unintentional fuel transfer between tanks
- Fuel leakage
- Compliance with regulations

Key fuel safety reference standards include:

1. FAA AC 25.981-1C, Fuel Tank Ignition Source Prevention Guidelines, 1 May 2002 (draft).
2. FAA AC 25-16 Electrical Fault and Fire Prevention and Protection.
3. SAE Aerospace Information Report (AIR) 1662, Minimization of Electrostatic Hazards in Aircraft Fuel Systems.
4. FAA AC 25-22, Certification of Transport Airplane Mechanical Systems, Section 25.863 Flammable Fluid Fire Protection, March 14, 2000.

Chemical/Biological Safety

Chemical/biological safety covers those safety aspects involved with the use of chemicals and biological agents in a system. Many chemicals and biological agents used in systems will fall into the HazMat category. One other aspect of Chemical/biological safety is that of protecting assets against the threat of or exposure to these potentially hazardous agents.

Chemical/biological safety-related hazards and safety concerns include:

- Chemical/biological exposure to individuals, workers, work areas, or facility equipment
- Moving chemical/biological materials within a facility
- Controlling areas contaminated with chemical/biological materials
- Fire protection
- Emergency procedures

Radiological Safety

Radiological safety covers those safety aspects involved with small amounts of radiological material in the system. Safety coverage includes the safe use, handling, storage, and disposal of radioactive material.

Radiological safety-related hazards and safety concerns include:

- An internal or external radiation dose to individuals
- Moving radioactive materials within a facility
- Controlling contaminated areas
- Radioactive materials management, including procedures and record system
- Dose monitoring
- Generation of radioactive waste and waste disposal
- Fire protection
- Emergency procedures
- Contamination of workers, work areas, or facility equipment
- Compliance with state and federal laws relating to radiation safety

Key radiological safety reference standards include:

1. U.S. Nuclear Regulatory Commission, CFR, Section 10.
2. U.S. Department of Transportation, CFR, Section 49.

Hydraulics Safety

Hydraulics safety covers those safety aspects involved with systems using hydraulics and involves hazard analysis for hydraulics-related hazards, and compliance with the appropriate hydraulics safety criteria.

Hydraulics safety hazards and concerns include the following:

- High pressure
- Fire
- Toxicity
- Failure to provide hydraulic pressure to critical subsystems
- Hydraulics pressure generator, valves, tubing, and so on
- Chafing, bending, and separation of hydraulic tubing
- Leakage and spills
- Storage

Pneumatics Safety

Pneumatics safety covers those safety aspects involved with systems using pneumatics and involves analysis for pneumatics-related hazards, and compliance with the appropriate pneumatics safety criteria.

Pneumatics safety hazards and concerns include the following:

- High pressure
- Fire
- Failure to provide pneumatic pressure to critical subsystems
- Chafing, bending, and separation of pneumatic tubing
- Pneumatics pressure generator, valves, tubing, and so on

Battery Safety

Battery safety involves ensuring that the system and system operators are safe when batteries are used in the system. Batteries have the capacity to explode and to overheat and generate a fire. Battery types fall into two categories: primary batteries and secondary batteries. A primary battery is one that cannot be recharged and is typically thrown away after the battery is completely discharged. Examples of these types of batteries include alkaline batteries and lithium metal batteries. A secondary battery is one that can be recharged and used again and again. They do eventually "die," but most can be charged and discharged many times. Examples of these battery types include nickel metal hydride, nickel cadmium (NiCAD), and lithium ion.

Battery safety hazards and concerns include the following:

- Explosion
- Fire from overheating
- High pressure of lithium battery
- Violent venting of lithium battery
- Lithium battery explosion with lethal projectiles (design must provide protection)
- If the lithium battery is not isolated from external power source, it may be inadvertently recharged
- Sulfuric acid in lead acid batteries
- Lead acid batteries under charge gives off the flammable gas hydrogen; therefore, keep away from naked flame or source of ignition
- Ensure there is adequate ventilation when charging lead acid to dissipate hazardous gasses that vent from the battery when charging

- If the following conditions are violated, the result may be a battery of lower capacity going into voltage reversal when discharged, causing the battery to vent:
  - Never mix primary (nonrechargeable) and secondary (rechargeable) batteries in equipment at the same time
  - Never mix different types of rechargeable batteries (i.e., lithium ion and NiCAD) batteries in equipment at the same time
  - Never mix new and used lithium ion batteries in the same equipment

Key battery safety reference standards include:

1. Duracell, Alkaline Manganese Dioxide Technical Bulletin, April 2, 2002.
2. Duracell, Lithium Manganese Dioxide Technical Bulletin, April 2, 2002.
3. Naukam, A.J., Wright, R.C., and Matthews, M.E., Safety Characterization of Li/MnO2 Cells, April 2, 2002.
4. U.S. Army Communications-Electronics Command, Technical Bulletin Number 7, Revision A, Battery Compartment Design Guidelines for Equipment Using Lithium-Sulfur Dioxide Batteries, October 1997.
5. ANSI C18.2M, Safety Requirements for Portable Rechargeable Cells and Batteries.
6. UL 2054, Safety Requirements for Household and Commercial Batteries.

Electrostatic Discharge (ESD) Safety

ESD safety covers those safety aspects involved with the hazardous effects of ESD. ESD is a charge of static electricity inadvertently discharged to the ground—after it has built up on a system, component, or human being. Personnel generate PESD; helicopters generate HESD. ESD damage is similar to damage associated with high-frequency radar pulses.

ESD-related hazards and safety concerns include:

- Damage to voltage sensitive electronics such as integrated circuits (ICs) and central processing units (CPUs).
- Ignition of electro-explosive devices (EEDs) and electronic safe and arm devices (ESADs)
- Refueling fires from static electricity
- Solder stations properly connected to ground
- Personnel using ground straps
- Cell phones as a source of static electricity

Key ESD safety reference standards include:

1. MIL-STD-331C, Test Method Standard, Fuze and Fuze Components, Environmental and Performance Tests For, January 5, 2005.

2. Technical Report (TR)-RD-TE-97-01, Electromagnetic Environmental Effects Criteria and Guidelines for Electromagnetic Radiation Hazard (EMRH), Electromagnetic Radiation Operational (EMRO), Lightning Effects, ESD, EMP, and EMI Testing of US Army Missile Systems.
3. DOD-HDBK-263, Electrostatic Discharge Control Handbook for Protection of Electrical and Electronic Parts, Assemblies, and Equipment (Excluding Electrically Initiated Explosive Devices).

Materials Safety

Materials safety covers those safety aspects involved with the safe use of materials in a system. A hazardous material is any substance or compound that has the capability of producing adverse effects on the health of humans. Many materials are inherently hazardous, while other classes of materials are contributing factors in hazards. For example, the fabric used in a system operator's chair may emit toxic fumes during a fire. The development of procedures for the control and use of HazMat, including explosives, is a responsibility of system safety. When appropriate, local hazardous material spill-response teams should be notified of activities, which may be potential spill or release situations requiring containment.

HazMat are solids, liquids, or gases that can harm people, other living organisms, property, or the environment. HazMat should be tracked within a HazMat database and handled within the framework of the safety program. The HazMat database should list the material, quantities, location, any required special procedures (e.g., firefighting), exposure levels, health hazard data, exposure times, disposal methods, and so on. The appropriate Material Safety Data Sheet (MSDS) must accompany each HazMat used in system manufacture, operation, and maintenance activities. Materials that are toxic (either in their normal state or when subject to burning, smoldering, etc.), carcinogenic, or which present other inherent category I or II hazards should not be used. Laws and regulations on the use and handling of HazMat may differ depending on the activity and status of the material. For example, one set of requirements may apply to their use in the workplace while a different set of requirements may apply to spill response, sale for consumer use, or transportation. Most countries regulate some aspect of HazMat, especially the transportation of HazMat.

Hazards and safety concerns associated with materials include the following:

- Flammability
- Toxicity
- Explosives
- HazMat
- Compatibility

- Environmental damage
- Pollution
- Personnel safety
- Disposal safety

Fire Safety

Fire safety covers those safety aspects involved with fire, fire prevention, fire protection, and fire suppression. This safety aspect crosses many different boundaries and interfaces of a system. For example, a particular system may have common fire safety concerns during facility storage, transportation, handling, operation, maintenance, and disposal. Fire is a potential common-cause failure mode that can cause the simultaneous failure of redundant equipment.

Fire safety hazards and concerns include the following:

- Ignition sources
- Personnel safety
- Toxicity
- Environmental damage
- Fire detection and suppression
- Pollution
- System loss
- Egress and escape
- Emergency preparedness procedures

Key fire safety reference standards include:

1. NASA-STD-8719.11, w/Change 1, Safety Standard for Fire Protection, August 2000.

Nuclear Weapon Safety

Nuclear weapon safety covers those safety aspects involved with a nuclear weapon system. A nuclear weapon system is a high-consequence system requiring an extensive SSP and Nuclear Safety Program. Nuclear safety must be planned in the conceptual phase, designed into components in the development phase, and continually examined throughout the test and operational phases of each device. The DoD has established four safety standards that are the basis for nuclear weapon system design and the safety rules governing nuclear weapon system operation. These standards require that, as a minimum, the system design shall incorporate the following positive safety measures:

1. There shall be positive measures to prevent nuclear weapons involved in accidents or incidents, or jettisoned weapons, from producing a nuclear yield.
2. There shall be positive measures to prevent deliberate prearming, arming, launching, firing, or releasing of nuclear weapons, except upon execution of emergency war orders or when directed by competent authority.
3. There shall be positive measures to prevent inadvertent prearming, arming, launching, firing, or releasing of nuclear weapons in all normal and credible abnormal environments.
4. There shall be positive measures to ensure adequate security of nuclear weapons.

Standard hazard analysis apply to nuclear weapon systems; however, because of the political and military consequences of an unauthorized or accidental nuclear or high explosive detonation, additional analyses are necessary to demonstrate positive control of nuclear weapons in all probable environments. The following analyses, in whole or in part, are performed:

1. A quantitative analysis to assure that the probability of inadvertent nuclear detonation, inadvertent programmed launch, accidental motor ignition, inadvertent enabling, or inadvertent prearming meets the numerical requirements specified in applicable nuclear safety criteria documents.
2. An unauthorized launch analysis to define the time, tools, and equipment required to accomplish certain actions leading to unauthorized launch. The results of this analysis are used by the nuclear safety evaluation agency in determining which components require additional protection, either by design or by procedural means.
3. A Nuclear Safety Cross-check Analysis of SW and certain firmware, which directly or indirectly controls or could be modified to control critical weapon functions. This analysis, by an independent contracting agency, must determine that the final version of SW or firmware is free from programming, which could contribute to unauthorized, accidental, or inadvertent activation of critical system function.
4. A safety engineering analysis of all tasks in modification or test programs at operational sites. This analysis is specifically oriented toward identifying hazards to personnel and equipment in the work area and is in addition to the analysis of the safety impact of the change to the weapon system.

Key nuclear weapon safety reference standards include:

1. Air Force Instruction AFI 91-101, Air Force Nuclear Weapons Surety Program, 2000

2. Air Force (AF) Manual 91-119, Safety Design and Evaluation Criteria for Nuclear Weapon Systems Software, 1999.
3. Department of Energy (DOE)-STD-3009-94, Preparation Guide For U.S Department Of Energy Nonreactor Nuclear Facility Documented Safety Analyses, July 1994.

Nuclear Power Safety

Nuclear power safety covers those safety aspects involved with a nuclear power plant. A nuclear power system is a high-consequence system requiring an extensive SSP.

Nuclear power safety-related hazards and safety concerns include:

- Personnel safety
- Backup safety systems
- Failure of emergency core cooling system
- Safety procedures
- Emergency preparedness (procedures, contacts, checklists, etc.)
- Human error
- Emergency cooling system
- Radiation containment
- Personnel qualifications and training
- SwS
- Safety culture

Key nuclear power safety reference standards include:

1. IEC 880, Software for Computers in the Safety Systems of Nuclear Power Stations, 1986.

Rail Transportation Safety

Rail transportation safety covers those safety aspects involved with rail-type transportation systems, such as railroads and airport people moving systems.

Rail safety-related hazards and safety concerns include:

- Personnel safety
- Safe headway (spacing) between vehicles
- Emergency preparedness (procedures, contacts, etc.)
- Emergency access/egress
- Emergency communication systems
- Safe vehicle/track switching systems
- Safe vehicle warning systems

- Properly maintained systems
- Elevated systems
- SW control of system operation

Environmental Qualification Safety

Environmental qualification safety involves ensuring that the correct environmental qualification tests related to safety are performed and successfully passed for system HW items. For example, electrical equipment that must operate in an explosive fuel environment must undergo explosive atmosphere testing to ensure it cannot cause accidental ignition of a fuel-rich compartment. System safety is responsible for ensuring the appropriate environmental tests are properly performed, passed, and documented for safety coverage.

Environmental qualification testing safety concerns and hazards include:

- The appropriate qualification tests are not performed
- Qualification tests are not performed correctly
- Qualification tests for commercial off-the-shelf items are unknown or not performed
- Suitable test evidence is available

Key environmental qualification reference standards include:

1. MIL-STD-810F, Test Method Standard, Environmental Engineering Considerations and Laboratory Tests, October 31, 2008.
2. MIL-HDBK-310, Global Climatic Data for Developing Military Products, June 23, 1997.
3. MIL-STD-2105C, Hazard Assessment Tests for Non-Nuclear Munitions, July 23, 2003 (includes environmental safety tests for weapons systems).

Flight Test Safety

Flight test safety covers those safety aspects involved with flight-testing of an aircraft. It includes the aircraft, special subsystems, test equipment, procedures, and personnel involved in the flight tests. Range safety is a major component of flight test safety.

Flight test safety-related hazards and safety concerns include:

- Personnel safety
- Weapons release
- Inadvertent weapons initiation
- Safety of test items and test equipment
- Safety of manuals and instructions
- Safety of flight (SOF) program

- Aircraft airworthiness certification
- Emergency preparedness

Range Safety

During the development or modification of a system it is often necessary to test the system on a dedicated test range. Since each test range is individual and unique, there are range safety requirements specific to each range. Overall, range safety is the responsibility of the range safety officer (RSO) at the host range. Range safety consists of facility and support equipment safety, as well as the safety built into the system being tested. The RSO is not trying to make the system safe—only the range and its collateral area. The RSO reviews all of the safety data on every system used at the range and examines and evaluates all interfaces between and among the systems being used at the range. Each of the system's safety assessment reports (SARs) is reviewed individually to assure that that system by itself is safe. Then, each of the interfaces will be examined to determine that no hazards are introduced when all these systems are integrated into one test system/range system.

Typical range safety concerns include the following:

- Site personnel safety (training, eyes, sound, physical barriers, etc.)
- Safety inspection for guarding of moving parts, safety hand rails, hazardous material review, and so on
- Lock out and tag out procedures
- Environmental impact
- Grounding system test (power, lightning, ESD, etc.)
- Power system test (all voltages)
- Emergency power-off system test
- Communication links functionality
- Fire department/emergency vehicle communication functionality
- SwS program
- Emergency lights, horns, fire alarms
- Noise level test (less than 85 dBA for an 8-h period)
- Radio frequency (RF) radiation areas (if applicable)
- Protection zones (e.g., launch gasses)
- Termination system (failure and/or premature operation)
- Exceeding flight zone
- Flying into populated area
- Weapons hazards
  - Veering off course
  - Exceeding flight zone
  - Flying into populated area

- Flying back to launch point
- Self-destruct (failure and/or premature operation)
- Failure of the self-destruct mechanisms
- Inadvertent detonation
- Premature detonation

Key range safety reference standards include:

1. Eastern and Western Test Range (EWR) 127-1, Eastern and Western Range, Range Safety Requirements, 1999.

Intrinsic Safety (IS)

IS refers to equipment and wiring that is inherently safe in a hazardous area (i.e., potentially flammable). In other words, an intrinsically safe system is one with energy levels so low they cannot cause an explosion. This is typically achieved through the use of barriers—either zener diode barriers or isolated barriers—that limit energy to a hazardous area. Hazardous area refers to any location with combustible material such as gases, dusts, or fibers that might produce an ignitable mixture. A hazardous area can be a sealed room filled with a volatile material or an area that is open to normal foot traffic, such as the area around a gasoline pump. An intrinsically safe component is an item that contains a hazardous energy source and operates in an explosive atmosphere, but the design is such that the energy cannot cause ignition of the explosive atmosphere. For example, intrinsically safe electrical lanterns are used in coal mines to prevent ignition of explosive vapors that may emanate from the mine.

IS-related hazards and safety concerns include:

- Separate intrinsically safe wiring from nonintrinsically safe wires by an air space, a conduit, or a partition
- Label wires to distinguish hazardous area wiring from safe area wiring
- Seal or vent conduit and raceways inside hazardous areas so they do not transfer the hazardous atmosphere to the safe area
- Compliance with local, state, and federal regulations for IS practices

Mine Safety

Mine safety involves all aspects of underground and surface mining.

Mine safety-related hazards and safety concerns include:

- Ingress and egress
- Noise

- Vibration
- Dust
- Poor lighting
- Fire/explosion
- Law violations
- Training
- Confined spaces
- Emergency preparedness
- Hazardous gases and vapors
- Elevator safety
- Rail safety (collisions, derailment, etc.)
- Safety culture
- Communications

Key mine safety reference standards include:

1. The Federal Mine Safety and Health Act of 1977 (Public Law 95–164), United States Code, Title 30, Mineral Lands and Mining, Chapter 22, Mine Safety and Health, November 9, 1977.

Fuze Safety

A fuze, or fuzing system, is a physical device in a system intended for the purpose of detonating an explosive device or munition when intended, and for preventing device initiation when not intended. It is typically designed to sense a target or respond to one or more prescribed conditions, such as elapsed time, pressure, or command, which initiates a train of fire or detonation in a munition. Safety and arming are primary roles performed by a fuze to preclude inadvertent ignition of a munition and intended ignition before the desired position or time.

Key fuze safety reference standards include:

1. MIL-STD-1316, Fuze Design, Safety Criteria for, July 10, 1998.
2. MIL-STD-331C, Test Method Standard, Fuze and Fuze Components, Environmental and Performance Tests For, January 5, 2005.

Explosives Safety

Explosives safety is the program and process used to prevent premature, unintentional, or unauthorized initiation of explosives and devices containing explosives and to minimize the effects of explosions, combustion, toxicity, and any other deleterious effects. Explosives safety includes all mechanical, chemical, biological, electrical, and environmental hazards associated with explosives or electromagnetic environmental effects. Equipment, systems, or procedures and processes whose malfunction would cause unacceptable mishap risk to manufacturing, handling, transportation, maintenance, storage, release, testing, delivery, firing, or disposal of explosives are also included.

Explosives safety-related hazards and safety concerns include:

- Premature initiation
- Unintentional initiation
- Unauthorized initiation
- Handling, transportation, and storage

Key explosives safety reference standards include:

1. DoD 4145.26M, Contractor Safety Manual for Ammunition and Explosives.
2. DoD 6055.9 STD, Ammunition and Explosives Safety Standards.
3. DoD Directive 6055.9, DoD Explosives Safety Board (DDESB) and DoD Component Explosives Safety Responsibilities.
4. MIL-HDBK-1512, Electroexplosive Subsystems, Electrically Initiated, Design Requirements and Test Methods.

Patient Safety

The objective of patient safety is to ensure the patient is safe while in the medical facility and that he or she leaves the medical facility in a healthier condition than when he or she entered. The application of patient safety is in medical facilities, primarily hospitals. Patient safety covers a wide area of concerns, such as surgery, facilities, medical devices, drugs/medications, pathogen control, training, and record keeping. A large part of patient safety involves human factors safety; human error is always a given, but it can be controlled. Human error does not mean negligence; it means natural human fallibility.

Patient safety-related hazards and safety concerns include:

- Wrong-site surgery errors (wrong arm, leg, eye, etc.)
- Incorrect dispensing of drugs (incorrect drug, amount, rate, etc.)
- Handling, transportation, and storage of medical equipment and drugs
- Laboratory errors
- Exposure to toxic chemicals
- Patient identification errors
- Infection control
- Adverse events
- Incorrect use of medical devices/equipment
- Faulty medical devices/equipment

- Errors in medical records
- Situation awareness (during surgery and recovery)
- Wrong treatment administered
- Faulty communications
- Inadequate supervision
- Fatigue and stress
- Training

CHAPTER 4

System Safety Acronyms

The following is a list of acronyms used in the system safety field, as well as acronyms that a system safety analyst must be familiar with while working in various technical domains and on various systems:

A/C	Aircraft
AC	Advisory Circular
ACAT	Acquisition Category
ADC	Air Data Computer
ADF	Automatic Direction Finder
ADI	Attitude Direction Indicator
AE	Architect and Engineering Firm
AEW	Airborne Early Warning
AF	Air Force
AFR	Air Force Regulation
AFRL	Air Force Research Laboratory
AFSRB	Army Fuze Safety Review Board
AGL	Above Ground Level
AGM	Air-to-Ground Missile
AGM	Aviation Ground Mishap
ALARA	As Low as Reasonably Achievable
ALARP	As Low as Reasonably Practicable
ANOVA	Analysis of Variance
ANSI	American National Standards Institute
AoA	Analysis of Alternatives
AoA	Angle of Attack
APGS	Auxiliary Power Generation System
API	Application Programming Interface
APS	Auxiliary Power System
APU	Auxiliary Power Unit
AR	Army Regulation

Concise Encyclopedia of System Safety: Definition of Terms and Concepts, First Edition. Clifton A. Ericson II.
© 2011 by John Wiley & Sons, Inc. Published 2011 by John Wiley & Sons, Inc.

ARAR	Accident Risk Assessment Report
ARC	Ames Research Center
ARINC	Aeronautical Radio, Inc.
ARP	Aerospace Recommended Practice
ARV	Armed Robotic Vehicle
ASAP	Aerospace Safety Advisory Panel
ASE	Aviation Safety Engineer
ASE-SW	Aviation Safety Engineer-Software
ASI	Aviation Safety Inspector
ASIC	Application-Specific Integrated Circuit
ASME	American Society of Mechanical Engineers
ASN	Assistant Secretary of the Navy
ASN (I&E)	ASN Installations and Environment
ASN (RD&A)	ASN Research, Development, and Acquisition
ASO	Aviation Safety Officer
ASQC	American Society for Quality Control
ASRS	Aviation Safety Reporting System
ASSE	American Society of Safety Engineers
ASTC	Amended Supplemental Type Certificate
ATC	Air Traffic Control
ATC	Amended Type Certificate
ATM	Air Traffic Management
ATP	Acceptance Test Procedure
ATV	All-Terrain Vehicles
AUR	All-Up-Round
AUV	Autonomous Underwater Vehicle
AUX	Auxiliary
AV	Aerial Vehicle
AV	Air Vehicle
AVO	Air Vehicle Operator
AW	Airworthiness
AWACS	Airborne Warning and Control System
AWC	Airworthiness Certification
BA	Barrier Analysis
BCSP	Board of Certified Safety Professionals
BDA	Battle Damage Assessment
BIT	Built-In Test
BITE	Built-In Test Equipment
BLOS	Beyond Line-of-Sight
BPA	Bent Pin Analysis
BUMED	Bureau of Medicine (Navy)
C2	Command and Control
C2P	Command and Control Processor
C3	Command, Control, and Communications
C4I	Command, Control, Communications, Computers, and Intelligence

SYSTEM SAFETY ACRONYMS

CAA	Civil Aviation Authority
CAD	Cartridge-Actuated Devices
CAD	Computer-Aided Design
CAE	Component Acquisition Executive
CAE	Computer-Aided Engineering
CAM	Computer-Aided Manufacture
CASE	Computer-Aided Software Engineering
CCA	Circuit Card Assembly
CCB	Change Control Board
CCD	Charge-Coupled Device
CCF	Common-Cause Failure
CCFA	Common-Cause Failure Analysis
CCP	Contamination Control Plan
CCP	Critical Control Point
CDR	Critical Design Review
CDRL	Contract Data Requirements List
CEC	Cooperative Engagement Capability
CFD	Computational Fluid Dynamics
CGI	Computer-Generated Images
CHENG	ASN (RD&A) Chief Engineer's Office
CIL	Critical Items List
CLIN	Contract Line Item Number
CM	Configuration Management
CMC	Commandant of the Marine Corps
CMC (SD)	Commandant of the Marine Corps (Safety Division)
CMF	Common Mode Failure
CMFA	Common Mode Failure Analysis
CMP	Configuration Management Plan
CNO	Chief of Naval Operations
CNR	Chief of Naval Research
COA	Course of Action
COL	Carrier Operating Limitation
COLD	Carrier Operating Limitations Document
CONOPS	Concept of Operations
CONUS	Contiguous United States
COTS	Commercial Off-the-Shelf
CP	Computer Program
CPCI	Computer Program Configuration Item
CPCR	Computer Program Change Request
CPDD	Computer Program Description Document
CPSCF	Computer Program Safety-Critical Function
CPU	Central Processing Unit
CRC	Cyclic Redundancy Check
CRT	Cathode Ray Tube
CSA	Code Safety Analysis

CSC	Computer Software Component
CSC	Critical Software Command
CSCI	Computer Software Configuration Item
CSE	Combat System Element
CSFP	Critical Single-Failure Point
CSI	Critical Safety Item
CSP	Certified Safety Professional
CSSQT	Combat System Sea Qualification Trial
CSU	Computer Software Unit
CTS	Clear to Send
CV	Carrier Vehicle
CV	Aircraft Carrier
CVN	Aircraft Carrier Nuclear
DA	Design Agent
DAL	Design Assurance Level
DAL	Development Assurance Level
DAR	Designated Airworthiness Representative
DARPA	Defense Advanced Research Projects Agency
DASN (S)	Deputy Assistant Secretary of the Navy (Safety)
DAU	Defense Acquisition University
DCN	Document Control Number
DD	Destroyer
DDESB	Department of Defense Explosive Safety Board
DEL	Data Element List
DER	Designated Engineering Representative
DEW	Directed-Energy Weapon
DFD	Data Flow Diagram
DIA	Defense Intelligence Agency
DID	Data Item Description
DIICOE	Defense Information Infrastructure Common Operating Environment
DM	Data Management
DOD	Department of Defense
DODD	Department of Defense Directive
DODI	Department of Defense Instruction
DOE	Department of Energy
DON	Department of the Navy
DOORS	Dynamic Object-Oriented Requirements System
DOT	Department of Transportation
DPA	Destructive Physical Analysis
DSF	Design Safety Feature
DSOC	Defense Safety Oversight Council
DT	Development Test
DT/OT	Developmental Test/Operational Test
E/DRAP	Engineering/Data Requirements Agreement Plan

E/E/PE	Electrical/Electronic/Programmable Electronic
E/E/PES	Electrical/Electronic/Programmable Electronic System
E3	Electromagnetic Environmental Effects
EA	Engineering Assessment
EASA	European Aviation Safety Agency
ECCM	Electronic Counter Countermeasures
ECM	Electronic Countermeasures
ECP	Engineering Change Proposal
ECS	Environmental Control System
EDEF	EDRAP Data Evaluation Form
EDRAP	Engineering/Data Requirements Agreement Plan
EED	Electro-Explosive Device
EEE	Electrical, Electronic, and Electromechanical
EEPROM	Electronically Erasable Programmable Read-Only Memory
EGI	Embedded GPS/INS
EHC	Explosive Hazard Classification
EHF	Extremely High Frequency
EIA	Environmental Impact Assessment
EID	Electrically Initiated Devices
EIL	Equipment Indenture List
EIS	Environmental Impact Statement
ELINT	Electronic Intelligence
EMC	Electromagnetic Compatibility
EMD	Engineering and Manufacturing Development
EMF	Electromagnetic Field
EMI	Electromagnetic Interference
EMR	Electromagnetic Radiation
ENG	Engineering
EO	Electro-Optical
EO	Engagement Order
EO	Engineering Order
EO	Executive Order
EOD	Explosives Ordnance Disposal
EPA	Environmental Protection Agency
EPROM	Erasable Programmable Read-Only Memory
ES	Electronic Surveillance
ESD	Electrostatic Discharge
ESD	Emergency Shutdown
ESF	Engineered Safety Feature
ESH	Environmental, Safety, and Health
ESOH	Environmental, Safety, and Occupational Health
ESSM	Evolved Sea Sparrow Missile
ET&E	Engineering Test and Evaluation
ETA	Event Tree Analysis
ETR	Eastern Test Range

EUC	Equipment Under Control
EW	Electronic Warfare
EWR	Eastern and Western Test Range
EWS	Electronic Warfare System
FAA	Federal Aviation Administration
FADEC	Full Authority Digital Electronic Control
FAR	Federal Acquisition Regulation
FAR	Federal Aviation Regulation
FBD	Functional Block Diagram
FCA	Functional Configuration Audit
FCO	Flight Clearance Officer
FCS	Fire Control System
FCU	Flight Control Unit
FDA	Food and Drug Administration
FDM	Fault Detection Monitoring
FDR	Flight Data Recorder
FFA	Free-Fire Area
FFD	Functional Flow Diagram
FHA	Facility Hazard Analysis
FHA	Fault Hazard Analysis
FHA	Functional Hazard Analysis
FHA	Functional Hazard Assessment
FI	Fault Isolation
FI	Final Inspection
FIFO	First In, First Out
FISTRP	Fuze and Initiation Systems Technical Review Panel
FLIR	Forward-Looking Infrared
FLS	Field-Loadable Software
FM	Field Manual
FMC	Flight Management Computer
FMEA	Failure Modes and Effects Analysis
FMECA	Failure Modes, Effects, and Criticality Analysis
FMET	Failure Modes and Effects Testing
FOC	Full Operational Capability
FOD	Foreign Object Damage
FOD	Foreign Object Debris
FOL	Flight Operational Limitation
FOLD	Flight Operating Limitations Document
FOSS	Free and Open Source Software
FOUO	For Official Use Only
FPGA	Field Programmable Gate Array
FQT	Flight Qualification Test
FQT	Final Qualification Test
FRACAS	Failure Reporting, Analysis, and Corrective Action System
FRB	Failure Review Board

FRD	Functional Requirements Document
FRR	Flight Readiness Review
FS	Fire Scout
FSCAP	Flight Safety-Critical Aircraft Part
FSD	Full-Scale Development
FSED	Full-Scale Engineering Development
FSIS	Food Safety Inspection Services (USDA)
FSMP	Facility Safety Management Plan
FT	Fault Tree
FTA	Fault Tree Analysis
FTP	File Transfer Protocol
FWA	Fixed Wire Antenna
G-48	System Safety Committee of Tech America (formerly GEIA)
GAO	General Accounting Office
GATM	Global Air Traffic Management
GCS	Ground Control System
GDT	Ground Data Terminal
GEIA	Government Electronics and Industry Association (old name)
GEIA	Government Electronics and Information Technology Association
GFE	Government-Furnished Equipment
GFF	Government-Furnished Facilities
GFI	Government-Furnished Information
GFM	Government-Furnished Material
GFP	Government-Furnished Property
GHA	Gross Hazard Analysis
GIA	Government Inspection Agency
GIDEP	Government–Industry Data Exchange Program
GIG	Global Information Grid
GOL	Ground Operating Limitation
GOLD	Ground Operating Limitations Document
GOTS	Government Off-the-Shelf
GPS	Global Positioning System
GPWS	Ground Proximity Warning System
GRC	Glenn Research Center (at Lewis Field)
GSA	General Services Administration
GSE	Government-Supplied Equipment
GSE	Ground Support Equipment
GSFC	Goddard Space Flight Center
GSN	Goal-Structured Notation
GUI	Graphics User Interface
HACCP	Hazard Analysis and Critical Control Point
HAR	Hazard Action Record

HAR	Hazard Action Report
HAR	Hazard Analysis Record
HAZCOM	Hazardous Communication
HAZOP	Hazard and Operability (Analysis)
HAZREP	Hazard Report
HAZWOPER	Hazardous Waste Operation
HCF	Hazard Causal Factor
HCI	Human–Computer Interface
HCR	Hazard Control Record
HEP	Human Error Probability
HERF	Hazards of Electromagnetic Radiation to Fuel
HERO	Hazards of Electromagnetic Radiation to Ordnance
HERP	Hazards of Electromagnetic Radiation to Personnel
HF	Human Factors
HFA	Human Factors Analysis
HFACS	Human Factors Analysis and Classification System
HFE	Human Factors Engineering
HFE	Human Failure Event
HHA	Health Hazard Assessment
HIS	High-Integrity System
HITL	Hardware in the Loop
HMI	Human–Machine Interface
HMMP	Hazardous Materials Management Plan
HMMR	Hazardous Material Management Report
HR	Hazard Report
HRA	Hazard Risk Assessment
HRA	Human Reliability Analysis
HRI	Hazard Risk Index
HRI	Human–Robot Interaction
HSC	Hardware/Software Configuration
HSCM	Hardware/Software Configuration Matrix
HSDR	Hardware Safety Design Requirement
HSI	Human Systems Integration
HSIP	Human Systems Integration Plan
HTDB	Hazard Tracking Database
HTS	Hazard Tracking System
HVPS	High-Voltage Power Supply
HW/SW	Hardware/Software
HWCI	Hardware Configuration Item
HX	Heat Exchanger
I/O	Input/Output
IAS	Indicated Airspeed
IAW	In Accordance With
ICAO	International Civil Aviation Organization
ICBM	Intercontinental Ballistic Missile

ICD	Initial Capabilities Document
ICD	Interface Control Document
IDD	Interface Design Description
IDD	Interface Design Document
IDS	Interface Design Specifications
IEC	International Electrotechnical Commission
IED	Improvised Explosive Device
IEEE	Institute of Electrical and Electronic Engineers
IFF	Identify Friend or Foe
IFR	Instrument Flight Rules
IG	Inspector General
IHA	Integrated Hazard Analysis
IHA	Interface Hazard Analysis
IIFF	Improved Identify Friend or Foe
ILS	Instrument Landing System
ILS	Integrated Logistics Support
IM	Insensitive Munitions
IMA	Integrated Modular Avionics
IMC	Instrument Meteorological Conditions
IMRB	Insensitive Munitions Review Board
IMU	Inertial Measurement Unit
IND	Inadvertent Nuclear Detonation
INS	Inertial Navigation System
INSRP	Interagency Nuclear Safety Review Panel
IOC	Initial Operational Capability
IOT	Interoperability Test
IPR	In-Process Review
IPT	Integrated Process Team
IPT	Integrated Product Team
IR	Infrared
IRIS	Incident Reporting Information System
IRM	Integrated Risk Management
IRS	Interface Requirement Specification
IS	Intrinsic Safety
ISA	Independent Safety Auditor
ISEA	In-Service Engineering Agent
ISO	International Standards Organization
ISR	Intelligence, Surveillance, and Reconnaissance
ISR	Interrupt Service Routine
IISAP	Integrated Interoperable Safety Analysis Process
ISSC	Internationals System Safety Conference
ISSPP	Integrated System Safety Program Plan
ISSS	International System Safety Society
IT	Information Technology
ITAR	International Traffic in Arms Regulations

IV&V	Independent Verification and Validation
JAA	Joint Aviation Authority (of Europe)
JAG	Judge Advocate General
JANNAF	Joint Army, Navy, NASA, Air Force
JATO	Jet-Assisted Takeoff
JCB	Joint Capabilities Board
JCD	Joint Capabilities Document
JCIDS	Joint Capabilities Integration and Development System
JHA	Job Hazard Analysis
JIT	Just In Time
JPL	Jet Propulsion Laboratory
JROC	Joint Requirements Oversight Council
JRP	Joint Robotics Program
JSA	Job Safety Analysis
JSC	Johnson Space Center
KCAS	Knots Calibrated Airspeed
KIAS	Knots Indicated Airspeed
KPP	Key Performance Parameter
KSA	Knowledge, Skills, and Abilities
KSC	Kennedy Space Center
KSLOC	1000 Source Lines of Code
LAN	Local Area Network
LaRC	Langley Research Center
LCC	Life Cycle Cost
LCCE	Life Cycle Cost Estimate
LCS	Littoral Combat Ship
LDP	Letter Data Package
LED	Light-Emitting Diode
LGB	Laser-Guided Bomb
LGM	Laser-Guided Missile
LGW	Laser-Guided Weapon
LIFO	Last In, First Out
LOA	Level of Autonomy
LOC	Lines of Code (Executable Source)
LOFI	Level of FAA Involvement
LOS	Line of Sight (Line-of-Sight)
LOT	Level of Trust
LOX	Liquid Oxygen
LRF	Laser Range Finder
LRIP	Low-Rate Initial Production
LRM	Line-Replaceable Module
LRU	Line-Replaceable Unit
LRV	Lightweight Reconnaissance Vehicle
LSO	Laser Safety Officer
LSRB	Laser Safety Review Board
LSSO	Laser System Safety Officer

LSSRB	Laser System Safety Review Board
LSSWG	Laser System Safety Working Group
LTD	Laser Target Designator
MA	Managing Activity
MA	Managing Authority
MA	Markov Analysis
MC/DC	Modified Condition/Decision Coverage
MCAS	Midair Collision Avoidance System
MCE	Mission Control Element
MDA	Milestone Decision Authority
MDA	Missile Defense Agency
MDAP	Major Defense Acquisition Program
MEL	Master Equipment List
MFHBMA	Mean Flight Hours Between Maintenance Actions
MFHBVF	Mean Flight Hours Between Verified Failures
MFHBR	Mean Flight Hours Between Removals
MH	Magnetic Heading
MIPS	Millions of Instructions per Second
MMI	Man–Machine Interface
MMP	Modular Mission Payload
MMU	Mass Memory Unit
MMU	Memory Management Unit
MNS	Mission Need Statement
MoD	Ministry of Defense
MooN	M out of N
MooND	M out of N with Diagnostics
MORT	Management Oversight and Risk Tree
MOTS	Military Off-the-Shelf
MOTS	Modified Off-the-Shelf
MOU	Memorandum of Understanding
MPE	Maximum Permissible Exposure
MRAL	Mishap Risk Acceptance Level
MRI	Mishap Risk Index
MSC	Marshall Space Center
MSDS	Material Safety Data Sheet
MSFC	Marshall Space Flight Center
MSL	Mean Sea Level
MTBF	Mean Time Between Failure
MTBOMF	Mean Time Between Operational Mission Failure
MTTR	Mean Time to Repair
MULE	Multifunction Utility/Logistics and Equipment Vehicle
N/A	Not Applicable
N/C	No Change
NA	Not Applicable
NAS	NASA Assurance Standard

SYSTEM SAFETY ACRONYMS 497

NAS	National Air Space
NASA	National Aeronautics and Space Administration
NATO	North Atlantic Treaty Organization
NATOPS	Naval Air Training and Operating Procedures Standardization
NAV	Navigation
NAVAIDS	Navigational Aids
NAVAIR	Naval Air Systems Command
NAVOSH	Navy Occupational Safety and Health
NAVSEA	Naval Sea Systems Command
NAWC	Naval Air Warfare Center
NBC	Nuclear, Biological, and Chemical
NBCC	Nuclear, Biological, and Chemical Contamination
NCIS	Naval Criminal Investigation Service
NDE	Nondestructive Evaluation
NDI	Nondestructive Inspection
NDI	Nondevelopmental Item
NEC	National Electrical Code
NEMA	National Electrical Manufacturers Association
NEPA	National Environmental Policy Act
NEW	Net Explosives Weight (TNT Equivalent)
NFPA	National Fire Protection Association
NFS	NASA FAR Supplement
NFS	Network File System
NFZ	No-Fly Zone
NGC	Northrop Grumman Corporation
NHB	NASA Handbook
NHS	NASA Health Standard
NHZ	Nominal Hazard Zone
NIOSH	National Institute for Occupational Safety and Health
NIST	National Institute of Standards and Testing
NMAC	Near Midair Collision
NNMSB	Non-Nuclear Munitions Safety Board
NOHD	Nominal Ocular Hazard Distance
NORAD	North American Aerospace Defense Command
NOSSA	Naval Ordnance Safety and Security Activity
NPD	NASA Policy Document
NRL	Naval Research Laboratory, Washington, DC
NSA	National Security Agency
NSC	National Safety Council
NSC	Not Safety-Critical
NSR	Not Safety-Related
NSRS	NASA Safety Reporting System
NSS	NASA Safety Standard
NSTC	NASA Safety Training Center

SYSTEM SAFETY ACRONYMS

NSTS	National Space Transportation System
NSWC PHD	Naval Surface Warfare Center Port Hueneme Division
NSWCCD	Naval Surface Warfare Center, Crane Division
NSWCDD	Naval Surface Warfare Center, Dahlgren Division
NTSB	National Transportation Safety Board
NWS	National Weather Service
NWSSG	Nuclear Weapons System Safety Group
O&SHA	Operating and Support Hazard Analysis
OA	Open Architecture
OACE	Open Architecture Computing Environment
OAT	Outside Air Temperature
OBA	Oxygen Breathing Apparatus
OCONUS	Outside Contiguous United States
ODS	Ozone-Depleting Substances
OEM	Original Equipment Manufacturer
OHA	Operating Hazard Analysis
OHA	Operational Hazard Assessment
OHEB	Ordnance Hazards Evaluation Board
ONR	Office of Naval Research
OOA	Object-Oriented Analysis
OOD	Object-Oriented Design
OOP	Object-Oriented Programming
OP	Ordnance Publication
OPEVAL	Operational Evaluation
OPNAV	Office of the Chief of Naval Operations
OPORD	Operation Order
ORD	Operational Requirements Document
ORM	Operational Risk Management
OS	Operating System
OSA	Operational Safety Assessment
OSD	Office of the Secretary of Defense
OSH	Occupational Safety and Health
OSHA	Occupational Safety and Health Administration
OSMA	Office of Safety and Mission Assurance
OSS	Open Source Software
OSSMA	Office of Systems Safety and Mission Assurance
OT	Operational Test
OT&E	Operational Test and Evaluation
OTD	Open Technology Development
P/N	Part Number
P2	Pollution Prevention
PAD	Pyrotechnic Actuated Devices
PAR	Program Action Request
PAR	Problem Action Record
PBXC	Plastic-Bonded Explosives (C)

SYSTEM SAFETY ACRONYMS

PBXN	Plastic-Bonded Explosives (N)
PCA	Physical Configuration Audit
PCB	Parts Control Board
PCB	Printed Circuit Board
PCMCIA	Personal Computer Memory Card International Association
PDA	Personal Digital Assistant
PDF	Probability Density Function
PDR	Preliminary Design Review
PDU	Power Drive Unit
PEL	Permissible Exposure Limit
PEO	Program Executive Officer
PES	Potential Explosion Site
PES	Programmable Electronic System
PESHE	Programmatic Environmental, Safety, and Health Evaluation
PETN	Pentaerythritol Tetranitrate
PFD	Probability of Failure on Demand
PFS	Principal for Safety
PHA	Preliminary Hazard Analysis
PHA	Process Hazard Analysis
PHL	Preliminary Hazard List
PHS&T	Packaging, Handling, Storage, and Transportation
PIDS	Prime Item Development Specification
PIP	Product Improvement Program
PITL	Pilot-in-the-Loop
PL	Programmable Logic
PLC	Programmable Logic Controller
PLD	Programmable Logic Device
PLOA	Probability Loss of Aircraft
PLOC	Probability Loss of Control
PLOM	Probability Loss of Mission
PM	Program Manager
PMO	Program Management Office
PNA	Petri Net Analysis
PO	Program Office
POA&M	Plan of Action and Milestones
POC	Point of Contact
POD	Probability of Detection
PPE	Personal Protective Equipment
PR	Problem Report
PRA	Probabilistic Risk Assessment
PRF	Pulse Repetition Frequency
PROM	Programmable Read-Only Memory
PSAC	Plan for Software Aspects of Certification

PSE	Program Safety Engineer
PSF	Performance-Shaping Factor
PSM	Program Safety Manager
PSSA	Preliminary System Safety Assessment
PTO	Power Takeoff
PTS	Position Tracking System
PTT	Push to Talk
PVCS	Professional Version Control System
PWB	Printed Wiring Board
QA	Quality Assurance
QD	Quantity-Distance
QRA	Quantitative Risk Assessment
R&D	Research and Development
R/T	Receiver/Transmitter
R3	Resource, Recovery, and Recycling
RA	Radar Altimeter
RAC	Risk Assessment Code
RADALT	Radar Altimeter
RADHAZ	Radiation Hazard
RAID	Redundant Array of Independent Disks
RAM	Random Access Memory
RAMS	Reliability, Availability, Maintainability, and Safety
RAMS	Reliability and Maintainability Symposium
RAN	Royal Australian Navy
RCC	Range Commanders Council
RCS	Radar Cross Section
RCVR	Receiver
RDTE	Research, Development, Test, and Evaluation
RDX	Cycoltrimethylene Trinitramine (Explosive)
RF	Radio Frequency
RFI	Radio Frequency Interference
RFID	Radio-Frequency Identification
RFP	Request for Proposal
RFR	Radio Frequency Radiation
RIA	Robotic Industries Association
RMB	Risk Management Board
RMP	Risk Management Plan
RMS	Root Mean Square
ROE	Rules of Engagement
ROM	Read-Only Memory
ROZ	Restricted Operations Zone
RPG	Rocket-Propelled Grenade
RSO	Range Safety Office
RTCA	Radio Technical Commission for Aeronautics
RTI	Real-Time Indicators

RTM	Requirements Traceability Matrix	
RTOS	Real-Time Operating System	
RVSM	Reduced Vertical Separation Minima	
SA	Situational Awareness	
SAD	Software Architecture Document	
SAE	Society of Automotive Engineers	
SAI	Safety Action Item	
SAM	Surface-to-Air Missile	
SAR	Safety Action Record	
SAR	Safety Assessment Report	
SAR	Synthetic Aperture Radar	
SAS	Safety Analysis Summary	
SAS	Software Accomplishment Summary	
SATCOM	Satellite Communications	
SAWE	Society of Allied Weight Engineers	
SC	Safety-Critical	
SCA	Safety Compliance Assessment	
SCA	Sneak Circuit Analysis	
SCA	Static Code Analysis	
SCBA	Self-Contained Breathing Apparatus	
SCC	Software Control Category	
SCCB	Software Change Control Board	
SCCB	Software Configuration Control Board	
SCCS	Safety-Critical Computer Software	
SCCSC	Safety-Critical Computer Software Component	
SCCSF	Safety-Critical Computing System Function	
SCD	Source Control Drawing	
SCF	Safety-Critical Function	
SCHC	Safety-Critical Hardware Components	
SCI	Safety-Critical Item	
SCI	Software Criticality Index	
SCM	Software Configuration Management	
SCMP	Software Configuration Management Plan	
SCN	Software Change Notice	
SCN	Specification Change Notice	
SCP	Software Certification Plan	
SCR	Safety-Critical Requirement	
SCUBA	Self-Contained Underwater Breathing Apparatus	
SDC	Signal Data Converter	
SDD	Software Design Description	
SDD	Software Design Document	
SDD	System Design Document	
SDP	Software Development Plan	
SDR	System Design Review	
SDRL	Subcontract Data Requirements List	

502 SYSTEM SAFETY ACRONYMS

SDZ	Surface Danger Zone
SE	Systems Engineering
SEA	Safety Engineering Agent
SEAL	Sea–Air–Land
SEB	Source Evaluation Board
SECDEF	Secretary of Defense
SEI	Software Engineering Institute
SEIT	Systems Engineering Integration Team
SEMP	Systems Engineering Management Plan
SEP	Systems Engineering Plan
SER	Safety Evaluation Report
SET	Safety Engineering Team
SETR	Systems Engineering Technical Review
SFAR	Special Federal Aviation Regulation
SFR	System Functional Review
SHA	System Hazard Analysis
SHAR	Safety Hazard Alert Report
SHRI	Software Hazard Risk Index
SI	System Integration
SIAM	Software Integrity Assurance Matrix
SIB	Safety Investigation Board
SIB	Software Implementation Board
SIL	Safety Integrity Level
SIR	Safety Investigation Report
SIR	Serious Incident Report
SIS	Safety Information System
SIS	Safety Instrumented System
SLAB	Sealed Lead Acid Battery
SLD	Single-Line Diagram
SLOC	Source Lines of Code
SME	Subject Matter Expert
SMS	Safety Management System
SOF	Safety of Flight
SOF	Special Operations Forces
SOLE	Society of Logistics Engineers
SOO	Statement of Objectives
SOOP	Safety Order of Precedence
SOP	Standard Operating Procedure
SoS	System of Systems
SOUP	Software of Unknown Pedigree
SOV	Shutoff Valve
SOW	Statement of Work
SPAWAR	Space and Naval Warfare Systems Command
SPF	Single-Point Failure
SPP	Safety Program Plan

SPR	Software Problem Report
SQA	Software Quality Assurance
SQAP	Software Quality Assurance Plan
SQMS	Software Quality Management System
SQT	Software Qualification Test
SQT	System Qualification Test
SR	Safety-Related
SRCA	Safety Requirements/Criteria Analysis
SRM	Safety Risk Management
SRR	Software Requirements Review
SRR	System Requirements Review
SRS	Software Requirements Specification
SS	System Specification
SSA	Software Safety Agent
SSA	System Safety Assessment
SSAR	System Safety Assessment Report
SSB	Single-Side Band
SSC	Stennis Space Center
SSDD	System/Subsystem Design Description
SSDR	Software Safety Design Requirement
SSE	System Safety Engineer
SSHA	Subsystem Hazard Analysis
SSI	Safety Significant Item
SSL	System Safety Lead
SSM	System Safety Manager
SSMP	System Safety Management Plan
SSP	System Safety Program
SSPP	System Safety Program Plan
SSR	Software Specification Review
SSR	System Safety Requirement
SSRA	Software Safety Requirements Analysis
SSRA	System Safety Risk Assessment
SSRP	System Safety Review Panel
SSS	System Safety Society
SSS	System Segment Specification
SSS	System Subsystem Specification
SSSTRP	Software System Safety Technical Review Panel
SSWG	System Safety Working Group
STA	System Threat Analysis
STANAG	Standardization Agreement (NATO)
STC	Supplemental Type Certificate
STE	Special Test Equipment
STP	Software Test Plan
STR	Software Trouble Report
SUGV	Small Unmanned Ground Vehicle

SVP	Software Verification Plan
SW	Software
SwAL	Software Assurance Level
SWAP	Size, Weight, and Power
SWG	Safety Working Group
SWHA	Software Hazard Analysis
SWIR	Shortwave Infrared
SWIT	Software Integration and Test
SwS	Software Safety
SwSA	Software Safety Analysis
SwSA	Software Safety Assessment
SwSWG	Software Safety Working Group
SyS	System Safety
SYSCOM	System Command
T&E	Test and Evaluation
TAA	Technical Assistance Agreement
TAAF	Test, Analyze, and Fix
TACAN	Tactical Air Navigation
TADIL	Tactical Digital Information Link
TAE	Technical Area Expert
TAS	True Airspeed
TBD	To Be Determined
TC	Type Certificate
TCAS	Traffic Collision Avoidance System
TCO	Total Cost of Ownership
TCTO	Time Compliance Technical Order
TD	Technical Directive
TD	Test Director
TDA	Technical Design Agent
TDP	Technical Data Package
TECHEVAL	Technical Evaluation
TEMP	Test and Evaluation Master Plan
TFOA	Things Falling Off Aircraft
TH	True Heading
THA	Threat Hazard Assessment
THERP	Technique for Human Error Rate Prediction
TIM	Technical Information Meeting
TLE	Top-Level Event
TLH	Top-Level Hazard
TLM	Top-Level Mishap
TM	Telemetry
TNT	Trinitrotoluene (Explosive)
TO	Technical Order
TPDR	Transponder
TQM	Total Quality Management

TR	Trouble Report
TRR	Test Readiness Review
TS	Technical Specialist
TSA	Test Safety Analysis
TSO	Technical Standard Order
TUAV	Tactical Unmanned Air Vehicle
TVC	Thrust Vector Control
TWA	Trailing Wire Antenna
TYCOM	Type Commands
UA	Unsafe Action
UAS	Unmanned Aerial System
UAS	Unmanned Aircraft System
UAV	Unmanned Aerial Vehicle
UBD	Underwater Breathing Device
UCAS	Unmanned Combat Aerial System
UCAV	Unmanned Combat Air Vehicle
UGV	Unmanned Ground Vehicle
UHF	Ultra-High Frequency
UL	Underwriters Laboratory
UML	Unified Modeling Language
UMS	Unmanned System
USACE	United States Army Corps of Engineers
USAF	United States Air Force
USB	Universal Serial Bus
USB	Upper Side Band
USD (AT&L)	Undersecretary of Defense for Acquisition, Technology, and Logistics
USDA	United States Department of Agriculture
USG	United States Government
USMC	United States Marine Corps
USN	United States Navy
USS	Unmanned Surface Ship
USSV	Unmanned Sea Surface Vehicle
USV	Unmanned Surface Vehicle
UUV	Unmanned Underwater Vehicle
UV	Ultraviolet
UVPROM	Ultaviolet Erasable Programmable Read-Only Memory
UXO	Unexploded Ordnance
V&V	Verification and Validation
VAC	Volts Alternating Current
VDC	Volts Direct Current
VHDL	VLSI (Very Large-Scale Integration) Hardware Design Language
VHF	Very High Frequency
VLS	Vertical Launching System

VLSI	Very Large-Scale Integration
VMC	Visual Meteorological Conditions
VMC	Vehicle Management Computer
VMS	Vehicle Management System
VOR	VHF Omni-Directional Ranging
VSM	Vehicle-Specific Module
VTOL	Vertical Takeoff and Landing
VTUAV	VTOL Tactical Unmanned Air Vehicle
WAAS	Wide Area Augmentation System
WAN	Wide Area Network
WBS	Work Breakdown Structure
WCA	Warning, Caution, Advisory
WCA	Warnings, Cautions, and Alerts
WESS	Web-Enabled Safety System (Navy/Marine Corps)
WFF	Wallops Flight Facility
WISE	WSESRB Interactive Safety Environment
WOW	Weight on Wheels
WP	White Phosphorous
WRA	Weapon Replaceable Assembly
WSESRB	Weapon Systems Explosives Safety Review Board
WSMR	White Sands Missile Range
WUC	Work Unit Code
WX	Weather
XMTR	Transmitter
XPDR	Transponder
ZA	Zonal Analysis
ZSA	Zonal Safety Analysis

INDEX

Abort, 16
Abnormal operation, 17
Above ground level (AGL), 17
Acceptable risk, 18
Acceptance test, 19
Accident, 19
   cause, 20
   class A, 54
   class B, 54
   class C, 54
   cost, 20
   investigation, 20
   scenario, 20
ASCII, 21
Action, 21
Actuator, 21
Ada, 22
Aerospace, 22
Airborne, 22
Aircraft airworthiness, 23
   authority, 23
Aircraft safety, 461
   card, 23
Airworthiness, 23
All-up-round (AUR), 24
Algorithm, 24
Ammunition, 24
   explosives (A&E), 25
Analysis of variance (ANOVA), 25
Analysis technique, 25
Analysis type, 26
AND gate, 26
Anomalous behavior, 26
Anomaly, 26
Anthropometrics, 27

Aperture, 27
Applications software, 27
Application-specific integrated circuit (ASIC), 27
Architecture, 27
Arm, 28
Arming device, 29
Artificial intelligence, 29
As low as reasonably practicable (ALARP), 29
Assembler, 30
Assembly, 31
   language, 30
Attribute, 31
Automatic test equipment (ATE), 31
Audit, 31
Authorized entity, 32
Autoignition, 33
Automatic mode, 33
Automatic operation, 34
Autonomous, 34
Autonomous operation, 34
Autonomous system, 34
Autonomy, 35
Availability, 35
Average, 35

Backout and recovery, 36
Barrier, 36
Barrier analysis (BA), 38
Barrier function, 39
Barrier guard, 39
BASIC, 39
Battery safety, 40, 474
Battleshort, 41

Concise Encyclopedia of System Safety: Definition of Terms and Concepts, First Edition.
Clifton A. Ericson II.
© 2011 by John Wiley & Sons, Inc. Published 2011 by John Wiley & Sons, Inc.

Bathtub curve, 42
Bent pin analysis (BPA), 43
Beyond-line-of-sight (BLOS), 45
Bingo, 45
Bingo fuel, 45
Biohazard, 45
Black box, 46
    testing, 46
Blasting cap, 47
Booster eexplosive, 47
Boundary, 47
Bow-tie analysis, 47
Build, 48
Built-in-test (BIT), 48
    equipment (BITE), 48
Burn, 48
Burn-in, 49
Burning, 49

C, 49
C++, 49
Calibration, 49
Capability maturity model (CMM), 49
    integration (CMMI), 51
Cascading failure, 51
Catalyst, 52
Catastrophe, 52
Catastrophic hazard, 52
Caution, 53
Certification, 53
Chain reaction, 53
Change control board (CCB), 53
Chemical/biological safety, 472
Class A accident, 54
Class B accident, 54
Class C accident, 54
Class desk, 54
Closed system, 55
Code coverage, 55
Collateral damage, 56
Collateral radiation, 56
Combustible liquid, 56
Combustion, 56
Combustion products, 57
Command mode failure sequence, 57
Commercial off-the-shelf (COTS), 57
    safety, 58
    software, 61

Common cause failure (CCF), 61
    analysis (CCFA), 63
Common mode failure (CMF), 67
Compiler, 69
Complexity, 70
Compliance, 70
Compliance based safety, 70
Component, 71
Computer-Aided Software Engineering (CASE), 71
Computer software
    component (CSC), 71
    configuration item (CSCI), 72
    unit (CSU), 72
Concept of operations (CONOPS), 72
Concurrent development model, 72
Condition, 72
Configuration, 73
    control, 73
    item (CI), 73
    management (CM), 73
Conflagration, 74
Conformance, 74
Construction/disposal safety, 468
Contamination, 74
Contingency, 74
    analysis, 75
Continuous wave (CW), 75
Contract, 75
Contracting process, 75
Contractor, 76
Contractor data requirements list (CDRL), 76
Control element, 76
Control entity, 77
Controlled area, 77
Controlled flight into terrain (CFIT), 77
Corrective maintenance (CM), 77
Correlation, 77
Corrosion, 78
Countermeasure, 78
CPU, 78
Crashworthiness, 79
Credible environment, 79
Credible event, 79
Credible failure mode, 80
Credible hazard, 80
Critical characteristic, 80
Critical design review (CDR), 80

Critical failure, 81
Critical few, 81
Critical hazard, 81
Critical item (CI), 81
 list (CIL), 82
Critical safety
 characteristic, 82
 item (CSI), 82
Culture, 83
Cut set (CS), 83
 order, 84
 truncation, 84

Damage, 84
 effects, 85
 mode and effects analysis (DMEA), 85
Danger, 85
 zone, 85
Data, 85
 element List (DEL), 86
 flow diagram, 86
 item description (DID), 86
 link, 87
 package, 88
Data base, 86
Deactivated code, 88
Dead code, 88
Debugging, 89
Deductive reasoning, 89
Deductive safety analysis, 90
Defect, 90
Defense-in-depth, 91
Deflagration, 91
Degradation, 91
Demilitarization (Demil), 92
Department of Defense Explosives
 Safety Board (DDESB), 92
Dependability, 92
Dependence (in design), 92
Dependent event, 94
Dependent failure, 95
Dependent variable, 95
Design load, 95
Deterministic process, 96
Derating, 96
Derived requirements, 96
Design, 96
 assurance level (DAL), 97

Designated engineering representative
 (DER), 97
Designated representative (DR), 97
Design
 diversity, 100
 for reliability (DFR), 100
 for safety (DFS), 100
 qualification test, 100
 requirement, 101
 specification, 103
Design safety
 feature (DSF), 101
 measure, 102
 mechanism, 102
 precept (DSP), 103
Destructive physical analysis (DPA), 103
Detectable failure, 103
Deterministic (analysis), 104
Detonation, 104
 velocity, 104
Development assurance level (DAL), 104
Developer, 106
Deviation, 106
Diagnostics, 106
Disease, 107
Discrepancy, 107
Dissimilar design, 107
Dissimilar software, 107
Diversity, 108
Dormant code, 108
Dormant failure, 108
Dormancy, 108
Downtime, 109
Dud, 109

Electrical/electronic/programmable
 electronic system (E/E/PES), 109
Electrically erasable programmable read
 only memory (EEPROM), 110
Electrical safety, 470
Electric shock, 110
Electrocution, 111
Electro-Explosive Device (EED), 111
Electromagnetic (EM), 111
Electromagnetic compatibility (EMC), 113
Electromagnetic environment (EME), 113

Electromagnetic field (EMF), 113
Electromagnetic interference (EMI), 113
Electromagnetic pulse (EMP), 114
Electromagnetic radiation (EMR), 114
 safety, 470
Electromagnetic susceptibility, 118
Electronic safety and arming dDevice (ESAD), 118
Electro-static discharge (ESD), 118
 safety, 118
Embedded system, 119
Emergence (emergent), 119
Emergency shutdown (ESD) system, 120
Emergency stop, 120
Emergent property, 120
Empty ammunition, 120
Enabling, 120
End-to-end tests, 120
End user, 121
Energetics, 121
Energetic materials, 122
Energy, 122
Energy barrier, 122
Energy path, 123
Energy source, 123
Engineering change, 123
 proposal (ECP), 123
Engineering critical, 124
Engineering/data requirements agreement plan (E/DRAP or EDRAP), 124
Engineering development model, 125
Entropy, 127
Environment, 127
Environmental impact statement (EIS), 127
Environmental qualification safety, 480
Environmental requirements, 128
Environmental validation program, 128
Equipment under control (EUC), 128
Erasable programmable read only memory (EPROM), 128
Electrostatic discharge (ESD) safety, 475
Evaluation assurance level (EAL), 129
Event, 131
 sequence diagram (ESD), 131
Event tree (ET), 132
 analysis (ETA), 132

Evolutionary acquisition, 134
Exclusive OR gate, 135
Exempted lasers, 135
Exigent circumstances, 135
Exit criteria, 135
Experimental design, 135
Explosion, 136
 proof, 136
 device, 136
 event, 136
Explosive material, 138
Explosive ordnance disposal (EOD), 138
Explosives, 138
 safety, 139, 483
 system, 141
Explosive train, 141
Exposure time, 142

Factor of safety, 142
Facilities safety, 467
Fail-safe, 142
 interlock, 144
Failure, 144
 cause, 145
 effect, 145
Failure mode, 145
 effect testing (FMET), 145
Failure modes and effects analysis (FMEA), 145
Failure modes and effects and criticality analysis (FMECA), 148
Failure reporting, analysis, and corrective action system (FRACAS), 149
Family of systems (FoS), 149
Fault, 150
Fault hazard analysis (FHA), 151
Fault injection, 152
Fault isolation (FI), 152
Fault tree
 analysis (FTA), 152
 symbols, 154
Fault-tolerant, 154
Feedback, 154
Feedforward, 156
Field-loadable software (FLS), 156
Field-programmable gate array (FPGA), 156
Final (type) qualified explosive, 156
Finding, 157

Firebrand, 157
Fire classes, 157
Fire point, 157
Fire door, 158
Fire resistive, 158
Fire retardant, 158
Fire safety, 477
Fire wall, 158
Firmware, 158
Fish bone diagram, 158
Flammable, 159
Flash arrestor, 159
Flash memory, 159
Flash point, 159
Flashover
  electrical, 160
  fire, 160
Flight acceptance, 160
Flight certification, 160
Flight clearance, 161
Flight-critical, 161
Flight-essential, 161
Flight operating limitation (FOL), 162
  document (FOLD), 162
Flight readiness review (FRR), 162
Flight safety critical aircraft part (FSCAP), 162
Flight test safety, 480
Foreign object
  damage (FOD), 163
  debris (FOD), 163
Formal qualification testing (FQT), 163
Formal methods, 163
Formal specification, 164
FORTRAN, 164
Fracture control program, 164
Fratricide, 164
Fuel, 164
Fuel safety, 471
Function, 165
Functional block diagram (FBD), 165
Functional configuration audit (FCA), 166
Functional hazard analysis (FHA), 166
Functional hierarchy, 169
Functional logic diagram (FLD), 169
Functional requirement, 169
Functional safety, 463
Functional test, 170

Fuse, 170
Fusible link, 170
Fuze/fuzing system, 170
  safety, 483

Government-furnished equipment (GFE), 174
Government off-the-shelf (GOTS), 174
Graceful degradation, 174
Graphical user interface (GUI), 175
Guard (safety guard), 176

Hang fire, 176
Hardware, 176
Hardware description language (HDL), 176
Harm, 176
Hazard, 177
  action record (HAR), 181
  analysis (HA), 182
    and Critical Control Point (HACCP), 184
  causal factor (HCF), 187
  checklist, 188
  components, 189
  control, 189
  countermeasure, 189
  description, 190
  elimination, 190
  identification, 190
  likelihood (hazard probability), 190
  log, 191
  mitigation, 191
  and operability (HAZOP) analysis, 185
  risk, 193
    index (HRI), 194
  symbol, 198
  tracking, 198
    system (HTS), 198
  triangle, 201
  typecast, 201
Hazardous, 192
Hazardous condition, 193
Hazardous function, 193
Hazardous material (HAZMAT), 192
Hazardous operation, 193
Hazard severity, 196
  levels, 197

Hazards of electromagnetic radiation
  to ordnance (HERO), 197
  to personnel (HERP), 198
Health hazard assessment (HHA), 201
Heat stress, 204
Heat stroke, 204
Hermetic sealing, 204
Hertz (Hz), 204
Heterarchy, 204
Hierarchy, 205
High intensity radio frequency (HIRF), 205
High-level language, 205
High-order language (HOL), 206
Holism, 206
Human engineering, 206
  safety, 463
Human error, 207
Human factors, 210
Human-machine interface (HMI), 210
Human reliability analysis (HRA), 210
Human systems integration (HSI), 210
Human-robot interaction (HRI), 212
Hydraulics safety, 473
Hypergolic, 212
Hyperthermia, 213
Hypothermia, 213

Illness, 213
Importance measure, 213
Improvised explosive device (IED), 213
Inadvertent functioning, 213
Inadvertent launch (IL), 214
Inadvertent arming, 214
Inadvertent release, 214
Incident, 214
Incremental development model, 214
Indenture level, 215
Indentured equipment list (IEL), 215
Independence, in design, 215
Independent event, 216
Independent failure, 216
Independent protection layer (IPL), 217
Independent safety feature, 218
Independent variable, 218
Individual risk, 218
Indoor air quality (IAQ), 218
Inductive reasoning, 219
Inductive safety analysis, 219

Informal specification, 220
Infrared, 220
Inherent hazard, 220
Initial risk, 220
Initiating event (IE), 221
  analysis (IEA), 221
Injury, 222
Insensitive munitions (IM), 222
Inspection, 223
Issue, 223
Instrument, 223
Interface, 224
  requirement, 224
Interlock, 224
Interoperability, 227
Interpreter, 227
Intrinsic safety (IS), 228, 482
Ionizing radiation, 228
Ishikawa diagram, 229

Java, 230
Jeopardy, 230
Job Hazard Analysis (JHA), 230
Job Safety Analysis (JSA), 233
Joule, 233

Label, 234
Laser, 234
Laser diode, 234
Laser footprint, 235
Laser safety, 235, 461
  officer (LSO), 239
  review board (LSRB), 239
Laser system, 239
  safety officer (LSSO), 239
Latent failure (or latency), 240
Layers of protection (LOP), 240
  analysis (LOPA), 241
Lead explosive, 241
Level of assembly, 241
Lightning safety, 471
Life cycle, 242
Life support item, 242
Light, 243
Likelihood, 243
Limited life items, 243
Line-of-sight (LOS), 243
Line replaceable unit (LRU), 243
Lockin, 244

INDEX   513

Lockout, 244
Lower explosive limit (LEL), 244
Lower flammable limit (LFL), 244
Low rate initial production (LRIP), 244

Main charge, 244
Maintainability, 245
Maintenance, 245
Malfunction, 245
Management oversight and risk tree (MORT) analysis, 245
Managing activity (MA), 247
Man-portable, 247
Man-transportable, 247
Manufacturing safety, 468
Marginal failure, 247
Marginal hazard, 248
Margin of safety, 248
Markov analysis (MA), 248
Master equipment list (MEL), 250
Master logic diagram (MLD), 250
Material safety data sheet (MSDS), 252
Materials safety, 476
Maximum permissible exposure (MPE), 252
May, 253
Mean time
    between failures (MTBF), 253
    to failure (MTTF), 253
    to repair (MTTR), 254
Microburst, 254
Milestone decision authority (MDA), 254
Mine safety, 482
Minimal cut set (MCS), 254
Misfire, 254
Mishap, 255
Mishap causal factor, 257
Mishap likelihood (mishap probability), 257
Mishap risk, 258
    index (MRI), 258
Mishap risk index (MRI) matrix, 259
Mishap risk analysis, 259
Mishap severity, 259
Mission critical, 260
Mission essential, 260
Mode, 260
Mode confusion, 261

Mode of control, 261
Modified condition/decision coverage (MC/DC), 262
Module, 262
Monitor, 262
Moral hazard, 263
Multiple occurring event (MOE), 263
Multiple-version dissimilar software, 264
Munition, 264

Near mishap, 265
Near miss, 265
Need not, 265
Negative obstacle, 265
Negligible hazard, 265
Net centric, 266
Net explosives weight (NEW), 266
Nit, 266
Nominal hazard zone (NHZ), 266
Nominal ocular hazard distance (NOHD), 266
Noise, 266
Noise pollution, 267
Non-developmental Item (NDI), 267
Nonconformance, 268
Non-functional requirement, 268
Non-ionizing radiation, 269
Non-line-of-sight (NLOS), 269
Normal distribution, 269
Normal operation, 270
Note, 270
Nuclear power safety, 479
Nuclear weapon safety, 270, 477
N-version programming, 271

Observation, 272
Offgassing, 272
Open architecture, 272
Open source software (OSS), 273
Open system, 273
Operating and support hazard analysis (O&SHA), 273
Operational environment, 276
Operational readiness review (ORR), 276
Operational risk management (ORM), 277
Operational safety precept (OSP), 279
Operator error, 279
Ordnance, 279

Organization, 279
OR gate, 280
Original equipment manufacturer (OEM), 280
Outgassing, 280
Override, 281
Oxidizer, 281
Ozone, 281

Paradigm, 281
Pareto principle, 281
Part, 282
Partial detonation, 282
Particular risk, 282
    assessment, 283
Partitioning, 283
Pascal, 283
Patch, 284
Patient safety, 484
Payload, 284
Perception, 284
Performance
    shaping factor (PSF), 285
    testing, 285
    validation, 285
Petri net analysis (PNA), 285
PHS&T safety, 467
Physical configuration audit (PCA), 287
Pitot tube, 287
Pivotal event, 287
Plasticizer, 288
Pneumatics safety, 474
Point of contact (POC), 288
Positive control, 288
Positive measure, 288
Potential hazard, 289
Power, 289
Preliminary design review (PDR), 289
Preliminary hazard analysis (PHA), 290
Preliminary hazard list (PHL), 292
Prescriptive safety, 294
Primary explosive, 295
Prime contractor, 295
Principal for safety (PFS), 295
Priority AND gate, 298
Probabilistic risk assessment (PRA), 298
Probability of failure on demand (PFD), 298

Probability of loss
    of aircraft (PLOA), 298
    of control (PLOC), 299
    of mission (PLOM), 300
Process, 300
Process safety, 469
Production readiness review (PRR), 301
Product safety, 301
Program, 301
Programmable electronic system (PES), 301
Programmable logic controller (PLC), 302
Programmable logic device (PLD), 303
Programmatic environmental, safety, and health evaluation (PESHE), 303
Programmatic safety precept (PSP), 304
Program safety engineer (PSE), 304
Program safety manager (PSM), 305
Propellant, 305
Pulsed laser, 305
Pyrophoric, 305
Pyrotechnic, 306

Qualification, 306
Qualified explosive, 306
Qualitative safety analysis, 306
Quantitative safety analysis, 307
Quantity-distance (QD), 308

Radiation hazard (RADHAZ), 308
Radio frequency (RF)
    identification (RFID), 309
    improvised explosive device (RFIED), 309
    radiation, 309
Radiological safety, 473
Radon, 309
Rail transportation safety, 479
Range safety, 309, 481
RAMS, 310
Random access memory (RAM), 310
Read only memory (ROM), 310
Red stripe, 311
Redundancy, 311
Reengineering, 312
Regression testing, 312
Refactoring, 313
Relationship, 314

INDEX **515**

Reliability, 314
  block diagram (RBD), 315
  growth, 315
Remote control, 315
Remotely guided, 316
Remotely operated vehicle (ROV), 316
Remote terminal unit (RTU), 316
Repair, 316
Repairable item, 317
Repair time, 317
Requests for deviation/waiver, 317
Requirement, 317
Requirements
  management, 321
  traceability, 323
Residual risk, 323
Resilience engineering, 324
Retinal hazard region, 325
Reverse engineering, 325
Rework, 325
Risk, 325
  acceptance, 328
    authority (RAA), 331
    process (RAP), 331
  analysis, 332
  -based safety, 332
  compensation, 332
  management, 333
    planning, 336
  mitigation, 336
    plan (RMP), 336
      implementation, 336
    priority number (RPN), 337
  tracking, 337
Robotic safety, 464
Runaway vehicle, 337

Safe, 337
Safe separation, 338
Safe software, 338
Safety, 338
  analysis
    technique, 339
    type, 341
  and arming (S&A) device, 342
  assessment report (SAR), 343
  audit, 345
  barrier diagram, 346
  case, 347

  -critical (SC), 350
  -critical function (SCF), 351
    thread, 351
  -critical item (SCI), 352
  -critical operation, 352
  -critical (SC) requirement, 352
  culture, 352
  device, 353
  factor, 353
  feature, 355
  of flight (SOF), 362
  instrumented function (SIF), 355
  instrumented system (SIS), 355
  integrity level (SIL), 357
  interlock, 358
  latch, 359
  management system (SMS), 359
  margin, 361
  measure, 361
  mechanism, 362
  order of precedence (SOOP), 362
  precept, 363
  -related (SR), 365
  requirements/criteria analysis (SRCA), 366
  significant item (SSI), 368
  of UMSs, 465
  vulnerability, 368
Safing, 368
Sampling, 369
Secondary failure, 369
Section, 370
Shall, 370
Sharp edges, 370
Shock hazard, 371
Shop replaceable unit (SRU), 371
Should, 371
Single point failure (SPF), 371
Situation awareness (SA), 373
Sneak circuit, 374
  analysis (SCA), 374
Societal risk, 376
Software
  capability maturity model (CMM), 376
  change control board (SCCB), 376
  criticality index (SCI), 377
  criticality level (SCL), 377
  development file (SDF), 381
  development library (SDL), 381

Software (cont'd)
  hazard risk index (SHRI), 381
  level, 382
  problem report (SPR), 384
  re-use, 384
  safety (SwS), 385, 459
    process, 387
    program (SwSP), 390
      plan (SwSPP), 391
Space, 393
Spiral development, 393
Specification, 393
  change notice (SCN), 393
State, 394
Staged photographs, 394
Statement
  of objectives (SOO), 395
  of work (SOW), 395
Sterilization, 395
Store, 395
Stress testing, 396
Subassembly, 396
Subject matter expert (SME), 397
Substantial damage, 397
Subsystem, 397
  hazard analysis (SSHA), 397
Suitability, 400
Supervisory control and data acquisition (SCADA), 400
Survivability, 401
Synergism, 401
Synergy, 402
Sympathetic detonation, 402
System, 402
  acceptance review (SAR), 404
  boundary, 404
  definition review (SDR), 404
  development model, 404
  environment, 405
  function, 405
  functional review (SFR), 405
  hazard analysis (SHA), 406
  hierarchy, 407
  interface, 409
  lifecycle, 409
    model, 409
  objective, 412
  requirements review (SRR), 413
  safety (SyS), 414, 458
    lead, 418
    management plan (SSMP), 419
    organization, 420
    process, 421
    program (SSP), 421
    program plan (SSPP), 423
    requirement (SSR), 425
    working group (SSWG), 427
    working group (SSWG) charter, 428
  of systems (SoS), 412
  type, 430
Systems engineering technical review (SETR), 429
Systems theory, 431

Tactical digital information link (TADIL), 431
Tailoring, 432
Technical data, 432
  package (TDP), 432
Technique for human error rate prediction (THERP), 433
Technology
  refresh, 433
  insertion, 434
Tele-operation, 434
Telepresence, 434
Temperature cycle, 434
Test readiness review (TRR), 435
Test witness, 435
Thermal balance test, 435
Thermal contact hazards, 435
Thermal-vacuum test, 435
Things falling off aircraft (TFOA), 436
Threat hazard assessment (THA), 436
Top level hazard (TLH), 437
Top level mishap (TLM), 437
Toxicity, 439
Track correlation, 439
Transponder, 440

Ultraviolet radiation, 440
Unattended system, 440
Undetectable failure, 440
Un-executable code, 441
Unexploded ordnance (UXO), 441
Unified modeling language (UML), 441

Unintended function, 441
Unit, 441
 testing, 442
Unmanned aircraft (UA), 442
 system (UAS), 442
 vehicle (UAV), 443
Unmanned ground vehicle (UGV), 443
Unmanned system (UMS), 443
Unsafe action (UA), 443
Use case, 444
User interface (UI), 444

Validation, 444
 by similarity, 444
Valid command, 445
Valid message, 445
Verification
Very large scale integration (VLSI)
 hardware design language (VHDL), 445
Vibroacoustics, 445
Visible radiation (light), 445

WASH-1400, 446
Warning, 446
Watt, 446
Wavelength, 447

Waypoint, 447
 navigation, 447
Weapon, 447
 replaceable assembly (WRA), 447
 system, 447
Weapons systems explosives safety review board (WSESRB), 448
Wearout, 448
Web-enabled safety system (WESS), 448
What-if analysis, 448
White box, 450
 testing, 450
Why-because analysis (WBA), 451
Will, 452
Wind shear, 452
Work breakdown structure (WBS), 453
Workmanship tests, 453
Worst case scenario, 453
Worst credible hazard, 453
Wrapper, 454

X-tree analysis, 454
X-rays, 454

Yoke, 454

Zonal safety analysis (ZSA), 455